THE
HANDY
PHYSICS
ANSWER
BOOK

About the Author

Charles Liu is a professor of astrophysics at the City University of New York's College of Staten Island and an associate with the Hayden Planetarium and Department of Astrophysics at the American Museum of Natural History. His research centers on colliding galaxies, starburst galaxies, quasars, and the star formation history of the universe. Among his many research projects, he is a founding member of the COSMOS collaboration, an international team of scientists studying the largest contiguous area ever surveyed with the Hubble Space Telescope.

(photo by Lillian M. Ortiz)

In addition to his academic research publications, Charles has written extensively for general audiences and appears often on television, radio, and podcasts, explaining and talking about scientific concepts and discoveries to the public. Together with co-authors Neil Tyson and Robert Irion, he received the American Institute of Physics Science Writing Award for their book *One Universe: At Home in the Cosmos.*

Charles has served as Councilor, Trustee, and Education Officer of the American Astronomical Society, and is a lifetime member of the International Astronomical Union. He is the current Chairman of the not-for-profit New York Astronomical Corporation, and he is the President of the Astronomical Society of New York. In 2020 he was elected as a Fellow of the American Astronomical Society.

Charles earned degrees from Harvard and the University of Arizona, and held postdoctoral positions at Kitt Peak National Observatory and Columbia University. He and his wife have three wonderful children.

THE HANDY PHYSICS ANSWER BOOK

PHYSICS ANSWER

BOOK

THIRD EDITION

Charles Liu, Ph.D.

VISIBLE
INK
PRESS

Detroit

ALSO FROM VISIBLE INK PRESS

The Handy Law Answer Book
by David L. Hudson, Jr., J.D.
ISBN: 978-1-57859-217-3

The Handy Literature Answer Book: An Engaging Guide to Unraveling Symbols, Signs, and Meanings in Great Works
by Daniel S. Burt, Ph.D., and Deborah G. Felder
ISBN: 978-1-57859-635-5

The Handy Math Answer Book, 2nd edition
by Patricia Barnes-Svarney and Thomas E. Svarney
ISBN: 978-1-57859-373-6

The Handy Military History Answer Book
by Samuel Willard Crompton
ISBN: 978-1-57859-509-9

The Handy Mythology Answer Book
by David A. Leeming, Ph.D.
ISBN: 978-1-57859-475-7

The Handy New York City Answer Book
by Chris Barsanti
ISBN: 978-1-57859-586-0

The Handy Nutrition Answer Book
by Patricia Barnes-Svarney and Thomas E. Svarney
ISBN: 978-1-57859-484-9

The Handy Ocean Answer Book
by Patricia Barnes-Svarney and Thomas E. Svarney
ISBN: 978-1-57859-063-6

The Handy Personal Finance Answer Book
by Paul A. Tucci
ISBN: 978-1-57859-322-4

The Handy Philosophy Answer Book
by Naomi Zack, Ph.D.
ISBN: 978-1-57859-226-5

The Handy Physics Answer Book, 3rd edition
by Charles Liu, Ph.D.
ISBN: 978-1-57859-695-9

The Handy Presidents Answer Book, 2nd edition
by David L. Hudson
ISB N: 978-1-57859-317-0

The Handy Psychology Answer Book, 2nd edition
by Lisa J. Cohen, Ph.D.
ISBN: 978-1-57859-508-2

The Handy Religion Answer Book, 2nd edition
by John Renard, Ph.D.
ISBN: 978-1-57859-379-8

The Handy Science Answer Book, 5th edition
by The Carnegie Library of Pittsburgh
ISBN: 978-1-57859-691-1

The Handy State-by-State Answer Book: Faces, Places, and Famous Dates for All Fifty States
by Samuel Willard Crompton
ISBN: 978-1-57859-565-5

The Handy Supreme Court Answer Book
by David L Hudson, Jr.
ISBN: 978-1-57859-196-1

The Handy Technology Answer Book
by Naomi E. Balaban and James Bobick
ISBN: 978-1-57859-563-1

The Handy Texas Answer Book
by James L. Haley
ISBN: 978-1-57859-634-8

The Handy Western Philosophy Answer Book: The Ancient Greek Influence on Modern Understanding
by Ed D'Angelo, Ph.D.
ISBN: 978-1-57859-556-3

The Handy Weather Answer Book, 2nd edition
by Kevin S. Hile
ISBN: 978-1-57859-221-0

PLEASE VISIT THE "HANDY ANSWERS" SERIES WEBSITE AT WWW.HANDYANSWERS.COM.

THE HANDY PHYSICS ANSWER BOOK

Visible Ink Press®
43311 Joy Rd., #414
Canton, MI 48187–2075

Visible Ink Press is a registered trademark of Visible Ink Press LLC.

Most Visible Ink Press books are available at special quantity discounts when purchased in bulk by corporations, organizations, or groups. Customized printings, special imprints, messages, and excerpts can be produced to meet your needs. For more information, contact Special Markets Director, Visible Ink Press, www.visibleink.com, or 734–667–3211.

Managing Editor: Kevin S. Hile
Art Director: Mary Claire Krzewinski
Typesetting: Marco Divita
Proofreaders: Christa Brelin and Shoshana Hurwitz
Indexer: Larry Baker

Cover images: Sir Isaac Newton (Institute for Mathematical Sciences, University of Cambridge); all other images are from Shutterstock.

ISBN: 978–1–57859–695–9

Cataloging–in–Publication data is on file at the Library of Congress.

Printed in the United States of America.

10 9 8 7 6 5 4 3 2 1

Contents

Acknowledgments

It is a privilege to have been passed the stewardship of this book from the distinguished authors of the two previous editions, P. Erik Gundersen and Paul W. Zitzewitz. Hopefully, my revisions and additions do justice to their work and legacies, which have made such a meaningful impact on physics, education, and research that extend well beyond this one title. To publisher *par excellence* Roger Jänecke and editor extraordinaire Kevin Hile: thank you for entrusting this project to me.

My parents, Frank Fu-Wen Liu and Janice Jui-Chi Liu, were my first teachers, and I am still learning from them today. Thanks to them, I love to learn, to ask questions, and to find answers; and through their work and sacrifice they gave me the freedom to study the universe with wonder and joy. After I started school, my many physics teachers did so much to shape me into the scientist and educator I am today. Here is a list of just a few of them in roughly chronological order: Walter Nazarenko, Sidney Coleman, George Rybicki, John Huchra, Peter Strittmatter, Rob Kennicutt, Bruce Woodgate, and Richard Green. Thank you all so much—without what you've taught me, I wouldn't have been able to write this book.

And to my wife, Dr. Amy Rabb-Liu, and our three wonderful children, Hannah, Allen, and Isaac: Thank you! Thank you! Thank you! Thank you! You are why things work in my world.

Photo Sources

Introduction

"It is indeed a grand time to be alive and asking questions." Harlow Shapley, the former director of the Harvard College Observatory, wrote that sentence in a book in which he describes how scientists use physics to find answers to the amazing way things work—not just in the world, but also on other planets, stars, galaxies, and even the entire cosmos. That sentence is as true today as it was then—and Dr. Shapley wrote it before I was even born.

If you think about it, when it comes to air, water, earth, and fire—sound, waves, heat, and light—space, time, matter, and energy—we all have more than a thousand questions. Some questions are simple, while others are complicated. Why is the sky blue? How does my cell phone work? Who *really* invented the light bulb? Are robots becoming intelligent? What happens when you fall into a black hole? Well, this book contains more than a thousand answers for you about these and many, many other fascinating topics in physics.

Physics explains how and why things work the way they do. Far from being dense or difficult, physics can be described by a set of simple laws; and just as the 26 letters of the alphabet can be combined in a myriad of ways to tell amazing stories, the laws of physics that govern the world and the universe around us produce powerful, beautiful, and even mind-blowing phenomena. Think of waves at the beach, music in your ears, the colors of the rainbow, or nuclear fusion in the Sun; all of these and so much more arise from physics. Understanding and applying physics has allowed us to improve and enrich our lives in both small and big ways—like heating our homes, hitting a baseball, watching a video, flying an airplane, and exploring outer space.

I invite you to open this book and have your questions answered. You can pick a topic that interests you, or choose a question at random, or even read the whole book from beginning to end. The chapters of this book are organized roughly in the order of how physics developed throughout the centuries; after starting off with "Physics Fundamentals," they will serve to guide you through the history of science. By experimenting with motion and force, the first physicists realized that quantities like momentum and energy are conserved when objects interact. Physics then expanded to explain the behavior not just of solid matter, but also of fluids like liquids and gases; and after that, physicists began to understand the thermodynamics of moving energy through matter as heat, and the carrying of mechanical energy through matter with waves. Studying a special type of high-speed wave led to insight into light and optics; and when physicists realized that light was an electromagnetic wave, they were able to see the connections between light and electricity and magnetism. Then, at the dawn of the previous century, examining the physics of sub-microscopic matter led to the revolution of atomic and quantum physics. The book concludes with the chapter "Physics Frontiers," covering the current century and beyond as scientists push the boundaries of what we know about the physical universe, matter, time, and even knowledge itself.

All the scientific terms on these pages are clearly described and explained, and in case you run into an unfamiliar symbol or word, there is a list of symbols and a glossary of terms at the end of the book, as well as a bibliography of books and websites to find out more. I hope every answer will entertain you, enlighten you, and inspire you to be even more curious and ask even more questions.

Enjoy your exploration of physics! Have a grand time.

PHYSICS FUNDAMENTALS

What is physics?

Physics is the scientific study of the structure, content, and activity of and in the world and universe around us. Physics seeks to explain natural phenomena in terms of a comprehensive theoretical framework in mathematical form. Physics depends on accurate instrumentation, precise measurements, and the rigorous expression of results.

Physics is often called the "fundamental science" because it is the core of many other sciences, such as astronomy, biology, chemistry, and geology. It is also the basis of many fields of applied sciences, such as aeronautics and astronautics, engineering, computer science, and information technology. A strong knowledge of physics is a powerful tool to have in today's high-tech world.

What is the origin of the field of physics?

The word "physics" comes from the Greek *physis*, meaning nature. Aristotle (384–322 B.C.E.) wrote the first known work called *Physics*, a set of eight books that presented a detailed study of motion and its causes. The ancient Greek title of the book is often translated as *Natural Philosophy*, or writings about nature. For that reason, those who studied the workings of nature were called "natural philosophers." They were educated in philosophy and called themselves philosophers.

One of the earliest modern textbooks that used the word "physics" in its title was published in 1732. It was not until the 1800s that those who studied physics were called physicists. In the nineteenth, twentieth, and twenty-first centuries, physics has proven to be a large and important field of study.

What are some of the areas of study within physics?

Due to the huge breadth of physics, physicists today often concentrate their work in one or a few areas of study within the field. Some of the most basic areas of physics include:

- *Mechanics,* which describes the effect of forces on the motion and energy of physical objects. Modern studies of mechanics often involve fluids (liquids and gases) and granular particles (like sand) as well as the motions of planets, stars, and galaxies.

- *Relativity,* which describes the motion of objects moving near the speed of light (special relativity) and the structure and effects of space, time, and gravity (general relativity).

- *Electromagnetism,* which describes how electric and magnetic forces interact with matter. Light, both visible and invisible to human eyes, is a form of electromagnetic wave, so the study of electromagnetism also includes the study of light.

- *Thermodynamics and statistical mechanics,* which describe how temperature and heat affect matter and are transferred. Thermodynamics deals with macroscopic (observable) objects, while statistical mechanics concerns the atomic and molecular motions of very large numbers of particles, including how those motions are affected by heat transfer.

- *Quantum mechanics,* which describes the interactions of elementary particles and fields—that is, the way the tiniest, most fundamental constituents of matter and energy in the universe behave on microscopic scales.

- *Nuclear, atomic, and molecular physics,* which describe the structure and phenomena related to the building blocks of the elements and compounds that make up everything in the world around us.

- *Physics education,* which investigates how people learn physics and how best to teach physics effectively.

What are some important specialized fields of physics?

As the "fundamental science," physics often connects with other sciences, resulting in important specialized fields of physics. These fields include:

- *Astrophysics,* which uses physics to examine the properties and behavior of the universe and its contents. The study of astrophysical origins, in particular the origin and early history of the universe itself, is known as *cosmology.*

- *Biophysics,* which examines the physical interactions of biological molecules and the physics of biological systems and processes.

- *Chemical physics,* which investigates the physical causes of chemical reactions between atoms and molecules and how light can be used to cause, measure, and interpret these reactions.

- *Geophysics,* which studies the physics of the planet Earth. It deals with the forces, interactions, matter, and energy found within and around Earth itself, including the study of tectonic plates, earthquakes, volcanism, the atmosphere, and the oceans.

- *Solid-state physics,* also known as condensed-matter physics, which studies the physical, electrical, and thermal properties of solid materials, including superdense, superhot, or supercold substances.

What are some of the specialized fields of applied physics?

The basic principles of physics are essential and present in just about every kind of activity, device, and structure you might do, use, and encounter in everyday life. Some of the fields of physics that relate to how the laws of physics are applied are:

- *Acoustics*: Musical acoustics studies the ways musical instruments produce sounds. Applied acoustics includes the study of how concert halls can best be designed. Ultrasound acoustics uses sound to image the interior of metals, fluids, and the human body.
- *Atmospheric physics* studies the atmosphere of Earth and other planets, including the active study of powerful storms and the causes and effects of the greenhouse effect, global warming, and climate change.
- *Materials science* investigates the properties of glasses, metals, and other materials and produces new compounds and mixtures of materials to be used in everything from cellular phones and computers to artificial hearts and airplanes.
- *Medical physics* investigates the use of physical processes to produce images of what's going on inside people as well as the use of radiation and high-energy particles to treat diseases such as cancer.
- *Optics* investigates ways to take clearer pictures and images of things we cannot ordinarily see from microscopes to eyeglasses to telescopes on Earth and in space.
- *Nanotechnology* examines the construction and operation of objects, materials, and machines on tiny scales like molecules that can contain and deliver medicines inside the human body or robots that can repair individual human cells.
- *Plasma physics* studies plasmas, which are gases composed of electrically charged atoms. Plasmas studied include those in fluorescent lamps, in large-screen televisions, in Earth's atmosphere, and in stars and interstellar space. Plasma physicists are also working to create controlled nuclear fusion reactors to produce electrical power.

MEASUREMENT

Why is measurement so important in physics?

As emphasized by the ancient Greek philosopher Aristotle (384–322 B.C.E.), who wrote the multivolume work *Physics*, the study of nature is based on observing what is here and what is happening around us. When you measure something, you are observing it as accurately and precisely as you can and then recording your observation so it can be examined later. Hence, measurement is fundamental to physics and to all scientific inquiry.

What is the difference between accuracy and precision?

"Accuracy" and "precision" are often used interchangeably in everyday conversation; however, each has a unique meaning. Accuracy defines how correct or how close to the

3

accepted result or standard a measurement or calculation has been. Precision describes how well the results can be reproduced. For example, a person who can repeatedly hit a bull's-eye with a bow and arrow is accurate and precise. If the person's arrows all fall within a small region away from the bull's-eye, then she or he is precise but not accurate. If the person's arrows are scattered all over the target and the ground behind it, then she or he is neither precise nor accurate.

What were the first clocks?

For thousands of years, all measurements of time were based on the rotation of Earth. The first method of measuring time shorter than a day dates back to 3500 B.C.E., when a device known as the gnomon was used. The gnomon was a stick placed vertically into the ground that, when struck by the sun's light, produced a distinct shadow. By measuring the relative position of the shadow throughout the day, the length of a day could be measured. In the Western Hemisphere, the gnomon was later replaced by the first hemispherical sundial about 2,300 years ago by the astronomer Berossus (born c. 340 B.C.E.).

What are the major limitations of gnomons and sundials?

Sun-based clocks like gnomons and sundials cannot be effectively used at night or other times when the sun isn't shining. To remedy this problem, timing devices such as notched candles were created. Later, hourglasses and water clocks (clepsydra) became popular. The first recorded description of a water clock is from the sixth century B.C.E. Ctesibius of Alexandria, a Greek inventor who lived in the third century B.C.E., used gears that con-

Sundials are a very old way to tell time. While accurate, they are limited by the fact that they only work when the sun is shining.

nected a water clock to a pointer and dial display similar to those in today's clocks. It wasn't until 1656, when a pendulum was used with a mechanical clock, that clocks began to keep accurate time.

What are the standards for measurement in physics today?

The International System of Units, officially known as the Système International d'Unités and abbreviated SI, was adopted by the 11th General Conference on Weights and Measures in Paris in 1960. Basic units are based on the meter-kilogram-second (MKS) system, which is commonly known as the metric system.

Most of the world uses the metric system for measuring quantities such as weight. Also known as the International System of Units (SI), the metric system was most recently refined at the 26th General Conference on Weights and Measures in 2018.

Do people in the United States use SI (the metric system)?

Although the American scientific community uses the SI system of measurement, the general American public still uses the traditional English system of measurement. In an effort to change over to the metric system, the U.S. government instituted the Metric Conversion Act in 1975. Although the act committed the United States to increasing the use of the metric system, it was on a voluntary basis only. The Omnibus Trade and Competitiveness Act of 1988 required all federal agencies to adopt the metric system in their business dealings by 1992. Therefore, all companies that held government contracts had to convert to metric. Although most American corporations manufacture metric products, the English system still is the predominant system of measurement in the United States in daily life.

How is time measured today?

Atomic clocks are the most precise devices to measure time. Atomic clocks, such as rubidium, hydrogen, and cesium clocks, are used by scientists and engineers when computing distances with global positioning systems (GPS), measuring the rotation of Earth, precisely knowing the positions of artificial satellites, and imaging stars and galaxies. The clock that is used to define the standard for the metric system is the cesium-133 atomic clock.

What is the standard unit of time?

The second is the standard unit of time in SI and the metric system. The measurement of the second is defined as the time it takes for 9,192,631,770 periods of microwave radiation that result from the transfer of the cesium-133 atom between lower-energy and higher-energy states. The second is currently known to a precision of 5×10^{-16}, or 1 second in 60 million years!

What is the standard unit of length?

The meter is the standard unit of length in SI and the metric system. In 1798, French scientists determined that the meter would be measured as 1/10,000,000th the distance from the North Pole to the equator. After calculating this distance, scientists made a platinum-iridium bar with two marks precisely one meter apart. This standard was used until 1960, when it was replaced with a measure of the wavelengths of light produced by glowing krypton atoms. Today, the meter is defined using the second and the speed of light in a perfect vacuum. One meter is the distance light travels in 1/299,792,458 second.

What are some of the metric prefixes of measurement, and what do they mean?

Prefixes in the metric system are used to denote powers of 10. The value of the exponent next to the number 10 represents the number of places the decimal should be moved to the right (if the number is positive) or to the left (if the number is negative). The following is a list of prefixes commonly used in the metric system:

femto	(f)	10^{-15}	deka	(da)	10^{1}
pico	(p)	10^{-12}	hecto	(h)	10^{2}
nano	(n)	10^{-9}	kilo	(k)	10^{3}
micro	(μ)	10^{-6}	mega	(M)	10^{6}
milli	(m)	10^{-3}	giga	(G)	10^{9}
centi	(c)	10^{-2}	tera	(T)	10^{12}
deci	(d)	10^{-1}	peta	(P)	10^{15}

What is the standard unit of mass?

The kilogram is the standard unit of mass in SI and the metric system. The kilogram was originally defined as the mass of 1 cubic decimeter of pure water at 4° Celsius. A platinum cylinder of the same mass as the cubic decimeter of water was the standard until 1889. A platinum-iridium cylinder with the same mass is permanently kept near Paris; for 130 years, the mass of this cylinder officially defined the kilogram. Copies exist in many countries. In the United States, the National Institute of Standards and Technology (NIST) houses the mass standard as well as the atomic clocks that define the second.

CAREERS IN PHYSICS

How can I become a physicist?

The first requirements to becoming a physicist are to have an inquisitive mind and the curiosity to ask questions. Albert Einstein (1879–1955), perhaps the most famous physicist in history, once said about himself, "I'm like a child. I always ask the simplest questions." It seems as though the simplest questions always appear to be the most difficult to answer and that the most important questions we have to answer are the ones we have yet to ask.

What kinds of classes should I take to become a physicist?

Usually, becoming a physicist requires quite a bit of formal schooling along with that inquisitive and curious mind. From elementary school through high school, a strong academic background, including mathematics, reading, writing, and science, is valuable

One type of job for a physicist is to work in a lab performing experiments using high-tech equipment. For example, these two experimental physicists are working with a laser deposition chamber to vaporize metal particles in a vacuum.

7

to enter college with a strong knowledge base. In college, you can study to become a physicist by taking courses such as mechanics, relativity, electricity, magnetism, thermodynamics, quantum mechanics, math, and computer science to obtain a bachelor's degree.

To become a research physicist, an advanced degree is usually required. This means attending graduate school, performing research, writing a thesis or dissertation, and eventually obtaining a Ph.D. (doctor of philosophy).

What does a physicist do?

Physicists teach other physicists and conduct research to make original discoveries. Research in physics usually is done in three general ways. Theoretical researchers create or extend mathematically based ideas or explanations of the physical universe and propose ways to test these theories. Experimental researchers develop and conduct experiments to observe the physical universe under controlled conditions to test theories, explore uses of new instruments, and investigate new materials. Computational researchers use computers and develop software to simulate—that is, reconstruct within a virtual environment—physical systems and conditions to explore and extend theories or to make observations and experiments that cannot yet be done directly in nature.

Where does a physicist work?

Physicists find employment in a wide variety of fields, and they frequently make more money than people who have not had an education in physics. Many physicists work in environments where they perform basic research such as colleges and universities, government laboratories, and astronomical observatories. Physicists who find new ways to apply physics to engineering and technology also often work at industrial laboratories for businesses and corporations. Physicists are also extremely valuable when they work in other areas; a few examples include computer science, information technology, economics and finance, medicine, communications, and publishing. Finally, many physicists who love to see young people get excited about physics become teachers in elementary, middle, or high schools or professors in colleges and universities.

What jobs do nonphysicists hold that use physics every day?

Every occupation has some relation to physics! There are many examples of jobs that people might not think of as being physics-intensive. Here are just a few:

- Athletes, both professional and amateur, use the principles of physics all the time. The laws of motion affect how balls are batted and thrown and what happens when athletes run, jump, block, and tackle. The more athletes and their coaches understand and use their knowledge of physics in their sports, the better the athletes perform.
- People who work for insurance companies often reconstruct automobile crashes, which are subject to the laws of physics, using physics concepts such as momentum, friction, and energy in their work.

- Modern electronic devices, from computers and tablets to mobile phones and music players, depend on the applications of physics. The internet and other telecommunication networks are often connected by electromagnetic waves or by fiber optics, and the professionals who develop and service them use the principles of the behavior of electricity and light to create signals and transmit them over thousands, millions, or even billions of miles.

- Medical doctors, nurses, and hospital technicians use all kinds of physics in their imaging and diagnostic equipment—X-rays, CT scans, PET scans, ultrasound, and MRIs, just to name a few. All these health care professionals must know how to use these methods to select the right device to use and interpret the results correctly.

FAMOUS PHYSICISTS

Who were the first physicists?

Although physics was not considered a distinct field of science until the early nineteenth century, people have been studying the motion, energy, and forces that are at play in the universe for thousands of years. The earliest documented accounts of serious thought toward physics, especially the motion of the stars and planets, date back to the ancient Chinese, Indian, Egyptian, Meso-American, and Babylonian civiliations. The Greek philosophers Plato and Aristotle analyzed the motion of objects but did not perform experiments to prove or disprove their ideas.

What contributions did Aristotle make?

Aristotle (384–322 B.C.E.) was a Greek philosopher and scientist of the fourth century B.C.E. He was a student of Plato and an accomplished scholar in the fields of biology, physics, mathematics, philosophy, astronomy, politics, religion, and education. In physics, Aristotle believed that there were five elements—earth, air, fire, water, and the quintessence or aether—out of which all objects in the heavens were made. He believed that these elements moved in order to seek out each other. He stated that if all forces were removed, an object could not move. Thus, motion, even with no change in speed or direction, requires a continuous force. He believed that motion was the result of the interaction between an object and the medium through which it moves.

The Greek philosopher Aristotle wrote the first known book about physics.

9

How did humanity's understanding of physics change after the time of Aristotle?

Through the third century B.C.E. and later, experimental achievements in physics were made in Alexandria and other major cities throughout the Mediterranean. Archimedes (c. 287–c. 212 B.C.E.) measured the density of objects by measuring their displacement of water. Aristarchus of Samos is credited with measuring the ratio of the distances from Earth to the sun and to the moon, and he espoused a sun-centered system. Erathosthenes (276–194 B.C.E.) determined the circumference of Earth by using shadows and trigonometry. Hipparchus (c. 190–c. 120 B.C.E.) discovered the precession of the equinoxes. And finally, in the first century C.E., Claudius Ptolemy (c. 100–170 C.E.) proposed an order of planetary motion in which the sun, stars, and moon revolved around Earth.

After the fall of the Roman Empire, a large fraction of the books written by the early Greek scientists disappeared. In the 800s the rulers of the Islamic Caliphate collected as many of the remaining books as they could and had them translated into Arabic. Between then and about 1200, numerous scientists in the Islamic countries demonstrated the errors in Aristotelian physics. Included in this group is Alhazen (Ibn al-Haitham), Ibm Shakir, al-Biruni, al-Khazini, and al-Baghdaadi, mainly members of the House of Wisdom in Baghdad. They foreshadowed the ideas that Copernicus, Galileo, and Newton would later develop more fully.

Despite these challenges, Aristotle's physics remained dominant in European universities into the late seventeenth century.

Who was the founder of the scientific method?

Ibn Al-Haitham (known in Europe as Alhazen or Alhacen) is considered to be the founder of the scientific method. He was born in Basra, Persia (now Iraq), in 965 C.E. and died in Cairo, Egypt, in 1038 C.E. He wrote 200 books, 55 of which have survived, including his most important work, *Book of Optics*, as well as books on astronomy, geometry, mechanics, and number theory. He is also known for his contributions to philosophy, medicine, and experimental psychology.

The scientific method, in which discoveries are made by the cyclical process of analyzing data, making hypotheses, and testing those hypotheses with experiments and observations, is the cornerstone of modern science. The scientific method was not developed in Europe until more than 500 years *after* Ibn Al-Haitham's time, when Englishman Francis Bacon (1561–1626) and Italian Galileo Galilei (1564–1642) started their work in the philosophy of science.

How did the idea arise that the sun was the center of the universe?

Aristotle and Ptolemy's view that the sun, planets, and stars all revolved around Earth was accepted for almost 18 centuries. Nicolas Copernicus (1473–1543), a Polish astronomer and cleric, was the first person to publish a book arguing that the solar system is a heliocentric (sun-centered) system instead of a geocentric (Earth-centered) system. In the same year as his death, he published *On the Revolutions of the Celestial*

Spheres. His book was dedicated to Pope Paul III. The first page of his book contained a preface stating that a heliocentric system is useful for calculations but may not be the truth. This preface was written by the German theologian Andreas Osiander (1498–1552) without Copernicus's knowledge. It took three years before the book was denounced as being in contradiction with the Bible, and it was banned by the Roman Catholic Church in 1616. The ban wasn't lifted until 1835.

Which famous scientist was arrested for agreeing with Copernicus?

Galileo Galilei (1564–1642) was responsible for bringing the Copernican system more recognition. In 1632, Galileo published his book *Dialogue Concerning the Two Chief World Systems*. The book was written in Italian and featured a witty debate among three people: one supporting Aristotle's system, the second a supporter of Copernicus, and the third an intelligent layman. The Copernican easily won the debate. The book was approved for publication in Florence but was banned a year later.

Galileo Galilei's *Dialogue Concerning the Two Chief World Systems* (1632) argued for the Copernican model of the solar system with the sun at the center and the planets circling the sun.

Pope Urban VIII, a longtime friend of Galileo, believed that Galileo had made a fool of him in the book. Galileo was tried by the Inquisition and placed under house arrest for the rest of his life. All of his writings were banned.

Galileo was also famous for his work on motion; he is probably best known for a thought experiment using the Leaning Tower of Pisa. He argued that a heavy rock and a light rock dropped from the tower would hit the ground at the same time. His arguments were based on extensive experiments on balls rolling down inclined ramps. Many scientists agree that Galileo's work started the science we now call physics.

Who is considered perhaps the greatest scientist of all time?

Many scientists and historians consider the Englishman Isaac Newton (1643–1727) one of the most influential people of all time. It was Newton who discovered the laws of motion and universal gravitation, made huge breakthroughs in light and optics, built the first reflecting telescope, and was one of the founders of the calculus. His discoveries published in *Philosophiae Naturalis Principia Mathematica*, or *The Principia*, and in *Optiks* are unparalleled and formed the basis for mechanics and optics. Both of these

Sir Isaac Newton, one of the most famous scientists of all time, discovered the laws of motion, developed calculus, and built the first reflecting telescope, among many other accomplishments.

books were written in Latin and published only when friends demanded that he do so, many years after Newton had completed his work.

Where did Isaac Newton conduct his scientific studies?

Newton was encouraged by his mother to become a farmer, but his uncle saw the talent Newton had for science and math and helped him enroll in Trinity College in Cambridge. Newton spent four years there, but he returned to his hometown of Woolsthorpe to flee the spread of the Black Plague in 1665. During the two years that he spent studying in Woolsthorpe, Newton made his most notable developments of calculus, gravitation, and optics.

What official titles did Isaac Newton receive?

Newton was extremely well respected in his time. Although he was known for being nasty and rude to his contemporaries, Newton became the Lucasian Professor of Mathematics at Cambridge in the late 1660s, the president of the Royal Society of London in 1703, and the first scientist ever knighted in 1705. He was famous as the Master of the Mint, where he introduced coins that had defined edges so that people couldn't cut off small pieces of the silver from which the coins were made. He is buried in Westminster Abbey in London.

Who was the most influential scientist of the twentieth century?

On March 14, 1879, Albert Einstein was born in Ulm, Germany. No one knew that this little boy would one day grow up to change the way people viewed the laws of the universe. Albert was a top student in elementary school, where he built models and toys and studied Euclid's geometry and Kant's philosophy. In high school, however, he hated the regimented style and rote learning. At age 16 he left school to be with his parents in Italy. He took, but failed, the entrance exam for the Polytechnic University in Zurich. After a year of study in Aarau, Switzerland, he was admitted to the university. Four years later, in 1900, he graduated.

He earned his diploma and began graduate studies at the University of Zurich. During the next three years while working at the Swiss Patent Office, he developed his ideas about electromagnetism, time and motion, and statistical physics. In 1905, his so-called *annus mirabilis,* or miracle year, he published four extraordinary papers. One was on the photoelectric effect, in which Einstein introduced light quanta, later called photons. The second was about Brownian motion, which helped support the idea that all matter is composed of atoms. The third was on special relativity, which revolutionized the way physicists understand both motion at very high speeds and electromagnetism. The fourth developed the famous equation $E = mc^2$. He earned his Ph.D. that year, and a few years later he was appointed a professor at the German University in Prague.

What discovery did Albert Einstein make that earned him worldwide fame?

By 1914 Einstein's accomplishments were well accepted by physicists, and he was appointed professor at the University of Berlin and made a member of the Prussian Academy of Sciences. Einstein published the General Theory of Relativity (also called the Theory of General Relativity) in 1916. Among its predictions was that light from a star would not always travel in a straight line but would bend if it passed close to a massive body like the sun. He predicted a bending twice as large as Newton's theory predicted. During a 1919 solar eclipse, these theories were tested, and Einstein's prediction was shown to be correct. The result was publicized by the most important newspapers in England and the United States, and Einstein became a world figure.

What was unusual about Einstein's Nobel Prize?

Albert Einstein won the 1921 Nobel Prize in Physics—the highest international scientific honor of that time. Einstein was a controversial person at that time, however, in part be-

Albert Einstein is most often remembered for his famous formula $E = mc^2$, but his Nobel Prize in Physics was awarded for his explanation of the photoelectric effect.

cause he was Jewish in an age of anti-Semitism and in part because he strongly supported pacifist causes. In addition, his approach to theoretical physics was very different from that of other physicists of that time. He was repeatedly nominated for the Nobel Prize, but members of the prize's selection committee, despite his public fame, repeatedly refused to grant him the prize, most likely for political reasons. The 1921 prize was not awarded to anyone that year. In 1922 the selection committee found a way to compromise. Einstein was awarded the previous year's prize not for his theory of relativity but "for his services to Theoretical Physics, and especially for his discovery of the law of the photoelectric effect"—one of his discoveries that had already been tested and proven experimentally.

How was Albert Einstein more than just a world-famous physicist?

Albert Einstein left a great mark on humanity not just for his remarkable scientific work but also for what else he wrote and did. Einstein supported a number of social and political causes, not all of them popular in his lifetime. The year he moved from Switzerland to Germany, he joined a group of people opposing Germany's entry in World War I. He joined pacifist and socialist causes. He was Jewish and opposed Nazism; when the Nazis came to wield political power in Germany in the 1930s, Einstein moved to the United States to take a position at the Institute for Advanced Study in Princeton, New Jersey, and eventually became a naturalized U.S. citizen.

In part at the urging of other physicists, Einstein wrote a letter to U.S. president Franklin D. Roosevelt (1882–1945) pointing out the danger posed by Germany's work on uranium that could lead to a new kind of bomb and recommending that America conduct research in this area. This letter helped launch the Manhattan Project, which led to creation of the world's first atomic bombs.

Although Einstein was never allowed to work on atomic bomb research himself, the development and use of those bombs at the end of World War II weighed heavily on his conscience. He remained in the public spotlight, spending time advocating nuclear disarmament and a peaceful world government. At one point, he was offered the presidency of the then-young nation of Israel; he refused the position.

Einstein also wrote and lectured extensively on issues of society, philosophy, education, and many other topics in addition to physics. At the end of the second millennium, he was named *Time* magazine's Person of the Century in 1999.

NOBEL PRIZE WINNERS IN PHYSICS

What is the Nobel Prize?

The Nobel Prize is one of the most prestigious awards in the world. It was named after Alfred B. Nobel (1833–1896), the inventor of dynamite; he left nine million dollars in trust, of which the interest was to be awarded to the person who has made a significant lifetime contribution to their particular field. The awards, given today in the fields of physics, chemistry, physiology and medicine, literature, peace, and economics, have prize money amounts of more than one million dollars each, and they recognize many of the most important scientists, writers, and social leaders in history.

Who received the first Nobel Prize in Physics?

The first Nobel Prize in Physics was awarded in 1901 to the German physicist Wilhelm Röntgen (1845–1923) for his discovery of X-rays. That same year, the first Nobel Prize in Chemistry was awarded to the Dutch chemical physicist Jacobus Hendicus "Henry" van't Hoff Jr. (1852–1911), considered to be one of the founders of modern physical chemistry, for his work on chemical thermodynamics.

Who among the Nobel Prize winners in physics have been women?

In 1903, Marie Curie (1867–1934) was the first woman to win the Nobel Prize in Physics for studies in spontaneous radiation. She later won another Nobel Prize, this one in chemistry. Two other women have since won: American Maria Geoppert-Mayer (1906–1972) in 1963 for discoveries concerning nuclear shell structure, and Canadian Donna Strickland (1959–) in 2018 for co-inventing the chirped pulse amplification laser (see below).

Which nation has the most winners of the Nobel Prize in Physics?

The United States of America has by far the most physics Nobel laureates of any nation. In the first 120 years that the prize was given, more than 90 Americans—many of them immigrants—won the Nobel Prize in Physics.

Who have been the winners of the Nobel Prize in Physics?

In 2019, the Nobel Prize for physics was shared by American astrophysicist James Peebles (1935–) for his pioneering discoveries about physical cosmology (the structure of the universe) and Swiss astronomers Michel Mayor (1942–) and Didier Queloz (1966–) for their discovery of exoplanets (planets orbiting stars other than the sun). In 2018, the prize was shared by Canadian Donna Strickland and Frenchman Gérard Mourou (1944–) for their invention of the chirped pulse amplification laser and American Arthur Ashkin for his invention of the laser tweezer.

MOTION AND FORCE

What is my position?

In physics, an object's location is called its position. How would you define your present position? Are you reading in a chair 10 feet from the door of your room? Perhaps your room is 20 feet from the front door of the house, or perhaps your house is on Main Street 160 feet from the corner of First Avenue. Notice that each of these descriptions requires a reference location. The separation between your position and the reference is called the distance—and if the direction of the separation is included, it's called the displacement.

What is displacement, and how is it different from distance?

The distance from a reference location does not include the direction to that location. Distance has only a magnitude, or size. In the example of a house, the magnitude of the distance of the house with respect to First Avenue is 160 feet. Displacement has both a magnitude and direction, so the displacement of the house from First Avenue is 160 feet west. Or, if you define west as the positive direction (because house numbers are increasing when you go west), then the house's displacement from the reference location, First Avenue and Main Street, could be written as +160 feet. A quantity like this, with both a magnitude and a direction, is called a vector.

How can I represent a vector quantity such as displacement?

A convenient way to represent a vector is to draw an arrow. The length of the arrow represents the magnitude of the vector; its direction represents the direction of the arrow. For example, you might create a drawing where 1.0 inch on the drawing represents 100 feet, and west points toward the left edge of the paper. Then, the displacement of the house from First Avenue would be represented as an arrow 1.6 inches long pointing toward the left.

17

How can I define displacement in more than one dimension?

You very often have to define a displacement in two or three dimensions. As an example, suppose you want to locate a house that is 160 feet west of First Avenue and 200 feet north of Main Street. The displacement is a combination of 160 feet west and 200 feet north. But how are they combined? You can't simply add them because they have different directions. Go back to the drawing with the arrow. Define north as the direction toward the top of the page. Then, add a second arrow 2.0 inches long in the upward direction. The tails of the two arrows are at the same place, representing the intersection of Main Street and First Avenue.

The two arrows are half of a rectangle 1.6 inches wide and 2.0 inches high. Draw lines completing the rectangle. The location of the house would be at the upper right-hand corner of the rectangle. Draw a third arrow, with the tail at the intersection of the other two vectors and the head at the upper right-hand corner. The length of the arrow can be either measured on your drawing or calculated using the Pythagorean Theorem: the square of the length (the hypotenuse of a right triangle) is equal to the sum of the squares of the other two sides. In this case: $1.6^2 + 2.0^2 = 6.56$. Then, length is the square root of that, or 2.56 inches. So, in real life (remembering that 1.0 inch on the drawing equals 100 feet), the displacement would have a magnitude of 256 feet.

What is GPS, and how does it find my location?

GPS, or the Global Positioning System, was developed by the U.S. Department of Defense and made operational in 1993. It consists of three parts: (1) a set of at least 24 satellites, each of which orbits Earth once every 12 hours and regularly broadcasts its location and the time the signal was sent; (2) a control system that keeps the satellites in their correct orbits, sends correction signals to their clocks, and updates their navigation systems; and (3) a receiver that might be mounted in a vehicle or installed in a handheld device. The receiver usually works with an app or other software to deduce your location and display a map of that location and the surrounding region.

How is GPS used?

One very important use of GPS is to send time signals that allow clocks to be calibrated to within 200 nanoseconds (200 billionths of a second). Who would need to know the time this accurately? Among others, businesses that use computers in many parts of the world can use this information to synchronize their computers so the precise time that transactions occurred are known.

GPS also provides navigation information. Hikers use GPS to replace maps (which can get outdated), compasses, and lists of landmarks. Automobile and truck drivers use GPS to replace paper maps and find local businesses such as banks, restaurants, and gas stations. Farmers use GPS to map precise locations in their fields. Not only does this information improve planting, it can be used to mark the locations of areas that need additional insect-control chemicals or fertilizers.

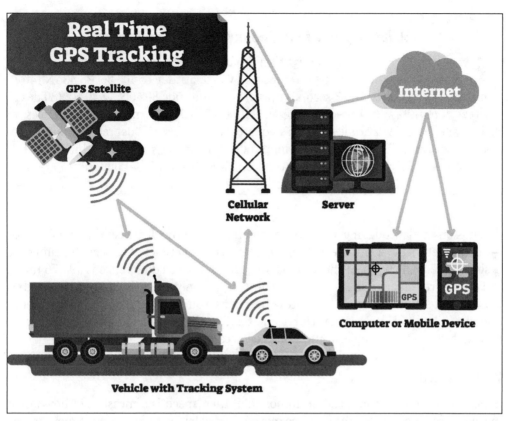

Your GPS device works by receiving signals from satellites that then calculate your position on the planet's surface. To do this effectively, your GPS must be able to receive signals from four satellites at any one time in order to make the needed calculations.

Self-driving vehicles rely on accurate GPS navigation to get from place to place safely. Engineers are also working on GPS systems to improve the flow of automobile traffic and reduce crashes. Each car would have a GPS receiver that would then broadcast its position. This information could be used to change red traffic lights to green if there was no opposing traffic. If two cars were equipped this way, the system could determine the distance between them and their relative speeds. If they were on a collision course, the system would apply the brakes to avoid a crash.

How can I convert latitude and longitude positions into a distance measurement?

A precise conversion is difficult because Earth is not a perfect sphere. Latitude is easier to convert. The circumference of Earth taken over the poles is 24,859.82 miles, which is equivalent to 360 degrees of latitude. Therefore, one degree is equal to 69 miles. So, the north–south distance between two cities 5 degrees of latitude apart would be 345 miles.

Longitude is more complicated. At the equator 360 degrees is Earth's circumference: 24,901.55 miles. But at the poles, it is zero! So, the conversion of longitude to distances

19

depends on the latitude. If you use trigonometry to find the distance, you will find that the circumference at a latitude of θ degrees is the circumference at the equator times the cosine of the angle θ. So, at the latitude 40 degrees, the circumference is 19,076 miles, and one degree of longitude is 53 miles. This result is only approximate because Earth is not a perfect sphere. For more accurate conversions, you can consult government science agencies (like the U.S. National Hurricane Center) that provide online calculators on their websites—for example, at nhc.noaa.gov/gccalc.shtml.

What is speed?

If something moves from position to position, then speed is a measure of how fast it moved. Speed is defined as the distance moved divided by the time needed to move (speed = distance/time). Both change in position (or distance) and time are measured quantities. Frequently, speed is called the time rate of change of distance. For example, if you drive 240 miles in 4 hours, then your speed is 60 miles per hour (abbreviated mph). It is unlikely that you drove the whole trip at a constant 60 mph; this example calculates your average speed. If you were pulled over for speeding and were told you were going 80 mph, you wouldn't be able to avoid a ticket by saying, "But officer, my average speed is only 60!"

What is instantaneous speed, and how is it measured?

If you reduce the time interval between measurements of position, both the distance traveled and the time elapsed are reduced. If the speed is constant, then the ratio of the two does not change. Instantaneous speed is defined as the limit of distance divided by time when the time interval is reduced until it approaches zero. In practice you can't reach the limit, but it is possible to measure positions every thousandth of a second. There are indirect methods of measuring instantaneous speed. For example, police use the Doppler shift (which will be discussed later in this book in the "Waves" chapter)—that is, the change in frequency that occurs when a radio or light wave is reflected from a moving object. Automobile speedometers often use the turning force (torque) on an aluminum disk produced by a magnet that is rotated by the turning car axle.

What units are used to describe speed?

In the English units used in the United States, speeds are usually given in feet per second or miles per hour. In the metric system, meters per second or kilometers per hour are more common. Here are some typical speeds in these four units of measurement.

Description	feet/sec	miles/hour	meters/sec	kilometers/hour
Walking	5.9	4.0	1.8	6.4
Sprinting	33	22	10	36
Fast car	103	70	31	113
Pitched baseball	139	95	42	153
Speed of sound	1,100	750	335	1,207
Space station (orbital speed)	25,667	17,500	7,823	28,164
Speed of light (in vacuum)	980,425,197	670,108,089	299,792,458	1,079,252,849

How was motion perceived in ancient and medieval Europe?

According to Aristotle in ancient Greece, motion was either natural (gentle) or violent. The four terrestrial elements—earth, air, fire, and water—sought their "natural" locations. Earth (rock and metal) and water fell down because they had a property called gravity. (And earth had more of it, so it sank in water.) Fire and air went up because they had an opposite property called levity. Objects in the heavens were made of "aether" and moved in large circles around the planet Earth.

Objects that were thrown or otherwise projected—for example, an arrow shot from a bow—were thought to move because they were given "violent" motion; that is, the bow transferred force to the arrow. Once the arrow was traveling through a medium like air or water, the medium pushed the object along. When the force ran out, the medium would now oppose the motion, and the arrow would fall or sink due to its gravity.

In the sixth century C.E. a philosopher named Philoponus expressed doubt about Aristotle's view on the role of the medium in motion. Another philosopher, Ibn Bajja (a Spanish Arab man known also as Avempace, who died in 1138), discussed the role of the medium as well; whereas Aristotle claimed that motion in a vacuum would be impossible because there would be no medium to provide a push, Ibn Bajja proposed that motion in a vacuum would continue forever because there would be nothing to oppose it.

In the 1300s, philosophers at Oxford University's Merton College and at the University of Paris began to develop ideas of instantaneous speed and acceleration, suggesting that the rate of motion could vary and be measured. Three centuries later, the work of Galileo Galilei and Isaac Newton solidified the concepts of inertia, acceleration, and force.

What is velocity, and how is it different from speed?

Just as you can add direction to a change in position and end up with displacement, you can also specify the direction of motion. The combination of speed and direction is called velocity. Velocity is the displacement divided by the time required to make the change, or the time rate of change of displacement. Velocity is a vector quantity, like displacement.

You might walk at 4 mph north, or a balloon might move at 5 feet per second up. If you assign the variable x to represent the north–south, y to east–west, and z to up–down position, then both the change in position and speed would be positive for movement to the north, east, or up. Your walking velocity would be $+4x$ mph. The balloon's would be $+5z$ ft/s.

Average speed is usually more useful than average velocity in daily life. For example, in a NASCAR race, the starting and finish lines are in the same position, so no matter how fast the cars go, their average velocity is zero because the beginning and ending positions are the same. Instantaneous velocity, however, is usually more useful than instantaneous speed.

How can I be moving even while I'm sitting still?

Suppose you are reading *The Handy Physics Answer Book* while sitting on a moving bus. To another passenger sitting on the bus, your velocity would zero. To a person standing on the sidewalk, however, your velocity would be equal to that of the bus. If you walk forward on the bus, then the person on the sidewalk would now see that your velocity is the sum of the velocity of the bus plus the velocity of your walking.

This example demonstrates the concept of relative motion. Earth itself is in motion—rotating on its axis, revolving around the sun, and moving with the entire solar system around the center of our Milky Way galaxy. Clearly, the frame of reference for any motion must be clearly specified. For most of our daily experiences, Earth's surface is used as the frame of reference, which means that its velocity is zero.

How do you add relative velocities?

In almost any example of motion in our everyday experience, relative velocities can just be added together. If, for example, you are riding on a bus with velocity V_1 and throw a ball with velocity V_2, then the velocity of the ball is simply $V_1 + V_2$.

As it turns out, however, this way of adding relative velocities is only an approximation. Albert Einstein showed that relative speeds actually behave in a more

You might think you are standing still, but Earth is in constant motion, spinning on its axis, orbiting the sun, and traveling around the galaxy and the universe.

complicated, very special way—which is why his theory on this topic is called the Special Theory of Relativity (also called the Theory of Special Relativity). Einstein explained that the speed of a beam of light is the same when measured by any observers, even if they are moving relative to one another; so if you shine a flashlight on the bus rather than throw a ball, the speed of the flashlight beam would be the speed of light for anyone on or off the bus.

How does velocity affect distance and time?

Albert Einstein's Special Theory of Relativity shows that both distance and time change with velocity. Einstein reached this conclusion by noting that one must define methods of measuring both distance and time. The result is that as objects move near the speed of light, their length (in the direction of motion) shrinks, and their internal clocks that measure time run more slowly. The amount of change is described by a factor called γ (gamma), which is always larger than one. Thus, the time given by a moving clock is γt, where t is the time shown by a fixed clock, and length is given by l/γ, where l is the length of the fixed object. The table below shows γ for a variety of velocities (note that c is the speed of light):

Velocity	γ
550 mph	1.00000000003
17,500 mph	1.00000003
0.5 c	1.2
0.9 c	2.3
0.99 c	7.1
0.995 c	10.0

How much does a moving clock slow down?

A clock on a jet plane ($v = 550$ mph) would lose 0.9 milliseconds per year, while one in a space shuttle ($v = 17,500$ mph) would lose 0.9 seconds per year. In 1971 atomic clocks were placed on planes, one of which flew around the world eastbound and the other westbound. The changes in time were measured and agreed with relativity theory. Clocks on GPS satellites must be adjusted for the loss of time. More conclusive tests have been done with fast-moving moving ($0.995c$) muons (a type of elementary particle). Muons, when at rest, decay in 2.2 μs (microseconds). The number of muons, produced high in Earth's atmosphere by cosmic rays, were measured at the peak and base of a high mountain. The ratio of numbers at the two heights showed that the muons lived at 22 μs, which agreed with their measured speed of $0.995c$. From the viewpoint of the muons, they decayed in 2.2 μs, but the height of the mountain was 10 times shorter from the muons' perspective than that measured by observers on Earth. Thus, the predictions of slower clocks and shorter distances have been tested and agree with Einstein's predictions.

ACCELERATION AND FORCE

What is acceleration?

Speed and velocity are seldom constant. Usually, they vary, and acceleration is a description of how that variation occurs. Acceleration is called the time rate of change of velocity. That is, it is the change in velocity divided by the time over which the change occurs. For example, a car might accelerate from 0 to 60 miles per hour. A sports car might do this in 5 seconds, while an economy car might require 9 seconds. Which car has the larger acceleration? Both have the same change in speed, but the sports car requires less time to make the change, so it has the larger acceleration.

What is the difference between acceleration and deceleration?

Like velocity, acceleration is a vector. That is, it includes both magnitude and direction. In speeding up, in which the change in velocity is positive, acceleration is a positive quantity. If the object is slowing down, then the acceleration is negative. If, however, the object is going in the negative direction—for example, a car going in reverse—then speeding up is a negative quantity because the final velocity is more negative than the initial velocity.

Therefore, in physics calculations, the term "deceleration" tends not to be used. Whether positive or negative, speeding up or slowing down, the term acceleration is always used.

What is a force?

Usually, a force is described as a push or a pull—an external agent that can cause or change the motion or velocity of an object.

What can exert a force?

The first answer that might come to mind is that a person can exert a force. A person can throw a ball, pull the string on a bow to shoot an arrow, push a chair across the floor, or pull a wagon up a hill. Many animals can do the same thing, so one might say that living organisms can exert forces.

What if you place a rock on a table? Does the table exert a force on the rock, or does it just block the rock's natural motion toward the floor? If you put a heavy rock in your hand, then your hand would sag downward because you would have trouble exerting the upward force on the rock.

In physics, it's all about the acceleration. Physicists don't think about deceleration but, rather, the velocity is either changing in a positive or negative direction.

The same happens with a table. If the table is made of thin wood, it will also sag. What would happen if the table were replaced with a sheet of paper? The paper might hold small stones, but with a heavy rock, it would tear because it could not exert a strong enough force to hold up the rock. The heavier the rock, the greater the force the table or floor exerts. In summary, inanimate objects can also exert forces.

Forces like these, exerted by humans or tables that touch the object, are called contact forces. What are other contact forces? You might think of rubber bands on slingshots, roads on the wheels of your car, or ropes pulling carts. Water and air can also exert a force. Think of a stick moving down a stream or what you feel when you stick your hand out the window of a fast-moving car.

What are some typical rates of acceleration experienced on Earth?

The table below shows some typical acceleration that might be experienced in a variety of units. The rightmost column, labeled "g," compares the acceleration to that of an object falling in Earth's gravitational field. This ratio is frequently used in both everyday and scientific writing.

Description	feet/ sec^2	miles per hour/sec	meters/ sec^2	km per hour/sec	g
Car from 0–10 mph	29.3	20.0	8.9	32.2	0.9
Sprinter first 1 s of 100 m dash	18	12	5	19	0.5
Commercial aircraft at takeoff	10	7	3	11	0.3
Car stopping from 60 mph	–30	–20	–10	–33	–0.9
Car crash from 35 mph	–1,585	–1,080	–483	–1,739	–49
Object falling due to gravity	32.2	21.9	9.8	35.3	1.0

How is acceleration related to force?

If you have a miniature toy car or even a smooth, hard ball that can roll on a smooth, level surface, you can explore the effects of force on motion. When the toy car or ball is motionless, give it a gentle tap with your finger. Note how it moves. Now, while it is moving, give it a second, third, or even fourth gentle tap. What happened?

You saw that when the toy car was at rest and a force was exerted on it, it started to move in the direction of the force. When a force was applied in the direction of its motion, it sped up. Each additional tap caused it to speed up more. What do you think would happen if you were able to exert a constant force on it while it was moving? It's difficult to do, but try it.

You can conclude from this exercise that a force applied in the direction of motion causes it to speed up or accelerate in the direction of the force. If the direction of motion is called the positive direction, then both the force and acceleration would also be positive.

Now, start the toy car moving, and give it a gentle tap in the direction opposite its motion. Don't tap it so hard that it stops or changes direction. See if you can tap it two or three times, again without stopping the car. What did you observe?

25

From your explorations you could draw the conclusion that when there is only one force exerted on an object, the larger the force, the larger the acceleration.

You should be able to conclude that a force applied in the direction opposite its motion causes it to slow down or accelerate in the direction of the force. If you had defined the direction of motion as positive, then the force and acceleration would both be negative.

What happens when no force is applied? You saw in the beginning that when the toy car started at rest, it remained at rest until you exerted a force on it. What happened while it was moving? It probably slowed down some, with the amount depending on the condition of the toy car and the hardness of the surface. But the amount it slowed was certainly much less than it was when you exerted a force in the opposite direction.

How does mass affect acceleration?

The best way to find out is to have two identical toy cars. Then, add some mass to one of them. For example, you could tape coins to the car. Then, line them up side by side, and use a pencil or ruler to apply the same force to both of them. Again, just give them a tap. Which one goes faster? You probably found that the one without the added mass sped up more.

Your observations tell you that for the same force, the greater the mass, the smaller the acceleration.

What happens if more than one force acts on an object?

You can explore this question with your toy car or ball. Try exerting two forces, like two finger taps in the same direction. You can see that the car or ball moves faster. The forces add together. What happens if the two forces are in opposite directions? You can try pushing each end of a motionless toy car. What happens if both forces are equal? If one is stronger than the other? If they're both equal, the car will remain motionless. That is, it will act as if there is no force on it. If one is stronger than the other, then it will move in the direction of the smaller force, but it will accelerate less. That is, the results will be the difference between the two forces. The combination, addition, or subtraction of forces produces a net force, and it is the net force that affects the acceleration.

What happens to an object's motion if there is no force?

If no force is exerted on an object, the object will experience no acceleration. That means that if the object's velocity is zero, then it will remain zero; and if the object is moving, it will continue to move with the same velocity.

NEWTON'S LAWS OF MOTION

What are Newton's Laws of Motion?

Sir Isaac Newton (1642–1727) summarized how the motions of objects anywhere in the universe relate to the objects' mass, acceleration, and force. The SI unit of force is called a newton in his honor. The three laws can be expressed as follows:

- Newton's First Law: If there is no net force on an object, then if it was at rest, it will remain at rest. If it was moving, it will continue to move at the same speed and in the same direction.
- Newton's Second Law: If a net force (F) acts on an object, it will accelerate in the direction of the force. The acceleration (a) will be directly proportional to the net force and inversely proportional to the mass (m). That is, $F = ma$.
- Newton's Third Law: For every action, there is an equal and opposite reaction.

What is inertia?

Newton's First Law of Motion is sometimes called the Law of Inertia. Inertia is the property of matter by which it resists acceleration. An object that has inertia will remain at rest or will move at constant speed in a straight line unless acted upon by a force.

If moving objects have inertia, why do they slow down and stop even if they are not pushed or pulled?

You may have noticed that the motion of your toy car or ball slowed down after a while, even when you didn't tap it. For Newton's First Law of Motion to be true, something must be acting to exert a force on the moving object to change its velocity. In this case,

In this simple pendulum game called Newton's Cradle, lifting one hanging ball and allowing it to drop and hit the second ball exerts a force on the next ball. The force on the ball at the other end causes it to swing out and back, according to Newton's laws.

that something is the surface on which the toy car or ball is moving, and the force it produces is caused by friction.

What does Newton's Third Law of Motion mean?

Forces are interactions between objects. You need more than one object to have a force. Therefore, if two objects interact, each exerts a force on the other. If you push on a friend's hand, his or her hand pushes back on you. If you stand on a floor, you exert a force on the floor, and the floor exerts a force on you.

Newton's Third Law of Motion states that these two forces are equal in magnitude but opposite in direction. If, for example, you exert an 800-newton force downward on the floor, the floor exerts an 800-newton force upward on you.

The Third Law describes forces on two different objects. Those forces can then be used with the Second Law to find out how the motions of the objects are changed.

How does Newton's Third Law of Motion affect us—when riding in a car, for example?

How do you accelerate your car? You press on a pedal called the accelerator. Does that cause the car to speed up? It cannot because the net force that causes the acceleration must be exerted on the car from outside it.

What does the car interact with? When it is not moving, it is touching the road and thus interacting with it. When you press on the accelerator, the engine causes the wheels to rotate in a direction that, because of friction between the tires and the road, pushes backward on the road. By Newton's Third Law, then, the road pushes forward on the tires, and the car accelerates in its forward direction. What happens if your car is on ice? Often, the friction between the tires and the ice is so small that the wheels cannot exert enough backward force on the ice for the ice to exert the force needed to accelerate the car. With too little friction, the car does not accelerate.

FRICTION

What are the properties of friction?

If you push a couch across a room, it will feel like someone is on the other end pushing back. If you pull on the couch, it will feel like someone on the other end is also pulling. The heavier the couch is, the larger the opposing force will be. The amount of force opposing you will also depend on the surface the couch is on. A carpet will usually have a stronger opposing force than a smooth surface, like wood or tile. These observations summarize the properties of friction between two objects that are in contact. Friction is always in the direction opposite the motion. Friction is greater if the moving object is heavier. Friction is greater if the surface is rougher. But friction does not increase if the speed of motion increases.

How can the friction between two surfaces be characterized?

In most cases, the frictional force is proportional to the force pushing the surfaces together (called the normal force, or N). This proportionality is usually written as $F_{friction} = \mu N$. The Greek letter μ (mu) is called the coefficient of friction. It is important to remember that the two forces, $F_{friction}$ and N, are not in the same direction.

Using the coefficient of friction is only an approximate characterization of frictional force, however. Friction is generally produced at the microscopic level, where atoms and molecules on two surfaces come into contact with one another. When two objects are not moving relative to one another, their surfaces "grip" each other more tightly; once they are moving, their surfaces "slip" across each other more easily. Physicists say that the coefficient of static friction is larger than the coefficient of kinetic (moving) friction, and you probably noticed this when it took more force to start the couch moving than it took to keep it moving.

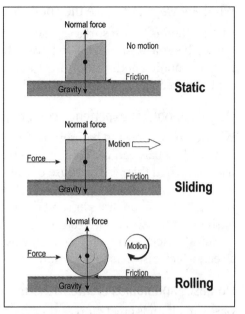

Static friction (top) is the resistance between two surfaces that are not moving against each other, and kinetic friction (bottom and middle illustrations) occurs with objects in motion that are also touching each other. Rolling a round object on a surface results in reduced friction versus two flat surfaces rubbing against each other.

What are some typical coefficients of friction for common surfaces?

Surfaces	Static Friction	Kinetic Friction
Teflon on teflon	0.04	0.04
Oak on oak (parallel to grain)	0.62	0.48
Steel on steel	0.78	0.42
Glass on glass	0.94	0.40

Do smoother surfaces always have less friction?

Perhaps surprisingly, if two metal surfaces are polished until they are extremely smooth, the coefficient of friction between them will actually increase. Experiments show that friction doesn't exactly depend on surface roughness at the atomic scale; rather, friction appears as well to be caused by chemical bonds between atoms on the two surfaces. The fundamental causes of surface friction are still a challenge in physics that has not yet been completely solved.

What are ways to reduce the coefficient of friction?

One method is to use surfaces that have less ability to form chemical bonds. Teflon is one such surface. Another is to have a thin film of oil between the surfaces. The oil will prevent bonding of the atoms in one surface with those of the other. Oil or other lubricants can reduce the coefficient of friction to about 0.1 to 0.2.

Do rolling objects experience contact friction?

Whether it is a bowling ball rolling down the alley or a wheel rolling on the road, there is no sliding between the ball or wheel and the flat surface, so there can be no contact friction. What, then, is the cause of rolling friction? Rolling friction is the result of deformations of either the rolling object or the surface. Think about how much faster a rolling ball stops on grass or sand than on a hard surface. In those cases, the ball has to push down the surface in front of it, which acts the same as contact friction. Or think of riding a bicycle with soft tires. In this case, the tire is squeezed when it contacts the street, which also acts like friction.

Do gases and liquids create friction?

Gases and liquids create forces that resist motion through them similar to, but not exactly like, contact friction. If you stick your hand out of the window of a moving car, you can explore the properties of "air drag." The faster you go, the stronger the force. If your palm is facing up or down, the drag is much smaller than when your palm faces forward or backward, showing that the shape of the object matters. A smaller hand experiences less drag than a larger one. So, drag depends on velocity, area, shape, and the density of the air. These properties are different from other contact friction forces, but the force is still exerted opposite the direction of motion, so it slows down the object. Liquids usually exert similar, but stronger, drag forces.

Is it always best to have as little friction as possible?

Friction is not always a bad thing. When you try to stop a car, for example, having a lot of friction between the car's tires and the road surface is a very good thing. If you try to open a jar of peanut butter, having a high coefficient of friction between your fingers and the lid of the jar helps, too.

How else is friction useful in accelerating things? Consider how you walk or run. Your feet interact with the ground. As long as there is sufficient friction, when you push your feet backward, the ground pushes you forward.

GRAVITY

What is gravity?

In our daily lives, gravity is quite simple: Isaac Newton (1643–1727) explained, using the Universal Law of Gravitation, that it is a force between two objects with mass that pulls the masses toward one another. On a cosmic scale, however, gravity is surprisingly complex: Albert Einstein (1879–1955) explained, using the General Theory of Relativity, that it is the curvature of space caused by massive objects, which makes ordinary objects behave exactly as if there were a force between them.

What is Newton's Universal Law of Gravitation?

Isaac Newton correctly explained that the gravitational force acts not only on objects close to Earth, like the apple that he famously saw drop from a tree. Rather, gravity also acts between any two objects—for example, celestial objects like Earth and the moon, Earth and the sun, or Mars and the sun.

According to Newton's Universal Law of Gravitation, the force of gravity between two objects is proportional to the product of the objects' masses, and it is inversely proportional to the square of the distance between the two objects' centers of mass. So, an object twice as massive as the moon would experience twice as much gravitational force from Earth as the moon, but an object twice the distance from Earth as the moon would experience only one-fourth as much gravitational force from Earth.

How does gravity work without two objects touching?

How exactly does a celestial object, such as the sun, reach out 93 million miles and hold Earth in its orbit? An early idea was that it was an "action at a distance" force; that is, the sun attracted all objects, like Earth, without anything between the two.

Later, when physicists began to understand that magnets create a magnetic "field" around them, the idea arose that massive objects create a gravitational field around them as well. In this model, objects like Earth and the moon exert gravity on one another through the gravitational fields that exist around them.

On what does the gravitational field of an object depend?

Newton demonstrated that the gravitational force on one object caused by another is proportional to the product of the masses of the two objects divided by the square of the distance between them. The gravitational field of an object would then be the force divided by the mass of the object on which the force is exerted. The symbol used for the gravitational field is g, and

The famous story about a falling apple giving Sir Isaac Newton the idea about gravity and how it is a force that acts upon any two objects.

31

G is the symbol for the universal gravitational constant. It is called universal because it is the same for objects made of any material and of any mass—from an apple to a galaxy.

The equation that describes the gravitational field at a specific location is

$$g = -GM/r^2$$

Here, M is the mass of the attracting object, and r is the distance from the center of this mass to the location in space. The gravitational field is a vector quantity, and its direction is toward the center of the attracting object.

What is the difference between a gravitational field and gravitational acceleration?

When Isaac Newton developed his Universal Law of Gravitation, nobody had yet understood the concept of a gravitational field. Instead, Newton found $g = -GM/r^2$ to be the acceleration of an object by another object due to the force of gravity. About 200 years after Newton's work, Albert Einstein developed the General Theory of Relativity; one part of that theory is called the Principle of Equivalence, which explains that you cannot tell the difference between being accelerated by a pulling force and being acted upon by gravity.

How does a gravitational field relate to force?

According to Newton's Second Law of Motion, the force exerted on an object is equal to the object's mass times its acceleration. Because of the Principle of Equivalence, it is appropriate to say that the force of gravity on an object is equal to the object's mass times the gravitational field strength or, written as an equation, $F = mg = -GMm/r^2$.

Has the Principle of Equivalence been tested and confirmed?

Mass defined as $m = F_{net}/a$ is called "inertial mass." Mass defined as $m = F_{gravitation}/g$ is called "gravitational mass." Many physicists, including the Hungarian Baron Roland von Eötvös (1848–1919), have conducted experiments to determine if the two kinds of mass are equal and if they are the same for all materials. Recent experiments have shown that if they are different, the difference is at most only one part in 10^{15}—a factor of a thousand million million! So, in this way at least, the General Theory of Relativity appears to be correct, and the laws of physics in an accelerating reference frame or a gravitating frame are indistinguishable.

What's the strength of gravitational fields of other astronomical objects?

This table shows the properties of the sun, planets, and Earth's moon. It also shows the strength of the sun's gravitational field at each planet and the strength of the planets' gravitational fields at their surfaces. The sun and Jupiter, Saturn, Uranus, and Neptune are gaseous and have no solid surfaces.

Which planet has the strongest gravitational field? Is it also the planet with the largest mass? Why do you think that the sun, which has a mass 1,000 times that of Jupiter, has a surface gravitational field only about 10 times that of Jupiter? As a hint, look at the equation that defines g and see what properties other than mass are involved.

Name	Distance from Sun (10^3 km)	Radius (km)	Mass (kg)	g (Sun) (N/kg)	g (surface) (N/kg)
Sun	0	695,000	1.99×10^{30}		274.66
Mercury	57,910	2,439.7	3.30×10^{23}	3.96×10^{-2}	3.70
Venus	108,200	6,051.8	4.87×10^{24}	1.13×10^{-2}	8.87
Earth	149,600	6,378.14	5.98×10^{24}	5.93×10^{-3}	9.80
Moon	384.4 (from Earth)	1,737.4	7.35×10^{22}		1.62
Mars	227,940	3,397.2	6.42×10^{23}	2.55×10^{-3}	3.71
Jupiter	778,330	71,492	1.90×10^{27}	2.19×10^{-4}	24.80
Saturn	1,429,400	60,268	5.69×10^{26}	6.49×10^{-5}	10.45
Uranus	2,870,990	25,559	8.69×10^{25}	1.61×10^{-5}	8.87
Neptune	4,504,300	24,746	1.02×10^{26}	6.54×10^{-6}	11.15

What is the strength of Earth's gravitational field?

Earth's mass is 5.9736×10^{24} kg, and the gravitational constant G is $6.674 \times 10^{-11} \text{m}^3$ $\text{kg}^{-1}\text{s}^{-2}$. At the surface of Earth, 6.4×10^6 m from the center, the gravitational field strength $g = 9.8$ N/kg. Thus, a kilogram of meat, for example, experiences a gravitational force of 9.8 N toward Earth's center.

What happens when we leave Earth's surface and go into space? The International Space Station orbits at about 320 kilometers above Earth's surface. How strong is the gravitational field at that altitude? Its distance from Earth's center is about 6.7×10^6 m, so $g = 8.9$ N/kg. That's not much different from the gravitational force at Earth's surface.

How about if we go even farther? The moon is 384,000 kilometers from the center of Earth. At that distance, $g = 0.0027$ N/kg. So, the force of Earth's gravity on the same kilogram of meat would be only three thousandths of a newton! How does this very small gravitational field keep the moon in its orbit? The answer is that the moon has a large mass, 7.2×10^{22} kg, so the gravitational force on it is 2.0×10^{20} N. This result is very useful when exploring how orbits work.

Does the gravitational force also follow Newton's Third Law of Motion?

All objects have gravitational fields surrounding them. So, just as the moon has a force exerted on it due to Earth's gravitational field, Earth has a force on it due to the moon's field. According to Newton's Third Law of Motion, these forces are equal in magnitude but opposite in direction. As a result, the moon does not orbit around Earth's center (or Earth around the moon's), but

Jupiter, the largest planet in our solar system, also has the strongest gravitational pull.

the moon and Earth orbit around a point roughly 2,900 kilometers from Earth's center. The sun and other planets also exert gravitational forces on Earth, and Earth does the same on them.

The gravitational fields of the planets, in particular Jupiter, cause the sun to orbit around a point near its surface. Thus, an observer on another planetary system who studied the motion of the sun carefully could determine that planets were orbiting the sun. This method has been used to detect hundreds of exoplanets, which are planets orbiting stars other than the sun.

How does the General Theory of Relativity explain a gravitational field?

Albert Einstein, in the General Theory of Relativity, showed that a gravitational field was actually a distortion of space and time—a four-dimensional structure called space-time—caused by the mass of an object.

Space-time has four dimensions that can be very difficult to visualize; it is possible, though, to see the effect with a two-dimensional model. See, for example, the illustration on the next page and imagine that it's a rubber sheet on which a bowling ball and a golf ball are placed. The sheet is pulled down by the bowling ball, which represents the sun. Earth is represented by the golf ball, which is placed on the sheet and given a push perpendicular to the direction of the sun. This ball "orbits" the sun a few times until friction causes it to speed up and spiral into the sun.

The interaction between mass and space-time has been well described by two sentences written by the American physicist John Wheeler (1911–2008):

Space-time grips mass, telling it how to move.
Mass grips space-time, telling it how to curve.

One can imagine gravity working as if heavy objects are sitting upon a stretchy fabric of space-time. The objects warp the space around them, which can draw other objects toward them—hence, creating a gravitational force.

How has the General Theory of Relativity been tested?

General relativity has been tested in many ways since Albert Einstein first developed it more than a century ago. One particular test is happening every day: the clocks in GPS satellites are adjusted every day according to the theory in order for all of us to find our locations on Earth accurately. The clocks in orbit run faster than those on Earth because they are at high altitudes, where the distortion of space-time is smaller. Both the effects of special relativity (clocks running slower) and general relativity (clocks running faster) must be used to keep the clocks running accurately.

MOTION IN A GRAVITATIONAL FIELD

How can the motion of an object in a gravitational field be described?

As long as the force exerted by a gravitational field, such as Earth's, is the only force on an object, then Newton's Second Law can be used in the form $a = F/m_{inertial}$. But the force due to the field is given by $F = m_{gravitational} \, g$. And, as has been tested by experiment and explained by Einstein's theory, $m_{inertial} = m_{gravitational}$, so $a = g$.

Here is another point to consider. The acceleration a is measured in meters per second squared, while the gravitational field strength is measured in newtons per kilogram. How can these two quantities be equal? The answer can be found by looking at Newton's Second Law again. In the form $F_{net} = ma$, you can see that the units of force, newtons, must be equal to the units in which m times a are measured. The mass, m, is measured in kilograms and the acceleration in meters per second squared. Therefore, a newton must be a kilogram times a meter per second squared. Thus, a newton per kilogram (N/kg) is a meter per second squared (m/s^2).

How do the speed and position of a dropped object vary with time?

As long as the only force is the gravitational force, then the acceleration is the acceleration due to gravity, g. At Earth's surface the value is 9.8 m/s^2. (In the English system, g = 32.2 ft/s^2 or about 22 mph per second.) If the object is dropped from rest at time t = 0, then the velocity at a future time t is simply $v = gt$. That is, the velocity increases by 9.8 m/s each second. If we measure the distance fallen from the position where it was dropped, then at time t, it has fallen a distance $d = 1/2\, gt^2$.

The following table shows velocity and distance fallen for some selected times.

Time	Velocity	Distance	Velocity	Distance
0.10 s	0.98 m/s	4.9 cm	3.2 ft/s	1.9 in
0.15 s	1.42 m/s	11 cm	4.8 ft/s	4.3 in
0.20 s	1.96 m/s	20 cm	6.4 ft/s	7.7 in
0.50 s	4.90 m/s	1.2 m	16.1 ft/s	4.0 ft
1.0 s	9.8 m/s	4.9 m	32.2 ft/s	16.1 ft
2.0 s	19.6 m/s	19.6 m	64.4 ft/s	64.4 ft

How can I use a table of velocity and distance fallen to conduct experiments?

The first three lines on the table above have been chosen so that you can explore your reaction time—the amount of time it takes you to move your hand once your eye has seen something.

To conduct this experiment, have another person hold a ruler or yardstick vertically by one end. Place your finger and thumb at the lower end of the ruler, but don't let them touch the ruler. Have the other person drop the ruler, and you use your finger and thumb to grab it. Note the distance the ruler has dropped in the time it takes you to react to it being dropped. Compare the distance with the first three distances on the table to see if your reaction time is between 0.1 and 0.2 seconds.

Does an object thrown upward have to stop moving before it starts to fall?

When the ball is at its maximum height, does it actually stop moving? At an instant in time, it is indeed at rest, but it doesn't stay motionless for any time interval because the gravitational force keeps acting on it, whether it is moving up or down.

What happens if an object is not dropped but is thrown up or down?

Physicists would say that the ball has been given an initial velocity. But this initial velocity doesn't affect the force of gravity on the ball, so the ball would still gain a downward velocity of 9.8 m/s each second it is in flight.

Suppose the ball is thrown down with a velocity of 2.0 m/s. Then, at time $t = 0$, the time it was thrown down, it would already have a velocity of 2.0 m/s downward. If you use the table above, you see that 0.10 seconds later, it would have a velocity of 2 + 0.98 m/s = 2.98 m/s. Half a second after launch, its downward velocity would be 2 + 4.9 m/s = 6.9 m/s. That is, the initial velocity is just added to the increasing velocity caused by the gravitational force.

Now suppose the ball is thrown upward with the same 2.0 m/s speed. Now we have to be very careful with the sign of the velocity. We have chosen the downward direction as positive. So, the initial velocity of the ball is actually –2.0 m/s.

The gravitational force acts in the downward direction. By Newton's Second Law, that means that the ball will slow down at a rate of 9.8 m/s each second. After 0.1 seconds, it would be going –2.0 + 0.98 = –1.02 m/s. After 0.20 seconds, it would be going –2.0 + 1.96 = –0.04 m/s. It is almost at rest! But the gravitational force is still acting on

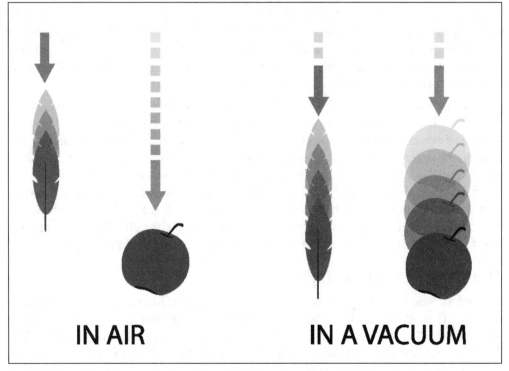

IN AIR

IN A VACUUM

Intuitively, it would make sense that an apple would fall faster than a light object, like a feather. If you remove any resistance caused, for example, by air, the two objects will reach the ground at the same time when dropped!

it, so it continues to accelerate downward. After 0.30 seconds, it would be going –2.0 + 2.94 = +0.94 m/s. That is, it would now be moving downward.

Does a heavier object fall faster than a lighter object?

The Italian scientist Galileo Galilei famously conducted this experiment four centuries ago. He dropped pairs of objects of different weight and mass—a pressing iron and a wooden ball, for example, and cannonballs of different sizes—from the Leaning Tower of Pisa, and they all took the same amount of time to fall.

You can try this experiment yourself: hold a heavy object (like a basketball) in one hand and a lighter object (like a tennis ball) in the other. Hold them at the same height and then let go of them simultaneously. They will hit the ground at the same time!

Why do objects of different mass fall at the same rate?

When you drop any object and let it fall to the ground, the force of gravity is the only force acting on it. All objects at the same distance from Earth's center will feel the same gravitational field strength, no matter how massive they are—so those objects will accelerate at the same rate and thus fall at the same speed.

How does gravity cause tides?

When a large object is in another object's gravitational field, different parts of the large object experience different field strengths. Earth, for example, is in the moon's gravitational field. Even though the moon is less massive than Earth, it still exerts gravitational effects on Earth—and the field strength is greater on the side closer to the moon than it is on the farther side. This is called a tidal effect.

Tidal effects are most clearly observable when there is liquid on the surface of a large object, and Earth has vast oceans across large parts of its surface. The moon's gravitational field exerts a force on that water. Because the field and force depend on distance, the force on the water closest to the moon is strongest. The force on Earth's center is smaller, and the force on the water on the far side of Earth is smallest. So, the water nearest the moon is pulled toward it, and Earth's center is pulled toward the moon more strongly than the water farthest from the moon. For that reason there is a tidal bulge in the water near the moon and another on the far side. The bulges, which are the high tides, are not directly under the moon because Earth's rotation drags the water along with it.

What are the ocean tides on Earth like?

As someone who lives near an ocean will know, there are two high tides and two low tides each day. They do not occur at the same time each day but depend on the phase of the moon. The heights of the tides vary over the seasons as well. Also, Earth's day is shorter than the lunar month, so high tides actually come about two hours before the moon is overhead.

Tides also vary greatly with location. Some of the world's largest tidal variations occur in the Bay of Fundy between the Canadian provinces of Nova Scotia and New

How does the air affect the motion of dropped objects?

The air we breathe—that is, Earth's atmosphere—creates a resistance to objects moving through it called viscous drag. The drag of air adds a force exerted on the object that is in the direction opposite its motion. The net force is the difference between the downward force of gravity and the upward force of air drag.

Brunswick, where the largest recorded range was 17 meters, or almost 56 feet. There have been many proposals to put a dam across the bay and use the tides to generate electricity, but environmental concerns have blocked construction in that bay.

The sun also affects the tides but not nearly as much as the moon does. When the sun, Earth, and the moon are aligned, which happens at both full and new moon phases, the tides are especially high.

What is terminal velocity?

The force of viscous drag on an object increases as the object's speed increases. Therefore, as an object falls through the air and gains speed, the upward (resistive) force from air drag increases. If that force gets strong enough, at some point it will equal the downward force, and there will be no net force on the object. So, according to Newton's First Law, the object's speed will then become constant. This constant velocity is called the terminal velocity.

On what does the air drag and terminal velocity of an object depend?

Air drag depends on the density of the air and the size, velocity, and shape of a moving object. Therefore, the terminal velocity will depend on all of these plus the object's mass.

You can test to see if the air drag depends on the velocity or the velocity squared by doing a simple experiment. You'll need only five filters for a drip coffee maker. The filters must be cup-shaped, with a flat bottom. If the drag depends on velocity, then the terminal velocity will be given by $kv = mg$, where k is some constant. Thus, v will be proportional to m. On the other hand, if the drag depends on the square of the velocity, then the terminal velocity will be given by $kv^2 = mg$, so v will be proportional to the square root of the mass.

The experiment depends on the fact that you can compare two velocities by dropping two objects at the same time, allowing them to fall different distances, and observing whether they hit the ground at the same time. If the drag depends on velocity, then doubling the mass will double the velocity. Stack two filters and hold them above your head. Hold a single filter in the other hand half that distance above the ground. Drop them at the same time and see if they hit the ground together, as they should if two filters fall twice as far in the same time. What did you observe? You probably found

39

that they didn't hit the ground at the same time. Now add two more filters to your stack so you have a total of four. The square root of four is two, so if the drag depends on the square of the velocity, the four filters will now fall twice as fast as the single filter. Try it. Do they hit the ground at the same time?

Air drag is more complicated overall than this simple experiment suggests. At slow speeds the drag constant, k, is larger than at higher speeds. Many studies have been done on the air drag of tennis balls, baseballs, and soccer balls. The fuzz, stitches, and panels have a strong effect on air drag, as anyone who has tried to hit a knuckleball will testify.

How do skydivers and parachutists use air drag to control their speed?

Skydivers can change their body shape, and thus their terminal velocity, by extending their arms or separating their legs. Parachutists can tilt the chutes to control the direction in which they're falling.

How do you measure weight?

Weight is defined as the force of gravity. If you stand on a scale, the scale measures the upward force of the scale on your feet. According to Newton's Third Law, that force is equal to the downward force of your feet on the scale. (Actually, there is a very slight difference because of the rotation of Earth.)

Interestingly, if you stand on a scale in an elevator, you'll see that your "weight" changes as the elevator accelerates. When it is going up and increases its speed, the scale will record a larger weight. The same will happen when you are going down and come to a stop. When it slows down while going up or speeds up while going down, the scale will show a smaller weight. Newton's Second Law can explain these changes using the fact that a net force is needed to accelerate you along with the elevator. (Note that you don't really need a scale to sense these accelerations. You can feel them in your stomach!)

What is weightlessness?

What would happen if the cable holding the elevator in the above example breaks? You and the elevator would both start falling freely, accelerating at g (the strength of the gravitational field). There would be no weight shown on the scale. Thus, in free fall, you and the elevator would both be in the state of "apparent weightlessness."

Strictly speaking, the way something can be weightless is to be so far from any massive object, like a planet or star, that

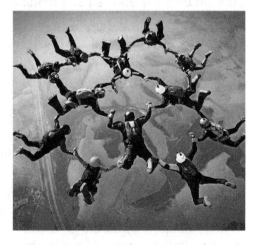

These skydivers can slow their descent somewhat by splaying their arms and legs wide, and then they can go faster by tucking in their arms and legs for less air resistance.

the gravitational field is zero! No human has yet achieved this state. In a satellite the gravitational field is about 80 percent as large as it is on Earth's surface, so the force of Earth's gravity on an astronaut in the satellite is not much smaller than it is on the astronaut on Earth. However, astronauts do experience apparent weightlessness—sometimes called "microgravity"—when they are in free fall or orbit.

CIRCULAR AND ANGULAR MOTION

How can you describe the path of a thrown ball?

You know what happens when a ball is dropped. It accelerates downward, gaining speed at a rate of 9.8 m/s each second. What would happen if the same ball rolled off a table?

To answer this question, you should define a coordinate system. One axis points down; the other, perpendicular to the first, points in the horizontal direction of the rolling ball. The force of gravity acts in the downward direction, causing the ball to accelerate downward, but there is no force in the horizontal direction, so the ball's horizontal speed would not change. Because only the downward acceleration affects its fall, it would hit the ground at the same time that a dropped ball would. The path the ball takes is in the shape of a parabola.

Galileo's principle of relativity (not to be confused with Einstein's special relativity or general relativity) is helpful in understanding this result. Galileo imagined a sailor dropping a ball from a high mast on a moving sailboat. The sailor would see the ball drop straight down, but an observer on the shore would see the ball having a horizontal velocity equal to the velocity of the boat. Therefore, this observer would see the ball's parabolic path. But both would see the ball hit the deck of the boat at the same time. Galileo said that the laws of physics are independent of relative motion. This statement is called the principle of relativity.

What happens when a ball is launched at an angle?

If a ball is thrown upward as well as outward, the ball will have both an initial horizontal velocity and an initial vertical velocity. Again, the horizontal velocity will be constant because there is no force in that direction. Its vertical velocity will be the same as it was when the ball was thrown either down or up. An initial upward velocity is much more interesting, so let's consider that.

We can specify the initial velocity two ways. First is the way we did before, by choosing the horizontal and vertical velocities separately. The second and more useful way is to specify the velocity and direction. Suppose a batter hit a baseball at a speed of 90 mph. This speed is 132 ft/s, or 40 m/s. The angle could be anything from 90 degrees, a vertical pop-up, down to an angle between 10 degrees and 30 degrees that might be called a line drive, to 0 degrees or even a negative angle that would be a ground ball.

The distance the ball travels before hitting the ground depends on both the speed and the angle. If air drag is very small, then the distance is maximum for an angle of 45 degrees. Air drag causes the angle for maximum distance to drop to around 35 degrees. During World War II, extensive tables were calculated so that gunners could find the angle for the gun to achieve the desired distance.

How can circular motion occur?

If an object moves in a circle, such as a planet, a satellite, or even a ball on a string twirled around, then there is no change in speed as it moves around the circle, so there can be no force in the direction of motion. Yet, there must be a force because the direction of the ball is changing, so its velocity is changing. The force must be perpendicular to the motion.

Try to make a ball move in a circle. This works best with a ball the size of a soccer ball or basketball on a smooth (wood or tile) floor. Start the ball moving; then kick it gently in the direction perpendicular to its motion. Try another kick or two. Note that its direction changes in the direction of the kick—the momentary force you placed on the ball. If you could exert a constant force that is always perpendicular to the motion, the ball would move in a circle. Note that the direction of the force is always toward the center of the circle.

Such a force is called "centripetal," or center seeking.

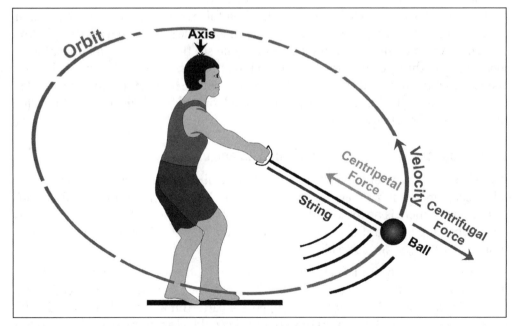

Centrifugal force is the force that pulls a spinning object away from a center axis, while centripetal force is the force pulling that object inward. If you have balance between these two forces, an object will have a stable orbit around another object.

The force required to keep an object moving in a circle depends on three quantities: the mass of the object, the speed of the object, and radius of the circle. The force must be larger for a larger mass, greater speed, and smaller radius of the circle.

How is centripetal force produced?

Centripetal force must be supplied by something that exerts force on the moving object. In a rotating drum ride at a carnival, for example, you stand with your back on the drum. As the drum speeds up, you can feel the force of the drum pushing on your back toward the center of the circle. Then the floor drops down, and only the force of friction between your back and the drum prevents you from dropping down with the floor.

When a car makes a turn, what supplies the centripetal force on it? The road supplies this force. The road is in contact with the tires, and friction between the tire and road is necessary for the force of the road to be exerted on the car. If the road is covered with ice, the friction often isn't large enough, and the car continues to go straight rather than along the curve. A racetrack is often banked so the tilt of the track can supply at least part of the inward force, reducing the need for friction.

If you're sitting in the car, there must be a sidewise force on you so that you stay with the car as it makes the turn. Usually, the friction between you and the seat is sufficient, but sometimes, you have to hold on to the door handle to exert more force.

Why do I feel an outward force when I am moving in a circle?

An object in circular motion is always being accelerated by the centripetal force. That means that this object is not in an "inertial reference frame"—that is, a reference frame where there is no acceleration. Newton's Laws of Motion are strictly correct only in inertial reference frames; in an accelerating reference frame, other effects that are like forces can be felt.

For example, a car that is speeding up, slowing down, or changing direction is accelerating. Therefore, a person in the car experiences other "forces." When the car is speeding up, you feel pushed back into the seat. When the car is slowing down, you feel yourself being pushed forward. When rounding a curve, you feel pushed outward. These are not real forces but are often called "inertial," "fictitious," or "pseudo" forces.

For objects in circular motion, there are two such forces: the centrifugal and Coriolis forces. The centrifugal force is the fictitious outward force you feel when your car is rounding a curve.

What are some effects of the Coriolis force?

Like the centrifugal force, the Coriolis force (named after the nineteenth-century French physicist Gaspard-Gustave de Coriolis) is an inertial or fictitious force that is experienced by objects in an accelerating frame of reference. In this case, the frame of reference is the surface of a rotating object.

An excellent example of the effects of the Coriolis force can be found on our planet Earth. A reference frame, or coordinate system, fixed on Earth is not an inertial frame.

The Coriolis force affects the rotation of winds around high- and low-pressure areas in Earth's atmosphere. Air flows into a low-pressure center. The Coriolis force in the Northern Hemisphere causes large flows of air to be deflected to the right, creating a counterclockwise flow around the center. In the Southern Hemisphere, the deflection and rotation are in the opposite direction. On the other hand, air flows away from a high-pressure center, so the deflection causes a clockwise rotation in the Northern Hemisphere and the opposite in the Southern Hemisphere.

How does gravity cause the planets to move around the sun?

If you twirl a ball on a string above your head, the string provides centripetal force that continually accelerates the ball toward your hand. As a result of this pulling effect, the ball goes around and around in a circle around your hand.

When an object such as a planet is moving in our solar system, the sun's gravity produces a gravitational field that continually accelerates the planet toward the sun. So, gravity provides a centripetal force on the planet, and just like a ball on a string, the planet goes around and around the sun. Unlike the ball, however, the strength of any gravitational field varies depending on the distance from the center of the object producing the field. As a result, different planets go around the sun at different speeds, and they take paths not always shaped like perfect circles.

How did scientists learn about the paths of planets moving around the sun?

In antiquity the planets (the name comes from the Greek word for "wanderer" because they seem to wander across the sky) were assumed to have circular orbits. To account for astronomers' observations, the planets' motion was thought to involve circles attached to other circles. In between 1600 and 1605, Johannes Kepler (1571–1630) made careful studies of the observations of Mars made by Tycho Brahe (1546–1601). He found that a circular path meant Tycho's observations were wrong by a few minutes of arc (much smaller than the moon's apparent size), but he knew that Tycho's work was more accurate than that. After some 40 failed attempts, Kepler finally discovered that the orbit could be described as an ellipse.

What are Johannes Kepler's Laws of Orbital Motion?

Johannes Kepler deduced three mathematical laws that describe the orbits of the planets around the sun. These laws, it

Johannes Kepler (pictured) studied the orbit of Mars using the measurements made by Tycho Brahe. He concluded that Mars's orbit is an ellipse and not a circle. The elliptical shape of the orbits of the planets in our solar system is part of Kepler's First Law.

What is the difference between a scientific law and a scientific theory?

In science, laws summarize observations. They describe phenomena. A theory, on the other hand, explains a large number of observations. Einstein's theory of gravity explains Newton's law of gravitation, which can be used to derive Kepler's three laws of orbital motion. A theory cannot become a law because they are essentially different things.

turns out, can be applied to any object orbiting any other object due to gravity, not just to our solar system.

Kepler's First Law is: Planets orbit the sun in a path shaped like an ellipse, with the sun at one focus of the ellipse.

An ellipse is not a circle, so the gravitational force of the sun is not always perpendicular to the motion. Therefore, the planet's speed changes as it moves around its orbit. When the planet is closer to the sun, it will move faster than it does when it is farther away.

Kepler's Second Law is: In equal times, a planet's orbital path will sweep out equal areas.

In 1595, some years after Kepler discovered the first two laws of orbital motion, he deduced the Third Law based on philosophical and mathematical arguments. It synthesizes two important properties of objects orbiting around a central massive object.

Kepler's Third Law is: The square of the orbital period of a planet is proportional to the cube of the radius of the orbit. The proportionality constant in this relation depends on the mass of the object about which the orbit occurs and the gravitational constant of the universe.

How does Kepler's Third Law relate to the motion of satellites around Earth?

Kepler's Third Law relates the period of a satellite to its distance to the center of Earth. The table below shows the mean radius of the orbit, the altitude above Earth's surface, and the period for some typical satellites.

Satellite	Type of Orbit	Altitude (km)	Mean Radius of Orbit (km)	Period
International Space Station	Equatorial	278–460	6,723	91.4 min
Hubble Space Telescope	Equatorial	570	6,942	95.9 min
Weather Satellite (NOAA 19)	Polar	860	7,234	102 min
GPS Satellite	Equatorial	20,200	25,561	718 min
Communications Satellite	Equatorial	36,000	42,105	1,436 min
Moon	Ecliptic	364,397–406,731	384,748	27.3 days

Most weather satellites have polar orbits, so the view of their cameras will sweep across the entire surface of Earth many times each day. The moon's orbit is aligned with Earth's orbit around the sun, not Earth's equator.

Kepler's Third Law is useful for satellites about other planets and for planets revolving around the sun. For very detailed calculations and descriptions of orbits, Newton's Laws of Motion and Einstein's General Theory of Relativity are useful as well.

STATICS

What is statics?

Statics is a branch of physics that describes the behavior and properties of objects that have volume, shape, and structure.

What does it mean to say an object is static?

The word "static" means "not moving." In the fields of engineering and physics, an object that is static is one that does not move. When static, all the forces acting on a body must sum to zero. That is, the net force on the body is zero, so the object does not move.

What is the name of the supporting force exerted by a chair?

Another term for a supporting force is "normal force." The normal force is always directed perpendicularly out of the surface. The normal force of a chair is straight up if the chair is on a level surface, while the normal force of an incline is perpendicular to the surface of the incline and not perfectly vertical. The term "normal" was derived from the geometrical name for a 90 degrees angle and is not the opposite of "abnormal."

What different kinds of forces can be exerted on an object?

Hold a book in your two hands. What kind of forces can you exert on it? You can pull on the book to try to make it longer, or you can push it to try to make it shorter. The name for the force on the book that pulls on it is a tension force, while the name for the force that pushes on it is a compressive force.

You can also try to twist the book, applying what is called a torsional force. Finally, place one hand on the front cover and the other hand on the back cover. Now push one hand to the right and the other to the left. This places a shear force on the book.

How do we characterize tension and compression?

Tension forces are common. If you hang from a chin-up bar, your arms experience a tension force. Pulling on a rope, wire, or cable also exerts tension on them. Materials experiencing tension will stretch, some more and some less. Materials that stretch easily are called pliant; those that do not are called rigid. Tensile strength refers to how well a material can withstand tension.

The compressive strength of a material is how well it can withstand compressive forces. Many materials will buckle if too great a compressive force is placed on them. Brittle materials will break. The compressive and tensile strength of your bones are almost exactly the same.

How do scientists measure compression and tension?

Scientists have measured the response of materials to tension and compression. The ratio of the applied pressure (called stress) to the change in length divided by the original length (called strain) is called Young's modulus. Its value varies from 10G N/m^2 (10 billion newtons per square meter) in wood to 200G N/m^2 in steel and cast iron. That means that for an equal force on it, steel will stretch (or compress) 1/20 as much as wood. If you would hang a 120-kilogram (264-pound) ball from a 6-meter- (19.7-foot)-long steel cable 2.5 millimeters (0.9 inches) in diameter, it would stretch 7 millimeters (0.3 inches).

Young's modulus also describes the change in length of an object when a compressive force is exerted on it. When you stand, you exert a compressive force on your leg bones, and they will shrink somewhat in length. They won't shrink much; a 70-kilogram (154-pound) person standing on one leg will compress it by only 0.01 percent of its length! The Young's modulus for cartilage, the material between the bones in all parts of your body, is one ten-thousandth as large as that of bone. So, putting the weight of a 70-kilogram (154-pound) mass on a piece of cartilage with same area as the femur in your leg would compress it by 10 percent of its original thickness.

What does shear do to structures?

Scissors, also known as shears, use their two blades to move the object they are attempting to cut in opposite directions. Thus, they exert a shearing force on the object to be cut. Earthquakes often cause land and roads

Think of torsion as twisting an object. When this happens to a building, it can stress the structure, causing it to crack or weaken.

to experience significant shearing forces. Pictures of torn-up roads after earthquakes show how one side of a street moved one way while the other side of the street moved in the opposite direction, tearing up the pavement as the parts moved past one another.

What does torsion do to structures?

A torsion force, usually from winds, is responsible for twisting structures. Buildings, bridges, and towers use cross supports to prevent such forces from damaging the structures. For example, the John Hancock Building in Chicago has cross supports that are easily visible.

CENTER OF GRAVITY

Why do hammers wobble end over end when tossed in the air?

If a baseball is tossed into the air, the ball follows a smooth parabolic path as described earlier in this book. If a hammer is tossed, its path appears much more complex. Why?

All objects on Earth are made of atoms. The mass of the object is the sum of the masses of all the atoms, and the force of gravity on each atom is proportional to the mass of that atom. In a baseball, the center may be made of different materials than the surface, but (if you ignore the stitches) the ball is made up of the same materials regardless what the direction is. That is, the ball is spherically symmetric. The center of gravity in any object is defined as the average location of its weight. Because the mass is distributed evenly throughout a baseball, the center of gravity is located in the center of the ball. However, for an object such as a hammer, with a metal head and a wooden handle, the center of gravity is not directly in the middle. Since more mass is located in the metal head of the hammer, the center of gravity is closer to that point.

According to the laws of motion, the center of gravity of any object follows a parabolic curve when tossed in the air. Indeed, although the ball and the hammer do not appear to have similar motions, their centers of gravity do. If you carefully watch both the center of a baseball and the center of gravity of a hammer, you will see that they both follow parabolic paths when thrown. Because the force of gravity is proportional to the mass of an object, the center of mass is at the same location as the center of gravity.

Where is the center of gravity of a person?

The center of gravity of a person depends on how that person's weight is distributed. The distribution is typically different in adult males and females. Males usually have more upper-body mass while females usually have more mass in the hip region. A typical male's center of gravity is about 65 percent of his height while that for a typical female is about 55 percent.

Try this. Stand facing a wall with your toes against the wall. Now try to rise to stand on your toes. Can you do it? Standing on your toes moves your point of support in front of your center of gravity, so you will tip backward. How do you stand on your toes when you're not against a wall? You naturally move your arms forward or bend forward to move your center of gravity forward.

If you stand with your back against a wall, can you bend forward so the trunk of your body is horizontal? If you're male, you will probably tend to fall forward, while if you're female, you are more likely to be successful. Again, the question is whether or not the center of gravity of your body is above your point of support—in this case your toes—or in front of it.

Why is it easy to tip over some objects?

If you want to tip something over, you'll have to rotate it. As you have seen, rotation means you need to apply a torque. An object sitting on a surface has several forces on it. First, there is the gravitational force that acts on its center of gravity. Second, there is the upward force of the surface. Third, there is friction between the object and the surface that exists only if the surface is on a slant.

Now suppose you exert a sideways force near the top of a box. That creates a torque, and the box begins to rotate. If you now let go, will the box continue to roll, or will it go backward? It depends on the relative locations of the center of gravity and the force of the surface, as seen in Figure 1 below.

If the center of gravity is above the bottom of the box, then the box will return to its upright position. If the center is directly above the corner of the box, then the box won't rotate at all. If the center is outside the bottom of the box, then it will tip over.

What is the best way to prevent something from tipping over?

As you can see, if the center of gravity is outside the base of the object, it will tip. That suggests a general rule: it will be stable if it is low and wide. That is, keep the center of gravity low and the base wide!

Not only is a larger torque needed to tip an object that is wide and low, but the center of gravity has to be lifted higher for the low and wide object. As a result, the work needed to tilt the object is larger.

Figure 2 shows how (on the left) a "tall and narrow" object can be more easily tipped, whereas (on the right) a "low and wide" object is more difficult to tip.

What are the applications of this rule? Autos and trucks that are narrow and have high centers of gravity are less stable against rollover than wider, lower vehicles. Football players and wrestlers are taught to crouch down low and spread their feet apart. This stance follows the "low and wide" rule, making their bodies more stable and difficult to knock over. If you are

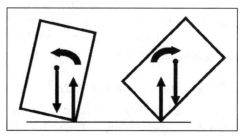

Figure 1.

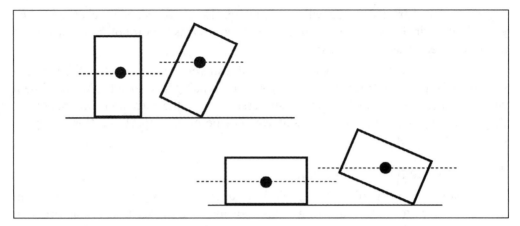

Figure 2.

crouched like this, then to knock you over, your opponent would have to exert a force to lift up your center of gravity and then push you over. If you simply stood upright with your feet close together, you would be "tall and narrow," and your opponent would have less difficulty pushing over you and your center of gravity.

MATERIALS FOR STRUCTURES

What materials were used by ancient peoples to make static structures?

The first materials used were made of stone, especially stones like flint that could be chipped to make sharp edges and points. Ancient peoples discovered metals, either in pure form or in ore, which is mixed with nonmetals. Copper, tin, gold, and iron were well known before 2000 B.C.E.

What is an alloy?

One method of changing the properties of metals is to mix two or more different metals, forming what is called an alloy. Copper and tin, for example, are very soft metals by themselves. Bronze is made by adding tin to copper, and it is a hard metal that can be melted and cast in various shapes, including statues, weapons, and tools. The technology to create bronze was known by about 4000 B.C.E. in what is now Iran and Iraq, with tin that might have come from southern England. Today, bronze is used for (among many other things) bells and cymbals.

What are some common alloys?

Pewter is a tin alloy that was developed in England for plates and cups. It adds copper, bismuth, and antimony to tin. Originally, lead was used, but because it is poisonous, it is no longer found in pewter.

Brass is typically 80 to 90 percent copper with zinc added. Actually, brass was made before metallic zinc was isolated! The zinc ore calamine was melted with copper. Brass is used as a decorative metal and in bullet casings. If aluminum and tin are added, it is resistant to corrosion by seawater.

Titanium is used in many alloys today because it is lightweight and corrosion resistant. Many alloys of both metals and nonmetals are used in electronics. Some LED lamps contain the metals gallium and aluminum and the nonmetal arsenic.

What is the difference between iron and steel?

Soon after bronze was developed, methods were found to convert iron ore, which is iron mixed with either oxygen or sulfur, to pure iron using a charcoal fire. Iron replaced bronze as a material for tools and weapons because it was cheaper and didn't require long trade routes to obtain tin. To keep iron from being brittle, the carbon must be removed. If the iron is heated to red heat and then hammered, the carbon is forced to the surface, where it can be removed. The result is steel, which contains about 1 percent carbon.

Steel is a versatile alloy of iron and carbon because additional materials can be added to change its properties. Very hard steel, called tool steel, contains tungsten, molybdenum, and chromium, among other minor additions. Its hardness can be changed by heating it and then cooling it very quickly by quenching it in water or other liquids. Stainless steel used in knives, forks, and spoons has 18 percent chromium and 10 percent nickel. Other stainless steels use different amounts of these two metals as well as molybdenum and magnesium. Stainless steel does not rust, but it isn't as hard as tool steel.

What determines how much a material will bend?

Try to bend a ruler. Hold one end tightly on your desk and push down on the other end. You are exerting a tension force on one surface and a compressive force on the other. The amount the ruler bends depends on the Young's modulus (Y) of the material as well as its length (L), width (w), and thickness (t). The bending (x) is proportional to the force applied (F) and the length cubed and is inversely proportional to its thickness cubed or, in the form of an equation:

$x = FL^3/(t^3wY)$

What is the difference between 24-karat gold and 18-karat gold?

Pure gold is a very soft metal; you could easily bend a gold rod that is the thickness of a pencil. It can be made harder by adding copper and silver in equal quantities. Gold alloys are measured in karats; 24-karat gold is pure gold, while 18-karat gold has 18 parts gold and 6 parts other metals, and 10-karat gold has 10 parts gold and 14 parts other metals.

That is why the joists supporting a floor are much thicker than they are wide. Typically, wood that is 1.5 inches wide but 10 or 12 inches thick is used. The larger thicknesses are used if the span between supporting walls is longer.

An I-beam is often used to support weight. The beam is in the shape of the capital letter I. The vertical member, tall and narrow, keeps it from bending, while the top and bottom members keep the vertical member from twisting. While I-beams are most often made of steel, wood beams are now used in houses because they are stronger, lighter, and cheaper than steel.

Reinforcing concrete blocks with rebar (steel bars) helps support the concrete and keep it from cracking. Composite materials, such as these blocks, are an example of how combining materials that have compressive strength with those that have tensile strength is an effective strategy in construction.

What are composite materials?

Composite materials have been used since ancient times, when straw was added to clay to make bricks. Brittle materials like concrete have larger compressive strength than tensile strength, so other materials are added to those brittle materials to increase tensile strength. For that reason, reinforcing bars (rebars) made of steel are embedded in concrete. The steel supports the concrete when it is under tension, reducing its tendency to crack and thus keeping it from falling.

Some modern composite materials use carbon fibers that are less than 0.01 mm in diameter, lightweight, and have a very high tensile strength but low compressive strength. They are added to plastics to make them more rigid (by increasing their Young's modulus) without adding weight. Golf club shafts are often made of a plastic filled with carbon fibers, giving the shaft great strength with low weight. By varying the amount of carbon fibers, the flexibility of the shaft can be changed.

How can glass be like a composite material?

Glass, although it isn't really a composite, can also be made less brittle using other materials. The key idea is to keep the surface under compression at all times so it won't crack. Glass sheets are made from the fluid state by cooling. If the surfaces are cooled quickly with strong blasts of air, the process produces a form of glass called case-hardened that was used for shatterproof lenses in eyeglasses. In the 1970s a chemical method was invented that could be used on cold glass. It involves putting the glass into a bath of potassium salts. The potassium replaces sodium in the surface layers of the glass. The larger potassium atoms expand the surface layers so that the interior portion of the glass compresses the surface, keeping it from developing cracks.

BRIDGES

What was the first type of bridge?

The first type of bridge ever used was probably a beam bridge. This bridge was probably just a fallen tree that was used to cross a ravine or a small stream; the tree was probably supported by the riverbed or by a group of rocks. Beam bridges consist of a horizontal roadbed supported by vertical piers on the shores that are planted in the ground. Beam bridges are limited by the resistance of the roadbed to bending.

How can you reduce the bending of a beam bridge?

The simplest way to keep the roadbed from bending is to use a king post. In Figure 3, the downward force of the center of the bridge pulls down on the vertical post. This places the diagonal braces under compression. They transmit the force to the piers. The upward force on the post makes the net force on the post zero. It is under tension.

While the king-post bridge can reduce the bending in the center of the bridge, it can do nothing about bending between the pier and the bridge center. One solution is to add a second vertical post and connect the two by a horizontal member, creating a queen-post bridge. But a method that allows much more support on a longer bridge is the truss.

The ancient Romans, who used stone and concrete in their construction, developed similar methods of translating downward forces to compression forces exerted on the piers on the ends of a bridge. They were famous for the arches used in their massive aqueducts. One such aqueduct, the Pont du Gard, was completed in 18 B.C.E. and was used to carry water a length of 270 meters (886 feet) over the Gardon river valley in southern France.

What is the difference between a dead load and a live load?

To remain static, bridges (and all structures, for that matter) must be able to withstand loads placed on them. A load is simply the engineering term for force. Dead load is the

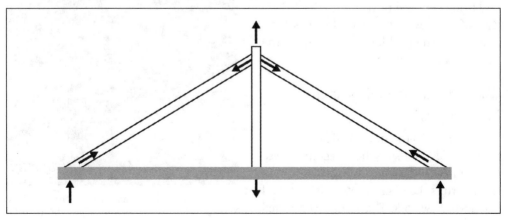

Figure 3.

weight of the bridge or structure itself. The live load is the weight and forces applied to the bridge as a result of the vehicles and people that move across the bridge at any given time. For safety, engineers account for much higher live loads than would normally occur.

What is a suspension bridge?

Suspension bridges are able to span huge distances without supports underneath them because the long cables suspending or holding up the roadbed are draped over a set of tall vertical towers called pylons. The pylons support the suspension cables from which vertical cables are attached that lift the deck. The ends of the suspension cables must be anchored into the ground at each end of the bridge to exert the tension forces on the cables.

The first known suspension bridge was constructed in the seventh century C.E. by Mayans at their capital, Yaxchilan, Mexico. It spanned 100 meters. Almost all of the longest bridges in the world are suspension bridges.

What happened to the Tacoma Narrows Bridge in 1940?

The Tacoma Narrows Bridge in Pierce County, Washington, was a suspension bridge that was almost 6,000 feet long and had a main span of 2,800 feet. When it opened to traffic on July 1, 1940, it was the third-longest suspension bridge in the world. Just four months later on November 7, 1940, it collapsed catastrophically as its main span broke and fell into the water below.

This bridge collapse is especially famous because a video of the collapse was taken, allowing physicists to study what happened very carefully. Eventually, it was concluded that the bridge was not properly built to prevent twisting of the roadbed, and a steady wind in just the right direction caused a torsional force that made the bridge vibrate, then flutter, and then fail completely.

A new Tacoma Narrows Bridge, built with the lessons of the old bridge in mind, replaced the destroyed one in 1950. A second bridge opened next to it in 2007. Neither of the new suspension bridges have had any of the problems that caused the first one's destruction.

What is the longest bridge in the United States?

The longest bridge in the United States, a suspension bridge, ranks as the sixteenth-longest bridge in the world. The Verrazzano-Narrows Bridge runs between Staten Island and Brooklyn, New York. This bridge, completed in 1964, spans 1,298 meters (4,260 feet).

At 4,260 feet, the Verrazzano-Narrows Bridge in New York is the longest suspension bridge in the United States.

The span between the towers of the Mackinac Bridge that links the upper and lower peninsulas of Michigan, at 1,158 meters (3,800 feet), is shorter than that of the Verrazzano-Narrows Bridge. When measured by the distance between the cable anchorages, however, it is the longest bridge in the western hemisphere. The length of the entire bridge, shore to shore, is five miles (eight kilometers).

What is the longest suspension bridge in the world?

The longest suspension bridge in the world is in Kobe, Japan. The Akashi Kaikyo spans a distance of 1,991 meters (6,532 feet). The total length of the bridge is 3,911 meters (12,831 feet). This bridge took 12 years and $3.3 billion to build and is designed to withstand major earthquakes and severe hurricane-force winds. In January 1995, the bridge weathered the 7.2-Richter-scale earthquake that killed 5,000 people; the only damage sustained by this tremendously well-engineered bridge was that one of the piers and anchorages had shifted by three feet. This high-tech bridge uses pendulums within the massive vertical towers to counteract dangerous bridge movement produced by seismic activity. These high-tech mechanisms move against the motion of the bridge, creating greater stability.

What is the newest kind of hybrid bridge?

One of the newest, prettiest, and most economical bridge designs is the cable-stayed suspension bridge. With its sleek lines and thin roadways, it is an excellent bridge for most midspan designs. The Russky Bridge in Vladivostok, Russia, completed in 2012, is the longest cable-stayed bridge in the world, with a span of 3,622 feet (1,104 meters). Cable-stayed bridges suspend the roadbed by attaching multiple cables directly to the deck supporting the roadbed. These cables are then passed through a set of tall, vertical towers and attached to abutments on the ground. Such engineering methods reduce the need for heavy, expensive steel and the massive anchorages that are needed to support suspension bridges.

SKYSCRAPERS

What makes a building a skyscraper?

The name "skyscraper" is an informal term for any strikingly tall building. The first skyscraper had load-bearing outer walls made of stone and concrete. Today, a skyscraper is typically supported by an internal iron or steel skeleton. Skyscrapers are economical in crowded cities because they take advantage of vertical space in areas where land is at a premium.

What challenges are there in building a skyscraper's foundation?

The first and perhaps most important challenge of any skyscraper is to design a foundation that can support the tremendous weight of a large building. The best way is to dig down to the bedrock. This can be as close as about 20 meters (70 feet) in New York City to about 60 meters (200 feet) in Chicago. If the distance is short, holes can be bored and concrete piers can be formed in the holes. More frequently, a caisson is required. This is a large, hollow, waterproof structure that is sunk through the mud, which is pulled into and then out the top of the caisson. A third method is to build a large steel and concrete underground pad that "floats" on the top of a hard, clay layer.

The load that the foundation must support includes the weight of the building, its furnishings and equipment, and the changing load of occupants. In addition to the loads, strong winds must also be considered.

How are the walls and frame of a skyscraper designed?

The walls of early tall buildings were constructed of masonry that supported the weight of the building. The 16-story, 65.5-meter (215-foot) Monadnock Building in Chicago, built from 1889 to 1891, required walls that were 1.8 meters (6 feet) thick at the base. It was so heavy that it sank, requiring steps to be constructed from the sidewalk down to the ground floor. The second half of the building used a steel frame on which masonry was attached, allowing much wider windows to be used.

A skyscraper's steel frames can be bolted, riveted, or welded together. When the fifty-nine-story, 279-meter (915-foot) Citigroup building was constructed in New York City from 1974 to 1977, the frame was bolted together, but later computer models showed that if hurricane-strength winds struck the building, it would be in danger of collapse. As a hurricane moved up the eastern seaboard in 1978, workers hurriedly welded plates over the bolted joints. Luckily, the hurricane moved out to sea, sparing the city.

What are some important safety considerations for skyscrapers?

One major effect of winds on tall buildings is to make them sway back and forth; earthquakes can also create dangerous swaying. While a variety of braces can reduce the sway, they add weight to the building. Another method is now used: tuned mass dampers. The 508-meter (1,667-foot) Taipei 101 skyscraper has a 728-ton damper, suspended from its

92nd floor to its 87th floor. The damper gently moves back and forth, opposing any sideways motion of the building and reducing sway. Dampers, both liquid and solid, are used in tall buildings, towers, offshore oil-drilling platforms, bridges, and skywalks. The 210-meter (690-foot) Burj Al Arab hotel in Dubai has 11 mass dampers.

Another consideration is the safety of occupants in case of fire. Some buildings have entire floors designed to be especially fire resistant so that people can gather there and be safer than on other floors. Transporting large numbers of people into and out of upper floors is a challenge to those who design the elevator systems; stairways can be used in emergency situations, but the simultaneous movement of occupants down and firefighters up the stairways of a skyscraper can cause severe traffic congestion and slow the movement of people.

MOMENTUM AND ENERGY

What is momentum?

In physics, momentum is defined as the product of mass and velocity. Momentum is a vector quantity—it has both magnitude and direction. Consider a football player—a lineman. Linemen are massive. If they can run fast, then their mass times their velocity is their momentum (mv). The momentum is in the direction the player is moving.

What is the relationship between momentum and force?

If a force is exerted on an object for a limited period of time, the product of that force and that time period is called the impulse of that interaction. Momentum and impulse are mathematically the same thing; to change the momentum of an object, you add an impulse. So, if you want to stop a lineman charging toward you with some momentum, you need to give the lineman an impulse in the opposite direction. The more time you have to push against the lineman, the less force you have to use to produce the same impulse.

How does a rocket increase momentum?

A rocket is an excellent example of how momentum is changed in a moving object. A rocket's engine expels gas backward at a high velocity. The gas, which is originally at rest in the rocket, is thus given a large momentum backward. This, in turn, provides a large impulse forward to the rocket, which is propelled forward because its forward momentum has been increased.

Why does a rocket work even if it's in space and can't push off the ground?

A common misconception about rockets is that they take off when their exhaust pushes off the ground. It only looks that way when, for example, a space-bound rocket rises off its launchpad. What the rocket exhaust gas does to the ground has no effect, however, on the rocket itself; it's what the gas does to the rocket that matters, a manifestation of Newton's

59

Third Law of Motion: for every action (the downward impulse of the gas), there is an equal and opposite reaction (the upward impulse on the rocket). The rocket rises slowly as it overcomes Earth's gravity, and it accelerates upward as its momentum increases.

How are cars designed to reduce momentum when they crash?

When large forces are exerted on a car, the chance of damage to the car's contents is much greater than if small forces are exerted. If a car hits a barrier or object, the momentum of the car changes rapidly, and the car could receive a large impulse in a very short period of time, causing a great deal of force to be exerted. As a safety feature, modern cars are designed so that the front, rear, and sides of the car will collapse when struck; this extends the amount of time over which the impulse is delivered, reducing the amount of force that is exerted and keeping the passengers inside the car safer.

How do the safety features inside a car reduce momentum?

Seat belts and airbags in cars are both designed to reduce the momentum of the passengers inside the car in case of a crash.

When a sudden, very large acceleration of the car is detected, the airbags are deployed. A chemical reaction within the bag rapidly fills the bag with gas. The front surface of the airbag speeds toward the passenger at up to 180 mph! But the momentum of the passenger is decreased much more slowly because the airbag compresses upon impact with the passenger and then decompresses, so the force of the collision is lessened. In addition, because the airbag has a large area, the force is not concentrated but spread out. This reduces the force on any single part of the body, further reducing the chance of injury.

Most recent cars also have side airbags to protect passengers from side collisions. However, airbags have caused injuries to small persons. Safer airbags inflate less forcefully, reducing the force on the passengers, who, because they have less mass, require less impulse to be stopped.

How is momentum useful in physics?

When physicists analyze the motion of more than just a few interacting objects at a time, using Newton's Laws of Motion to figure out the nature of those motions be-

With nothing to technically "push against" in space, it might seem odd that rockets could accelerate in space. The explanation is that the force of the combusting rocket fuel itself pushes the rocket, accelerating it forward.

What is an example in sports of impulse and change of momentum?

If you catch a baseball or softball, especially without a mitt, you should move your hand in the direction the ball is moving. If you don't want your hand to hurt when making the catch, you certainly don't move your hand toward the moving ball. No matter how you catch the ball, the change in momentum is the same, and therefore, the impulse your hand gives the ball is also the same. What changes? When you move your hand in the direction of the ball's motion, you increase the time the force of your hand acts on the ball. That means that the force is less.

comes very complicated. Using momentum, it is possible to focus on the beginning and end of a physical interaction, and thus, it is much easier to solve complex problems without becoming bogged down in the details of the interaction.

CONSERVATION OF MOMENTUM

What is the conservation of momentum?

If a group or system of objects has no outside forces acting on it, then the total amount of momentum in all the objects combined will stay the same no matter how those objects interact. Physicists say that the momentum in the system is "conserved." This is the conservation of momentum, and it is one of the most fundamental principles in physics.

How is the conservation of momentum related to Newton's First Law of Motion?

Newton's First Law of Motion states that an object will move at a constant speed in a straight line unless acted upon by an outside force. So, in the absence of outside forces, that object's mass, speed, and direction stay the same—which means its momentum, which is its mass times its velocity, stays the same, too. In other words, the conservation of momentum is like an extension of Newton's First Law of Motion from just one object to any number of objects moving and interacting.

If two objects interact, what might happen to each object's momentum?

Suppose you are sitting in a desk chair with wheels on a smooth floor, and you catch a heavy ball thrown by someone. In this case, you and the chair will roll backward. In other words, the momentum of you and the chair will increase as the momentum of the ball decreases. As long as external forces are zero, then in any interaction, the total momentum of the system—in this case, the ball, you, and the chair—will be constant. The sum of the decrease in momentum of the ball and the increase in momentum of you and the chair will be zero.

If I catch a ball in my hand but I don't move, what happened to the momentum of the ball?

According to Newton's Third Law of Motion, if your hand exerts a force on the ball, the ball exerts an equal and opposite force on your hand. Thus, there must be an impulse on your hand. Does the momentum of your hand change? Of course, your hand is just a part of your body, and if you are standing firmly on the ground, your body is unlikely to move. But your hand does move a little bit, as do the skin, muscles, and blood in your hand, absorbing the impulse of the ball. You may not even notice if your hand and its constituents spring right back after the ball is caught, but momentum was definitely

Why do you brace a shotgun butt against your shoulder when firing in it? To brace for the recoil, which is the result of Newton's Third Law.

transferred. Then, your body produces external force to restore the parts of your hand to their original positions—so after that, the momentum will not have been conserved.

How does shooting a rifle exemplify the conservation of momentum?

If there is only one object in the system, then with no external forces, Newton's First Law says that its velocity will not change, and conservation of momentum says that its momentum will not change. If the momentum was zero, it will remain zero.

If you shoot a rifle or shotgun, you are often told to hold the gun tightly against your shoulder. What's the physics explanation behind this advice? When the gun is fired, the bullet's momentum changes. Its new momentum is in the forward direction. So, according to the law of conservation of momentum, the gun must gain momentum in the opposite direction. It will recoil. If the gun isn't held tightly to your shoulder, its mass is relatively small, so its recoil velocity will be large. When it hits your shoulder, it could cause injury. If, on the other hand, the rifle is tight against your shoulder, then the mass is that of the rifle and your body. The recoil velocity will be much smaller.

ANGULAR MOMENTUM

Does momentum apply to objects that rotate?

Momentum applies to objects moving from one location to another. Since such motion is typically described in straight lines, physicists sometimes call this linear momentum. When objects (such as a spinning top) or systems of objects (such as planets in a solar system) rotate around a central point or axis, quantities apply that are similar to those that describe linear motion. Position is replaced by angle; velocity is replaced by angular velocity; acceleration is replaced by angular acceleration; and force is replaced by torque. The rotational analog to momentum is called angular momentum, and in any system of objects, in the absence of external torques, the angular momentum in the system is conserved.

How does torque take the place of force when measuring rotational motion?

You have had experiences that illustrate how torque works. Suppose you want to push open a door that rotates about its hinges. You know that the speed with which the door opens depends on how hard you push. It also depends on how far from the hinges you push—the farther, the faster. It also depends on the angle at which you push. Pushing at a right angle to the door is much more effective than pushing at a smaller or larger angle. If you push at a right angle, then torque equals the force times the distance from the axis of rotation.

What takes the place of mass when measuring rotational motion?

Mass is defined as the net force on an object divided by its acceleration. By analogy, then, the property that takes the place of mass should be the torque divided by angular acceleration. The property is called rotational inertia or the moment of inertia. It depends not only on mass but on how far the mass is from the axis of rotation. The further the mass is from the axis, the larger the moment of inertia. If you sit on a swiveling stool or chair while holding heavy weights, the further you extend your arms, the more difficult it is for someone to start you rotating. That is, it will require more torque to achieve the same angular acceleration.

How does angular momentum work in rotational motion?

The angular momentum of a rotating object is proportional to the product of its moment of inertia and its angular velocity. If there is no external torque on the object, then its angular momentum does not change.

An object with linear momentum that has no external forces on it cannot change its mass, so its velocity is constant. But a rotating object can change its moment of inertia, so, even without external torques, its rotational speed can be changed.

What is the significance of the conservation of angular momentum?

In the absence of external torques, if the moment of inertia of a rotating object gets smaller, its rate of rotation speeds up and vice versa. This applies especially to astrophysical systems, where strong external torques are rare. For example, when a big star collapses into a tiny white dwarf or neutron star, it can start spinning around dozens or even hundreds of times per second, creating a pulsar. In the other direction, the length of a lunar month has increased as the moon's distance from Earth has gradually increased over billions of years.

How is angular momentum used in sports?

Let's explore two different sports. In platform diving, a diver pushes off the tower, and thus, the platform exerts a force on her (Newton's Third Law). If she isn't standing straight up, the force also exerts a torque on her, which starts her rotating. If she pulls her arms and legs in, then her mass comes closer to her axis of rotation, and her speed of rotation increases. To slow this rotation, she can extend her arms and legs. With good timing, she can hit the water with a bare minimum of rotation.

Put a toy gyroscope on a stand and spin the wheel. Gravity pulls down on the center of gravity of the gyroscope, creating a torque on the axis of rotation, causing it to rotate downward.

Next, let's consider a figure skater. She can start spinning on the point of one skate by pushing on the ice with the second skate. Again, the force of the ice exerts a torque, and so her rotational speed increases. She can extend her arms to slow the rotation or pull them in as close as possible to attain a very high spin rate.

What happens when an axis of rotation tilts?

A toy gyroscope contains a rotating wheel. If you put it on a stand, the axis moves in a circle. Why? The gravitational force pulls down on the center of mass of the wheel. Thus, the gyroscope begins to rotate downward. The effect of this new torque is to cause the axis to change direction. This kind of tilting is called precession.

Precession is also important for bicycles and motorcycles. If the cycle starts to tip to the right, then the rotating front wheel's axis will rotate, and the wheel will turn to the right, helping to keep the cycle from tipping over.

ENERGY

What is energy?

An object with energy can change itself or its environment. That's a pretty abstract definition; energy is a challenging concept to define because it can take so many different forms, and objects can have energy in many different ways.

What are some of the ways an object can have energy, and what changes can it cause?

A speeding car has energy; think what damage it can do if it hits a wall or another car. The energy of motion is called kinetic energy. A rotating wheel also has energy; if you try to stop a spinning bicycle wheel with your hand, it may hurt you. This kind of energy is called rotational kinetic energy.

A compressed spring or a stretched rubber band can cause a stone to move. The energy in the squeezed spring or stretched band is called elastic energy. A variety of other forms of energy can also be stored in a material. The random motion of the atoms that make up the material is kinetic energy. A measure of the amount of the kinetic energy in the random motion of the atoms is called temperature; the more energy, the higher the temperature. Kinetic energy in the random motion of atoms in a material is called thermal energy. If you charge or discharge a battery, like the one in your cell phone, you change the chemical composition of the battery materials. When you charge it, you increase its chemical energy. You can also increase the chemical energy of your body by eating. Even mass has stored energy; splitting the nucleus of a uranium atom results in elements that have smaller mass but a large amount of kinetic energy.

How can energy be transferred?

Any energy transfer involves a source, whose energy is reduced; a means of transferring the energy; and an energy receiver, whose energy is increased. It's convenient to use diagrams to keep track of the source, the transfer, and the receiver. (See Figure 4.)

For example, if a moving pool ball collides with another ball, it can transfer all or part of its kinetic energy to the other ball. Transfer of energy by this kind of mechanical interaction is called work. The moving ball does work on the stationary ball, and its kinetic energy is reduced. The kinetic energy of the stationary ball is increased by the work done on it.

When a slingshot does work on the stone, its stored elastic energy decreases, and the kinetic energy of the stone increases.

When you throw a ball, your stored chemical energy is reduced, work is done on the ball, and the ball's kinetic energy is increased.

What energy transfers are involved when a ball is tossed?

If you lift an object like a ball, you increase the energy in Earth's gravitational field. Energy is transferred from you to the field, resulting in a decrease in your stored chemical energy (see Figure 5). Suppose you toss the ball up. When you toss it, you do work on the ball, transferring energy from your body to the ball, increasing its kinetic energy as well as the gravitational field energy. Once you let go of the ball, it continues to rise, but its velocity decreases as the gravitational field energy increases and the ball's kinetic energy decreases. It reaches its maximum height; at that instant, the kinetic energy is zero—all energy is in the field. On its way back down, it speeds up, so its kinetic energy increases, but the energy stored in the gravitational field decreases. While you are not touching it, the sum of the kinetic energy of the ball and the gravitational field acting on it is a constant. Energy changes from one form to the other and back again.

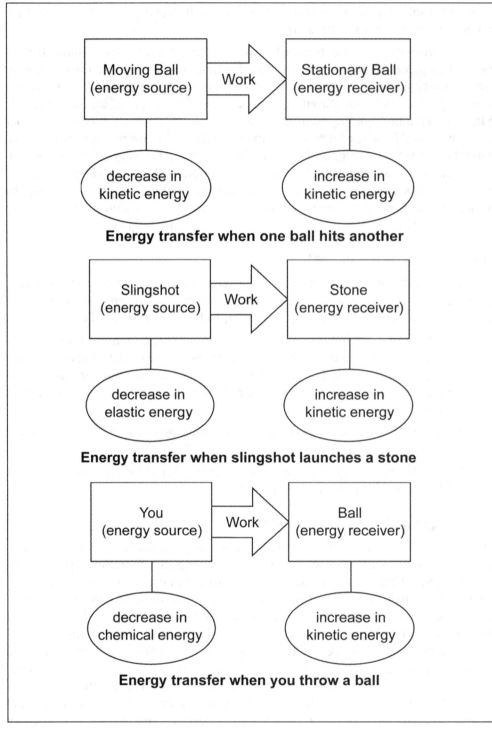

Energy transfer when one ball hits another

Energy transfer when slingshot launches a stone

Energy transfer when you throw a ball

As you catch it, stopping its motion and thus reducing its kinetic energy to zero, the ball does work on you, but your stored chemical energy does not increase.

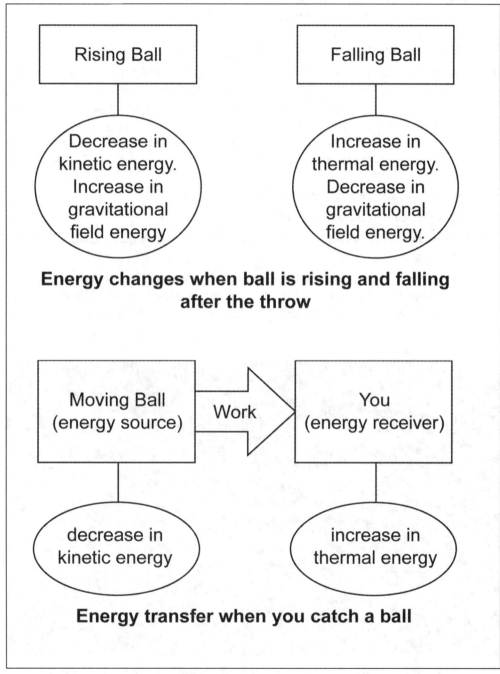

Figure 5.

Does all energy transfer involve doing work?

Physicists refer to the transfer of energy through mechanical interaction as "doing work" when the object receiving the energy is displaced. It is possible to transfer energy without doing work but, rather, with heat.

How does a hot object transfer thermal energy?

Rub a pencil eraser on the palm of your hand. Then quickly put the eraser against your cheek. You probably found that both the eraser and your hand became warmer—that is, their thermal energy increased. What happens to the increased thermal energy of the eraser and your hand? A short time later, they will have cooled. They have transferred their thermal energy to the cooler surroundings. Energy transfer that results from a difference in temperature is called heat.

What are the mechanisms to transfer thermal energy?

Thermal energy can be transferred in three ways:

1. *Conduction* occurs when two objects are in contact, like when you put your hand in warm water.

2. *Convection* is the motion of a fluid, usually air or water. The fluid is heated by the hotter object and then moves until it contacts a colder object, and it heats that object. You can think of convection as two instances of conduction in sequence.

When a moving pool ball collides with another one, all or part of its kinetic energy will be transferred to that ball, which is what makes this fun game possible.

3. *Radiation* is electromagnetic waves. When these waves are emitted by hotter objects and absorbed by colder ones, thermal energy is transferred. You can feel radiation if you bring your hands near (but not touching) a hot radiator or electric burner on a stove. The kind of radiation that we humans can feel as heat is called infrared radiation.

How does heat move?

Heat always flows from hotter objects (energy sources) to cooler objects (energy receivers). If an object is warm, it has an increased amount of thermal energy compared to its surroundings, so it transfers this energy to a colder object or its colder surroundings. The sun, for example, heats our planet Earth by radiation. Earth, in turn, is warmer than the space around it, so it radiates energy into space. When that energy reaches still other objects, they are made warmer, and so on through the entire universe.

Is there a difference between work and energy or between heat and thermal energy?

Energy, whether kinetic, stored, or thermal, is a property of an object. Gravitational field energy, for example, is a property of the gravitational field. Work and heat are means of energy transfer. Work is transfer by mechanical means. Heat is transfer between two objects with different temperatures. Examples of work include throwing a ball, a slingshot launching a stone, and ball being caught in a mitt. Examples of heat include your hand getting warm in warm water, a bottle of soda being cooled in a refrigerator, and Earth's surface being warmed by sunlight.

CONSERVATION OF ENERGY

What is the conservation of energy?

We often hear the term "energy conservation" to mean that we should use less electricity, gasoline, or other energy sources in our daily lives to help conserve resources and protect the environment around us. Physicists, on the other hand, refer to the "conservation of energy" in the same sense that we refer to the conservation of momentum: it is a fundamental principle of physics and nature.

The principle of conservation of energy states that as long as no objects are added to or removed from a system or group of objects, and as long as there are no interactions between the system and the rest of the universe, then the total energy of the system does not change.

How is a block of wood on a table an example of the conservation of energy?

Think about a block of wood on a table. If you push the block along the table, you are doing work on the block and transferring energy from your body to the block. The block

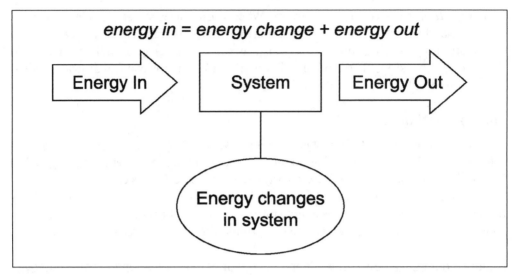

Figure 6. While energy in a system can change in nature, the total energy of a system will always remain the same.

starts moving, but when you stop pushing, it quickly slows and comes to a stop. Where did its kinetic energy go?

It turns out that, although the temperature change was probably too small for you to detect, the friction between the block and table changed the kinetic energy into thermal energy. Thus, the decrease in kinetic energy of the block was accompanied by increased thermal energy in both the block and the table. The energy was conserved; it just changed forms.

Scientists have made careful measurements of energy in a variety of forms and have always found that energy is neither created nor destroyed. In other words, the energy put into a system always equals the energy change in the system plus the energy leaving the system.

What is the difference between conservation of energy and conservation of kinetic energy?

Kinetic energy, which is the form of energy contained in moving objects, is not always converted to other forms of energy when objects collide. In those cases, kinetic energy is conserved, and the collisions are called "elastic." These are the kinds of collisions where objects bounce off one another without sticking or experiencing much friction. When objects do stick together or rub against one another for a period of time, kinetic energy can be transformed into thermal energy, and the collisions are called "inelastic."

What are examples of elastic and inelastic collisions?

The toy Newton's Cradle has a set of steel balls that swing on strings. In these elastic collisions, kinetic energy is almost totally conserved. The incoming ball stops when it hits a second ball, and the second ball moves away with the same speed, and thus kinetic energy, as the incoming ball had. Billiard balls (or pool balls) colliding on a pool table provide another case where kinetic energy is conserved.

An example of an inelastic collision is throwing a snowball into a snowman. The snowball doesn't bounce but sticks to the snowman. Another example would be dropping a golf ball into a mud puddle; rather than bouncing back up, the golf ball lands into the soft mud and just stays there.

How are conservation of energy and conservation of momentum used in everyday life?

These two physics principles are most often used to examine what happens when two or more objects collide. Momentum is conserved if there are no external forces, and it changes very little if the forces during the collision are much greater than the external forces. For example, if two cars collide, the two cars are the system. The forces between them are much larger than the forces on the wheels that come from outside the system, so momentum is conserved. Is energy conserved? While total energy is conserved, the kinetic energy before the collision is much greater than the energy afterward. Much of the energy goes into bending metal and breaking glass and plastic. Automobile crash reconstruction is a way of using conservation of momentum to figure out the speeds of one or both of the cars before the collision.

Who further developed the idea of conservation of energy?

The physicist Benjamin Thompson (1753–1814) was born in Massachusetts, but because he opposed the American Revolution, he left for England and was knighted by King George III and given the title Count Rumford. While in America, he spied for the British.

Who first investigated the ideas of conservation of momentum and conservation of energy?

Isaac Newton (1642–1727), considering collisions, first described momentum as the product of mass and velocity, but he called it "the quantity of motion." It took 150 years from the first statement of principles until the terminology was worked out. Collisions also inspired the Dutch physicist Christiaan Huygens (1629–1695), who wrote that in the collision of two perfectly elastic spheres, the sum of what we today call kinetic energy would not be changed by the collision. The German scientist Gottfried Wilhelm Leibniz (1646–1716) gave the name *vis viva* in 1695 to kinetic energy. But how could the conservation of *vis viva* be extended beyond elastic collisions? Finding the answer to this question took more than 150 years!

While in England, he spied for the French and was a counterspy for the British. He moved to Bavaria, now part of Germany, and became minister of war, among other duties. Because he ran an orphanage and wanted to save money, he studied heat and invented many items, like an efficient stove and a coffee percolator. A long series of experiments led Rumford to conclude in 1798 that thermal energy was nothing more than the vibratory motion of what we know today as the atoms that make up the material.

About 20 years before Rumford's work, the French scientists Antoine-Laurent de Lavoisier (1743–1794) and Pierre-Simon Laplace (1749–1827) showed that heat produced by a guinea pig after eating was very close to the heat produced when the food was burned. The development of steam engines by James Watt (1736–1819) and others simulated studies of the relationship between work done and heat produced and how to make engines more efficient. Around 1807 the word "energy" was used with its modern meaning.

In 1842 a German physician, Julius Robert von Mayer (1814–1878), proposed that all forms of energy are equivalent and that the sum of all forms is conserved. He wrote in general, qualitative terms, although in later essays, he included quantitative evidence based on the work done when a gas was heated. But his work resulted in little recognition until the end of his life.

How did James Joule and Hermann von Helmholtz contribute to understanding the conservation of energy?

About the same time that Julius Robert von Mayer was proposing his ideas about energy, a British amateur of science, James Prescott Joule (1818–1889), began a series of experiments designed to determine the relationship between work done and thermal energy increase that resulted in heat transmitted to the outside. He explored electric generators, the compression of gases, and stirring water. His experiments lasted 18 years. As he continued to publish his results, they were taken more and more seriously. Today, the metric system's unit of energy is named the joule in his honor.

The German physicist and physiologist Hermann von Helmholtz (1821–1894) developed a mathematical description published in 1847 that showed precisely how energy was conserved in many fields, including mechanics, thermal energy and heat, electricity and magnetism, chemistry, and astronomy. With his results, the scientific community recognized the great

Julius Robert von Mayer believed in the conservation of energy and that all types of energies were really all the same.

achievement of Rumford, von Mayer, Joule, and others and fully accepted the physical principle of the conservation of energy.

What does energy efficiency mean?

Energy efficiency refers to useful energy compared to waste energy. In most cases, thermal energy is not a useful form of energy for a system. You want the energy content in the gasoline in your auto to give it kinetic energy, not to make it hotter. The cooling system uses a mixture of water and antifreeze to cool the engine and warm the radiator, where air flowing through it is heated, thus cooling the fluid. The thermal energy in the heated air is often called rejected or waste energy. An auto is about 20 percent efficient. That is, only one-fifth of the energy in the gasoline is converted into the kinetic energy of the auto. In addition to the hot air from the engine, tires get warm from flexing, and the brakes get hot when they are applied. All this thermal energy is rejected or waste energy.

What is the energy efficiency of the human body?

Your body also uses only a portion, about 25 percent, of its food energy to move your limbs, control your organs, and maintain the processes that keep you alive. Your body is cooled by contact with the air or by evaporating liquid—either perspiration or the humid air expelled by your lungs.

What are some units of energy?

The SI unit for energy is the joule (J), but energy is often measured in other units. The calorie (cal) and the kilocalorie (kcal) or food calorie are used both for chemical energy stored in foods and to measure heat. The British thermal unit (BTU) is most often used to measure heat in homes and industries. The kilowatt-hour (kWh) is used to measure electrical energy. A completely unofficial but useful unit is the jelly doughnut (JD), the energy in a medium-sized, jelly-filled doughnut. It helps relate all these units to a tasty treat.

Why is it important to increase energy efficiency?

Means of increasing the efficiency of auto and home appliances is an active area of research as nations try to conserve as much of the produced energy as they can.

The furnace in your home, for example, converts the chemical energy in oil, gas, or electrical energy into thermal energy, either of air or water, depending on whether you have forced-air heat or hot-water radiators. But not all the energy goes into heating the house; some leaves through the chimney as rejected or waste energy. Heating systems used to be about 60 percent efficient. Newer systems can be as much as 95 percent efficient.

How do different units of energy relate to one another?

The table below illustrates conversions among the units. To use it, read across. For example, 1 J = 0.239 cal, 1 cal = 4.186 J, 1 JD = 239 kcal.

	joule (J)	calorie (cal)	food calorie (kcal)	British thermal unit (BTU)	kilowatt-hour (kWh)	jelly doughnut (JD)
J	1	0.239	0.000239	0.000949	0.000000278	0.0000001
cal	4.186	1	0.001	0.00396	0.00000116	0.00418
kcal	4,186	1,000	1	3.96	0.00116	4.18
BTU	1,055	253	0.253	1	0.293	0.001055
kWh	3,600,000	859,000	859	3.41	1	3.6
JD	1,000,000	239,000	239	949	0.278	1

How much energy is there in commonly used fuels?

In comparing fuels used to heat a home (natural gas, electricity, fuel oil, and wood), more has to be considered than just the cost of the fuel per MJ of energy. The furnaces that distribute the heated air or water have quite varied efficiencies themselves.

The following table gives the energy content and cost of some common fuels in various units: million joules (MJ) per liter, per kilogram, and per dollar.

Fuel	MJ/liter	MJ/kg	Common Units	Cost (2019)	$/MJ	MJ/$
Gasoline	35	47	121 MJ/gal	$2.60/gal	0.021	47
E85 (ethanol)	22	38	76 MJ/gal	$2.50/gal	0.033	30
Diesel	39	48	135 MJ/gal	$2.70/gal	0.020	50
Natural gas	0.039		1,100 MJ/mcf (1,000 ft^3)	$3/mcf	0.003	370
Electricity			3.6 MJ/kWh	$0.13/kWh	0.036	28
Fuel oil	38		148 MJ/gal	$2.75/gal	0.019	54
Wood		18	36,000 MJ/cord	$250/cord	0.007	140
Coal		12	12,000 MJ/ton	$40/ton	0.003	330
Uranium (not enriched)		540,000				
Candy bar		2.1	0.12 MJ/1 oz. bar	$1/bar	8.4	0.12

POWER

What's the difference between energy and power?

Power is the term used to describe the rate at which energy is produced or consumed over a given period of time. Suppose you climb the stairs to the second floor. Whether you run or walk, because you have gone up the same distance, the increase in the gravitational field energy will be the same. The difference is the rate at which the energy has changed. The rate—that is, the change in energy divided by the time taken—is the power.

What is a watt?

A watt is a unit of power, named after Scottish inventor James Watt (1736–1819), that is commonly used in household electrical devices. One watt is one joule per second. A kilowatt is 1,000 watts, or 1,000 joules per second.

What are some typical power levels of physical phenomena?

The following table shows some typical power levels.

Unit	Example
femtowatt (10^{-15} watt)	10 fW — approximate lower limit of power reception of digital cell phones
picowatt (10^{-12} watt)	1 pW — average power consumption of a human cell
microwatt (10^{-6} watt)	1 μW — approximate consumption of a quartz wristwatch
milliwatt (10^{-3} watt)	5–10 mW — laser in a DVD player
watt (10^0 watt)	20–40 W — approximate power consumption of the human brain 70–100 W — metabolic rate used by the human body 500 W — power output of a human physically working hard 909 W — peak output power of a typical healthy human adult during a 30-second cycle sprint
kilowatt (10^3 watts)	1.4 kW — approximate power received from the sun on one square meter at the distance of Earth's surface 1–2 kW — approximate peak power output of a human athlete 2.7 kW — approximate power usage per person in the world in 2018 10.3 kW — approximate power used per person in the United States in 2018 2 kW — rate of heat output of a large electric teakettle 40–200 kW — approximate range of power output by typical automobiles
megawatt (10^6 watt)	1.5 MW — peak power output of a large wind turbine 2.5 MW — peak power output of a blue whale 3 MW — mechanical power output of a diesel locomotive 16 MW — average power consumption of a commercial jumbo jet liner 140 MW — electrical power output of a typical nuclear power plant
gigawatt (10^9 watts)	2.1 GW — peak power generation of Hoover Dam 4.1 GW — installed capacity of the world's largest coal-fired power plant 18.3 GW — approximate peak electrical power generation of China's Three Gorges Dam, the world's largest hydroelectric power plant
terawatt (10^{12} watts)	50–200 TW — rate of heat energy released by a strong hurricane

Unit	Example
petawatt (10^{15} watts)	4 PW — estimated total heat flux transported by Earth's atmosphere and ocean away from the equator toward the poles
	170 PW — approximate total power received by Earth from the sun
yottawatt (10^{24} watts)	380 YW — approximate luminosity of the sun
higher	5×10^{36} W — approximate luminosity of the Milky Way galaxy
	10^{40} W — approximate luminosity of a quasar
	10^{45} W — approximate luminosity of a gamma-ray burst

Why do the labels of some light bulbs give different numbers of watts?

In the twentieth century, most of the light bulbs used in homes were incandescent light bulbs. They were so common that people started thinking of the electrical power used by these bulbs to represent how brightly the bulbs shined. As compact fluorescent bulbs and light-emitting diode (LED) bulbs became more widely used in the twenty-first century, the light bulb manufacturers showed both the amount of electricity used by the bulbs and the number of watts an incandescent bulb would use to produce the same amount of light. A modern LED bulb, for example, uses only about 10 watts of electric power to produce the same amount of light as a 60-watt incandescent bulb—a significant savings of electric energy and cost.

What is a horsepower?

Horsepower is a unit of power created by James Watt. The value for a unit of horsepower was determined after Watt made an extensive study of horses pulling coal. He originally determined that the average horse was able to lift 33,000 pounds of coal one foot in one minute. The conversion between watts and horsepower (hp) is that 1 hp = 746 W = 0.746 kW.

In the United States, automobile engines are usually rated in hp, while in the rest of the world, kilowatts are used. In a hybrid or electric car, the power of the gasoline engine is usually measured in hp, while the electric motor is measured in kW.

What are the biggest sources of energy in the United States?

In 2018, according to information from the U.S. Energy Information Administration, the United States consumed 101 quadrillion BTU of energy. About 33 percent of that energy was produced with natural gas, 24 percent with petroleum (oil), 16 percent with coal, 9 percent with nuclear power, 6 percent with natural gas liquids, and 12 percent with renewable sources such as wind, water, solar, geothermal, and biomass.

How much energy is used—and wasted—in the United States?

In 2018, the United States used 101 quadrillion BTUs of energy, which is about the equivalent of burning 840 million gallons of gasoline. Much of that energy, however, is wasted. Electrical generation is very inefficient, as only 32 percent of the energy

Scottish inventor James Watt came up with the term "horsepower," which is equal to 746 watts, the amount of energy it takes an average horse to pull 33,000 pounds of coal one foot in one minute.

from the source (such as coal, natural gas, oil, or nuclear) is transformed into electrical energy. The transportation industry is especially wasteful in terms of energy, with 75 percent of the energy (mostly petroleum products, like gasoline) it uses being wasted as heat exhaust. About 20 percent of the energy generated for residential and commercial use is also wasted.

What might be the consequences of automobiles powered by electricity rather than gasoline?

As the use of hybrid electric and all-electric cars and trucks increases, the use of petroleum products will decrease, reducing the need for oil drilling and refining and reducing wasted energy and global carbon emissions into the atmosphere. However, to reduce the overall environmental impact of automobiles, the fuels used for electrical generation need to change as well to those that produce less carbon emission.

What are fossil fuels?

Fossil fuels are carbon-rich substances found on or near Earth's surface that produce large amounts of energy when they are burned. They include coal, petroleum (oil), and natural gas. Almost all of these fuels were made by or made of fossilized organisms that lived many millions of years ago; as a result, they cannot be easily created from raw materials today, are limited in amount, and must generally be mined or retrieved. Once fossil fuels are used up, they cannot be replaced within a few years or even a few centuries.

What is green energy?

When fossil fuels are burned to generate energy, they produce substantial amounts of carbon dioxide, which contributes extensively to global warming and climate change, as well as other waste products that contribute extensively to pollution and environmental degradation. People often use the term green energy to refer to energy sources that do not produce pollutants as fossil fuels do; these primarily include solar, wind, and water (hydroelectric) power.

Is nuclear energy green energy?

Nuclear power plants do not produce greenhouse gases the way fossil-fuel power plants do. However, they produce nuclear waste, which is very dangerous and can last for thousands of years, and the amount of nuclear fuel (usually uranium, which is mined, and its by-products) is limited. Also, when nuclear power plants fail, they can cause devastating impacts; for example, the nuclear disasters of Chernobyl, Ukraine, and Fukushima, Japan, have each rendered entire cities uninhabitable.

What does sustainable energy mean?

Sustainable energy sources, primarily wind, water, and solar energy, ultimately receive their energy from the sun and therefore will be available for billions of years; that is, their use can be sustained for the foreseeable future. The use of biomass (for example, wood or garbage) as a fuel is also considered in a way to be sustainable because more can be continually created; however, burning biomass typically adds substantial amounts of carbon dioxide and other pollutants into the atmosphere, whereas wind, water, and solar energy typically do not.

What is modern wind energy generation like?

The largest windmills in commercial use today are up to 60 stories high. Their rotor blades are more than 60 meters (200 feet) long, and at maximum power, they spin around once every five seconds, generating 7,000 kilowatts—enough to provide electricity to about 2,000 homes. The largest wind farms in the United States today are in California, Oregon, and Texas, each with hundreds of windmills. The states of Oklahoma, Kansas, and Iowa each generate more than one-third of their electricity with wind. Many of the world's largest wind farms are offshore in seas and oceans, especially off the coast of many European countries.

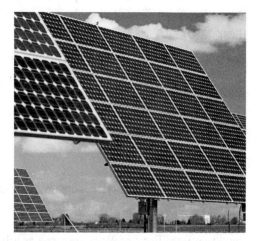

Because we will eventually run out of fossil fuels, we need to explore other, more sustainable forms of energy such as solar power.

What is modern water-powered energy generation like?

The nine largest electric power plants currently operating in the world are all hydro-electric (water-powered) plants. They are all located on major rivers around the world with very powerful water flows—for example, the Three Gorges Dam on the Yangtze River in China, the Itaipu Dam on the Parana River in Brazil, and the Grand Coulee Dam on the Columbia River in the United States. The number of large rivers in the world is limited, so the future expansion of hydroelectric power generation will probably be more limited than wind or solar power.

What is modern solar-powered energy generation like?

Sunlight is plentiful in the world, and if it can be effectively and economically harnessed, it could supply far more electricity than coal, oil, natural gas, or any other fuel source. Although solar energy currently produces a smaller amount of electric power than other sustainable energy sources, it has been the fastest-growing method of energy generation over the past several years. There are now more people employed in the solar power industry in the United States than in either the coal industry or the oil and gas industry.

Solar power is commonly generated today both from large solar-power plants and by small, individual solar-power systems such as those on the roofs of houses and other buildings. In some parts of the United States, such as Hawaii and Southern California, so many homes now have solar-power systems that the utility companies in those places can hardly keep up with connecting those homes to their local utility grids.

What are the challenges to using sustainable energy sources?

Wind power is highly variable. Water energy from traditional dams and reservoirs causes environmental problems. The technology to generate energy profitably from ocean waves and tides has not yet been fully developed. Solar energy can be directly converted to electricity using photovoltaic cells, but smaller solar-power systems are still quite expensive. Large solar farms, at which solar energy heats a fluid so it can boil water to use with steam turbines that drive electrical generators, usually have to be located far away from population centers.

Perhaps most importantly, the power grids that bring electricity from power plants to homes and businesses must be upgraded, most likely at great expense, to use sustainable energy sources effectively.

For these and other reasons, the present use of sustainable energy sources is low compared to the use of fossil fuels overall. In 2018, only about 12 percent of the energy produced in the United States was provided by sustainable sources. Scientific analyses show, however, that with concerted effort, sustainable wind, water, and solar energy can eventually replace most or even all fossil fuels used to produce electricity.

SIMPLE MACHINES

What are simple machines?

Simple machines are devices that match the human capability to do work to tasks that need to be done. They can reduce the force required to do a task or reduce the distance or change the direction that an object must be moved.

How does a simple machine work?

Suppose you want to lift a heavy object. If you use a machine, you will use the stored chemical energy in your body to do work on the machine. The machine, in turn, does work on the object, which increases the object's energy. If the force is constant and in the direction of motion, then work is the product of force and distance moved (as shown by the formula $W = F \times d$). The drawing below illustrates this process.

What is mechanical advantage?

The mechanical advantage of a simple machine is the output force divided by the input force. To make the output force larger than the input force, you must choose a machine that has a mechanical advantage larger than one. If, for example, you exert a small input force over a large distance into a machine, the machine could then exert a larger output force over a smaller distance. The brakes on a car or bicycle work on this principle.

How can a machine with mechanical advantage less than or equal to one be useful?

If a simple machine has a mechanical advantage less than one, you exert a larger force on it than it puts on whatever object it contacts. The output force is less than the input

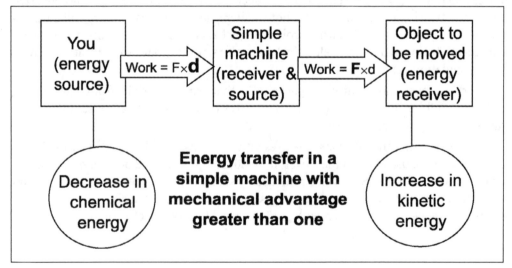

Figure 7.

force, but that can be very useful because the output distance (the distance the object moves) will then be larger than the input distance (the distance you move). Baseball bats, brooms, and golf clubs are all examples of this kind of simple machine. Figure 7 illustrates this process.

A machine with mechanical advantage of one, where the input force and distance are both equal to the output force and distance, can be useful for changing the direction of the target object's motion. A seesaw is an example of this kind of machine.

What are the limitations on simple machines?

Simple machines must, like everything else, obey the law of conservation of energy. That means the work done on the machine equals the work the machine does plus the heat the machine produces because friction has increased its thermal energy. Some machines are highly efficient, meaning input and output work are almost the same. Other machines put out only a small fraction of the work put in.

The drawing below shows a simple machine where the machine has warmed up—that is, its thermal energy has increased. It is now hotter than its surroundings, so it transmits heat to its surroundings (see Figure 8). The output work of the machine is smaller than the input work, and the machine is thus inefficient.

What types of simple machines are there?

There are four major groups of simple machines: levers, wheels and axles (which are kind of like levers), pulleys (including gears), and inclined planes (including wedges and screws).

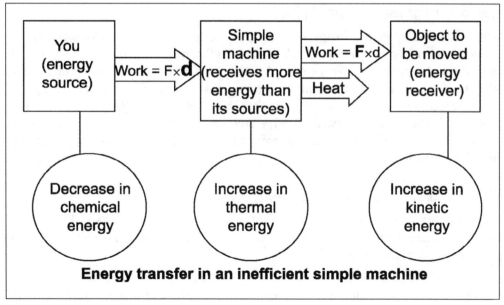

Energy transfer in an inefficient simple machine

Figure 8.

81

What is an inclined plane?

A ramp is a simple example of an inclined plane. Instead of lifting an object to the height at the end of the ramp, you move it along the surface of the ramp. The distance is longer, but it requires much less force. The ramp has a larger output force than input force, so it has a mechanical advantage greater than one. That is, $MA = F_{output} / F_{input} =$ (length of ramp)/(height of ramp).

Here is an example calculating the forces and distances of an object on an inclined plane. According to the Americans with Disabilities Act, a wheelchair ramp slope ratio must be 1:12, or one inch of rise for every foot of ramp. A ramp that must rise 30 inches would therefore have to be 30 feet long, for example. The input work is $F_{input} \times L$, where L is the length of the ramp. The output work is $F_{output} \times d$, where d is the rise and F_{output} is the weight of the person plus the wheelchair. If there is no sliding or rolling friction, then $F_{input} \times L = F_{output} \times d$. The mechanical advantage is L/d. So, $F_{input} = F_{output}/MA$, and the force needed to push a wheelchair up the ramp is given by $F_{input} = F_{output}/(L/d)$. If the weight of the person plus the wheelchair is 200 lbs., then the force needed to push the person up the ramp is $F_{input} = $ 200 lbs./(30 ft./2.5 ft.) = 17 lbs.

How can inclined planes be used to do more than just lift or move objects?

Inclined planes, when manufactured with sharp edges, can be used to create cutting tools. Look at the cutting edge of a pair of scissors. It's an inclined plane where the input is the force and motion of the closing blades, and the output is the outward movement of the paper after it is cut.

How is a wedge like an inclined plane?

A knife is one example of a wedge. Look carefully at the sharp edge of a kitchen knife. It looks like two inclined planes put back-to-back. As the knife blade moves down through food, the wedge pushes the pieces apart. A hatchet or axe is another example of a wedge. Wedges are also used to split wood. In this case, the flat end of an axe is often used to drive the wedge into the end of a log, which is then forced apart. Wedges are inefficient machines because there is usually a large amount of friction between the wedge and the material, which leads to increased temperature of both the material and wedge and, thus, heat transfer.

What kind of simple machine is a screw?

You can think of a screw as an inclined plane wrapped around an axle. The ancient Greeks used a screw to lift water. Screws and bolts are also used to fasten together two pieces of wood, plastic, or metal.

Screws have a large mechanical advantage, and thus, the distance the screw thread moves in its circular motion is much larger than the distance the screw moves into the material. Therefore, the force the screw or bolt exerts to hold the material together is large in comparison to the torque used to turn the screw. Screws and bolts are inefficient machines because of the friction between the screw and material or between the metal of the bolt and nut.

What is a lever?

A lever is a bar that rotates around a fulcrum or pivot. The locations where the input and output forces are exerted relative to the location of the pivot determines the class of the lever, as shown on the diagram below. The width of the arrows illustrates the amount of force, while the length of the arrows shows the distance moved.

How efficient are levers?

Unlike wedges or screws, which are rather inefficient, levers can be very efficient simple machines. The only place where friction can occur with a lever is at the pivot point. So, as long as friction is minimal there, the lever can approach 100 percent efficiency!

What Earth-moving statement did Archimedes make about a lever?

The ancient Greek natural philosopher Archimedes (c. 287–212 B.C.E.) first described the properties of the lever in 260 B.C.E. While conducting his research, he stated, "Give me a firm spot on which to stand, and I will move the Earth." Archimedes was referring to the use of levers; theoretically, with a long enough level and a non-Earthbound place on which to rest a fulcrum, one could indeed move our planet. Which class of lever would you use for this task?

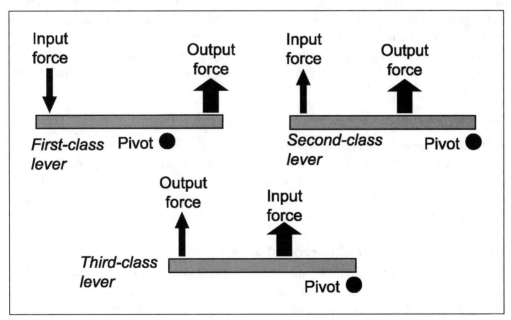

Figure 9.

What are some examples of levers in everyday life?

The first-class lever (Figure 9) has a mechanical advantage greater than one. How could you arrange the location of the pivot to make the mechanical advantage less than one? Look closely at a pair of scissors. Note that you can adjust the mechanical advantage by moving the region of the scissors you are using to cut. Where would you put the material to be cut if more force is needed to make the cut? You would put it nearer the pivot. On the other hand, if the material is easily cut, cutting near the tip of the blades provides enough force and speeds up the cutting. What class lever is a can or bottle opener? Locate the pivot point and compare to the three drawings in Figure 9.

Your forearm is a lever, with the elbow joint being the pivot. Where does the bicep muscle attach? The attachment is close to the elbow, so the forearm is a third-class lever. Are other muscles and bones in your body so easily characterized? Most are not because tendons that transmit the force from the muscle to the bone are long and go through several bends.

Consider sports equipment like baseball bats, tennis racquets, and golf clubs. They are often used as extensions of your arms, so the person plus the bat or club has to be examined together. But note that in every case, the system is a third-class lever, where a large distance moved (and therefore greater speed) is favored over an increased force.

How is a wheel and axle similar to a lever?

A wheel and axle consists of a disk (the wheel) attached to a thin rod (the axle) so that the two rotate together (see Figure 10). Typically, the input force is applied to the outer edge of the wheel, and the output force is exerted on something, like a rope, attached to the outer edge of the axle. If the two forces are exerted in the same direction, then it is like a third-class lever. If the two are exerted in the opposite direction (for example, a person pushing down on one side of the wheel while a rope is pulled up on the other side), then it is like a first-class lever. One may also have the input force exerted on the axle, in which case it is like a second-class lever.

Are levers and wheels and axles really force multipliers?

Both the lever and the wheel and axle are really torque, not force, multipliers. Recall that if the force is at right angles to the line from the axis of rotation to the point where the force is applied, then the torque is given by Fr. Therefore, if the radius of the axle is a and the radius of the wheel is w, then if the input force is applied to the wheel, the output force is given by $F_{output} = F_{input} (w/a)$.

What are some examples of wheels and axles?

A screwdriver is one common example of a wheel and axle machine. The larger the diameter of the handle, the greater the torque that can be applied to the screw. Many examples have a rope or chain wrapped around the axle. A sailing ship's steering wheel could be represented by either the left or the center drawing (Figure 10), depending on

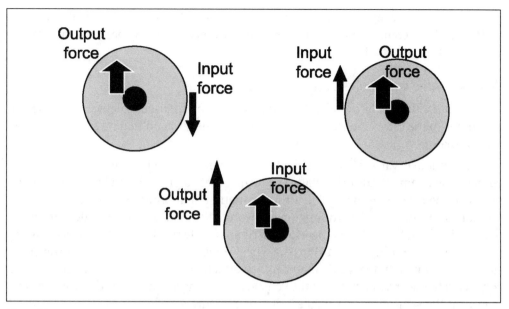

Figure 10.

whether the helmsman pushes down or pulls up on the edge of the wheel. The rope around the axle is connected to the rudder, which is then turned to the right or left.

A similar machine would be a device to lift a bucket from a well. The wheel is then replaced by a crank, but the operation of the device is exactly the same. Again, the crank can be either pulled up or pushed down to exert a force on the rope to pull the bucket up.

The rear wheel of a bicycle can also be thought of as two wheels on an axle. The large wheel has a rubber tire and exerts a backward force on the road, while the smaller wheel is the sprocket that the chain turns. Figure 10 shows that the force of the chain is larger than the wheel's force on the road.

How are pulleys used as simple machines?

A fixed, or unmovable, single pulley can be considered a wheel and axle where both have the same radius (Figure 11). The mechanical advantage is one, but the pulley changes the direction of the force (Figure 11 shows how pulleys are used).

If the pulley is allowed to move and the output force is exerted by the axle of the pulley, then the input force is shared by the two ends of the rope. If you fasten one end and pull up on the other, then you achieve a mechanical advantage of two.

A combination of fixed and movable pulleys is called a block and tackle. Archimedes is said to have pulled a fully loaded ship using a block and tackle more than 2,200 years ago. Energy can be lost in a block and tackle, however, from friction between the axle and its holder as well as the stretching of the rope and rolling friction between the rope and pulley.

Two or more pulleys, on fixed axles, can be connected together with a belt. If the two pulleys are of different diameters, then the one with the smaller diameter will turn faster, and thus, it can exert a larger torque.

What are some ways pulleys are used in cars?

In an automobile, one or more pulley-and-belt systems are used to deliver torques from the engine to the valve crankshaft, the water pump, the air-conditioning compressor, and the alternator.

Continuously variable automobile transmissions are used in a few modern cars to connect the engine to the driveshaft. The torque that can be delivered by an engine depends on the rotational speed. The torque is highest at an intermediate speed. A transmission is designed to allow the engine to revolve at a speed where it can deliver torque to the driveshaft, the axle, and the wheels that are revolving at a variety of speeds. When the auto is accelerating from a stop, the wheel rotation speed is slow, so the transmission needs to match a large-diameter pulley attached to the engine to a smaller one connected to the driveshaft. When the vehicle is traveling at a high speed, the engine can revolve at the same or even a smaller rate than the driveshaft.

One method of creating a pulley with a variable diameter is to use a V-shaped belt and a pulley with a groove that can be adjusted in width. This kind of transmission is often used with electric hybrid cars, and it is being developed as a means of decreasing fuel use.

Who developed gears?

As early as 2600 B.C.E. in India and other parts of southern Asia, gears were used to open and close doors in temples and to lift water. In the fourth century B.C.E. Aristotle described how gears were used. Archimedes described worm gears around 240 B.C.E., and in 40 B.C.E. Vitruvius showed how gears could convert motion from a horizontal axle to a vertical one. In the 1300s C.E. gears were used in clocks in bell towers and on churches.

What kind of simple machines are gears?

Gears are toothed wheels that transmit torque between two shafts. The teeth prevent them from slipping, so they are called positive drives. The smaller gear in a pair is called a pinion and the larger one a gear.

Because the teeth mesh, in a given amount of time the number of teeth engaged on the pinion equals the number on the gear. The number of teeth is proportional to the circumference of the gear, which, in turn, is proportional to the radius r of the wheel. Therefore, if the pinion makes one turn, the gear will make (r_{pinion} / r_{gear}) turns. Because energy is conserved, the torque exerted by the gear equals the torque applied to the pinion (r_{gear} / r_{pinion}).

What are some examples of gears around us?

Gears are plentiful in cars and trucks—for example, in an automatic transmission, converting the high-speed, small-torque output of the engine to the low-speed, large-torque

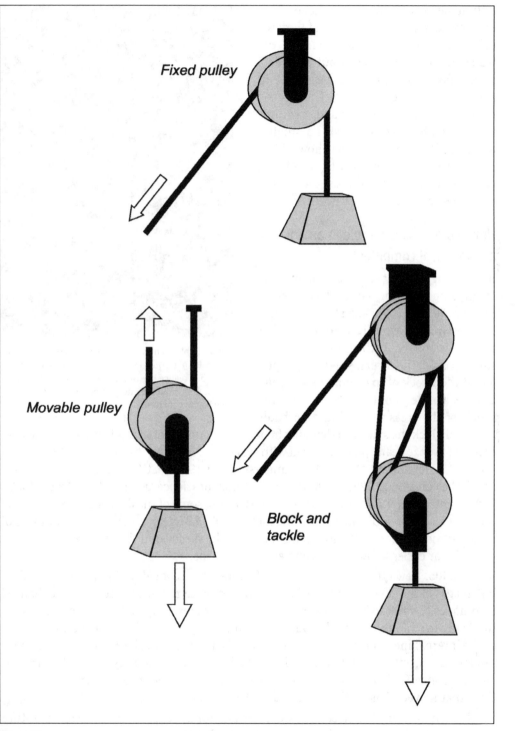

Fixed pulley

Movable pulley

Block and tackle

Figure 11.

output needed to turn the wheels. This kind of gear drive is called a step-down drive. By contrast, in a windmill, the blades turn at a low speed but provide a large amount of torque. The gear drive in a windmill is a step-up drive. The electric generator, on the other hand, requires high speed but can work with a smaller amount of torque. Gears also allow a bicyclist to match the speed with which she can rotate the pedal sprocket with the speed needed to drive the wheels.

How did gears play a role in an ancient astronomical computer?

In the first century B.C.E., a Roman merchant ship carrying Greek treasures to Rome sank. In 1900 C.E. a storm caused a party of Greek sponge divers to seek shelter on the island of Antikythera. After the storm passed, the divers, seeking sponges, found the wreck of the Roman ship. Over the next nine months, they recovered much of the treasure, including a badly

Windmill blades may move slowly, but they can produce a lot of torque. Gears increase the rotational speed but reduce the torque, driving an electric generator. Almost all the energy from the wind is delivered to the generator.

corroded block the size of a telephone book. A few months later, it fell apart, showing remains of bronze gears, plates covered with scales, and inscriptions in Greek. The German scientist Albert Rehm (1871–1949) understood in 1905 that the device was an astronomical calculator. In 1959 the American historian of science Derek de Solla Price (1922–1983) wrote a *Scientific American* article describing some of the details of the device and the calculations it could do. By 1974 he had described 27 gears, including the number of teeth on most of them. By analyzing the teeth, he discovered that the ratio could describe lunar cycles known by the ancient Babylonians.

In 2005, Hewlett-Packard and X-Tek Laboratories teamed up to produce astonishing images of the device. HP contributed a camera system that used computer-enhanced techniques to show inscriptions that were otherwise invisible. X-Tek brought an eight-ton X-ray machine that could produce extremely high-resolution 2-D and 3-D images of the internal mechanisms. Using the X-ray machine, archaeologists discovered thousands of Greek letters that described details of the mechanisms. They found 30 gears, but analysis suggests that there must have been at least five more to perform the calculations that moved dials on the front and back of the box.

The Antikythera device can predict solar and lunar eclipses and the dates of future Olympic games, and it can show the complicated motion of the moon. The precision with which the gears were made was greater than any made in the world for the next thousand years. But who made it and where? It is uncertain but likely to have been made in one of the colonies of Corinth on the island of Sicily, perhaps by a student of Archimedes, decades after he was killed in 212 B.C.E.

How are clocks important to the development of gears?

As clocks improved, so did the gears used within them. The pendulum clock uses a type of gear called an escapement to drive the pendulum, which regulates the time marked out by the clock. Precision gears allow clocks to use less power and have greater accuracy.

How are gears used today?

In addition to automobiles and clocks, gears are used in washing machines, electric mixers and can openers, and electric drills as well as computer hard drives and CD/DVD drives. Today, much of the development of gears is associated with improvements in the materials used. New metal alloys increase the lifetime of gears used in automobiles and industries. Consumer electronics often contain plastic gears that require no lubricant and are comparatively quiet.

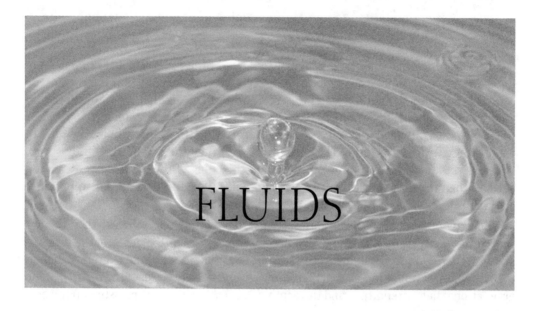

FLUIDS

What is a fluid?

Solid objects retain their shape because strong forces hold the atoms in their places. In a liquid, the forces keep the atoms close together, but they are free to move. In a gas, the atoms are about ten times farther apart than in either solids or liquids, and forces between them are very weak. As a result of weaker forces, a liquid or gas can flow freely and assume the shape of its container. Both liquids and gases are called fluids. The study of fluids in a state of rest (static fluids) or in motion is called fluid mechanics, and the study of the movement of fluids is called fluid dynamics.

WATER PRESSURE

What is pressure?

Pressure is a measure of force exerted divided by the area over which it is exerted. You can exert pressure, for example, by pressing your hand against a surface, like a table or wall. You feel water pressure when you go into a swimming pool. Your body is experiencing air pressure all the time, although you might not feel it. One very important physical principle is that any volume of liquid, such as water, exerts equal pressure in all directions.

What is Pascal's Principle, and how does it work?

In 1647 the French scientist Blaise Pascal (1623–1662) recognized that liquids, like water, exert the same pressure in all directions. This statement is known as Pascal's Principle.

To understand Pascal's Principle, think of a small cube of water, as shown in Figure 12. The gray arrow in the middle is the force of gravity on the cube. As a result of gravity, the total downward force of the cube is larger than the upward force.

Now by Newton's Third Law, the outward force of the water is equal to the inward force on the water. What exerts this force on the water? If the cube is at the edge of the container, then the container exerts the force. You can check that for yourself. Use an empty fluid container—say a juice box or milk carton. Punch a hole in one side. Pour water into the container. To keep the water from leaving through the hole, you have to exert a force with your thumb. The same would be true if the hole were in the bottom. What exerts the downward force on the top of the cube? If the cube is not at the top surface, then the force is exerted by water above it. If it is on the top, then air pressure exerts this force.

Because of gravity, the downward force exerted by the fluid is greater than the upward force, so the force, and hence the pressure, increases as the depth of the fluid increases. Therefore, the pressure at the bottom of a container of fluid is greater than at the top. The amount of increase in pressure is given by the product of the density, ρ, the gravitational field strength, g, and the depth, h, or ρgh. That is, the pressure on the bottom, $P_{bottom} = P_{top} + \rho gh$.

What does it mean to say water "seeks its own level"?

The surface of water placed in a single container (a glass or a bathtub or a lake) will remain at the same level relative to Earth on both sides of the container. Adding water to only one side will make the entire level uniformly rise; there can never be one section of the glass or tub or lake that is at a higher elevation than another section. To understand this fact, consider adding the small cube of water on top of the surface at one location. It would exert a downward force on the water under it. But because water can flow, water under it would flow outward, raising the level elsewhere in the container until the pressure is equal everywhere.

Water also seeks its own level in other containers. If you fill a hose or tube with water and hold the tube in a "U" shape, the water level will be at the same locations in the two ends. You can use the "U" tube to make a device to show you equal heights at two different locations. You may have a coffee maker that has a water height indicator on the side. This is a small tube that connects to the water reservoir at the bottom. The water level in the narrow tube and the wide reservoir is the same.

What are the units to measure pressure?

Pressure is force divided by area, so in the metric system, pressure is measured in newtons per square meter, called a pascal (Pa). One pascal is a very small pressure, so usually, the kPa, or 1,000 Pa, is used. In the English system, pounds per square

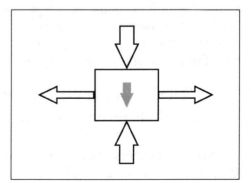

Figure 12.

inch (psi) is often used. Another unit used is a measurement of the height of mercury in a glass tube that would create a pressure that balances the pressure of the fluid. That unit, which used to be called millimeters of mercury, is now called the torr. Here is how these units compare: 760 torr = 14.7 psi = 101.3 kPa.

Why are water towers often located on tall buildings?

A typical home requires water pressures of 50 to 100 psi. City water systems use pumps to maintain that pressure in the pipes. Vertical pipes are needed to supply the upper floors with water. Each foot of height reduces the pressure by 0.443 psi. Auxiliary pumps at various floors can provide the needed increase in pressure. An alternative is to put a large storage tank on the roof and use pumps to fill it. It then supplies the building with water under pressure due to the height of the tank. It also allows the pumps to be run to fill it at night when electricity rates may be cheaper. In addition, it provides a backup source of water in case of fire.

Why are water tanks placed on high towers?

The height of the water determines the pressure. Since a holding tank is up high, it places a lot of pressure on the water in the rest of the water network. The tank should be as large across as possible so that, for the same amount of water, the vertical dimension of the tank can be as small as practicable. This design limits the variation in pressure as the tank empties or is filled.

Why are water towers particularly important in small towns?

Small towns, which often use wells as a water source, use water towers to store water in case there is an interruption in electrical service. It also allows the town to use smaller

pumps because the tower can supply the pressure during peak water demands. Typical daily water use in a small town is about 500 gallons per minute, but this can rise to 2,000 gallons a minute in peak times. A tower typically stores one day's worth of water.

Why might my ears hurt when I dive to the bottom of a swimming pool?

Just as the weight of the air above us creates atmospheric pressure, the weight of water creates liquid pressure. Close to the surface of a pool, there is very little water that can push down and increase water pressure. The further a person dives below the surface, however, the greater the water

Water towers, such as these in New York City, are often placed on top of tall buildings. This way, the force of gravity supplies the water pressure needed to deliver water throughout the building.

93

pressure. The eardrums are especially sensitive to the increased pressure, for they do not have the reinforcement that the diver's skin has. In fact, your eardrums can usually feel pressure when diving just 1.5 to 3 meters (5 to 10 feet) below the surface of the water.

Where is the water pressure greater, in the ocean at a certain depth or in a small lake at twice that depth?

The ocean contains a lot more water than a small lake. Nevertheless, it is the depth that determines the weight of water directly over a diver—and that determines the amount of pressure a diver experiences. Therefore, divers 20 meters below the surface of a lake will experience more pressure than ocean divers 10 meters below the surface of the ocean—twice the pressure, in fact, because they are at twice the depth. Saltwater is denser than freshwater, but the increase in pressure caused by the increased density is small compared to that caused by the difference in depth.

Why are dams thicker at the bottom than at the top?

Dams hold back bodies of water, and water pressure increases with the depth of the body of water, so the pressure from the water pushing horizontally on the dam is greater at the bottom than at the top. If holes were bored near the bottom, middle, and top of a dam, the longest horizontal stream of water would fire out through the bottom hole because the water pressure is greatest there.

BLOOD PRESSURE

What is blood pressure, and why is it important?

Blood pressure is the pressure your blood exerts on the walls of your arteries. Human blood is mostly water, so the rules that govern blood pressure are very similar to those that govern water pressure. The fluid dynamics of blood play a major role in blood pressure. The heart is the pump that moves the blood throughout the body, with vessels carrying the blood to different sections of the body. If a person's blood pressure is too high, there is a risk of damage to blood vessels and of strokes (which can result from broken or blocked blood vessels). If a person's blood pressure is too low, there might not be enough blood supplied to key organs, like the brain, which could cause fainting.

How is blood pressure usually measured?

The most common device used to measure blood pressure is the sphygmomanometer. It is placed around your upper arm, inflated, and then deflated, while a meter measures the pressure passing through that section of the arm. Either a person using a stethoscope or an electronic sensor then detects your pulse. The cuff is inflated until no pulse can be heard. It is then slowly lowered. As the pressure falls below the systolic pressure (pressure in the arteries caused by the heart contracting), the pulse can be heard. When it's below the diastolic pressure (the lessening of pressure as the heart fills with blood), the pulse gets weaker. The report "120 over 70" means that the systolic pressure is 120 torr and the diastolic pressure 70 torr.

ATMOSPHERIC PRESSURE

How is gas pressure similar to or different from liquid pressure?

The pressure from a gas acts the same way liquid pressure does. One difference between gaseous and liquid pressure is that gases are about 1/1,000 as dense as liquids and therefore apply less pressure. The second difference is that gases can be easily compressed, while the compressibility of liquid is very small.

What is Earth's atmospheric pressure?

Earth's atmosphere extends approximately 100 kilometers (328,000 feet) above the ground, but it gets less and less dense as the altitude increases. The amount of force on an area of one square meter is about 101,300 newtons. That is, the pressure is 101.3 kPa. This atmospheric pressure varies with temperature and other conditions. Our weather is mostly influenced by high- and low-pressure regions, which can deviate by about 5 percent from normal.

How does atmospheric pressure change with altitude?

Imagine yourself rising through the air in a hot-air balloon or on an airplane. The atmospheric pressure exerting force on you decreases as your altitude increases because the amount of air above you creating the pressure is smaller. At an altitude of 3 kilometers (10,000 feet) above sea level, the atmospheric pressure is about 70 percent of what it is at sea level. At 100 kilometers, it is 3 millionths of the sea-level pressure! If Earth's atmospheric density were constant, then instead of 100 kilometers, the atmosphere would be only about 8 kilometers high (26,246 feet; close to the height of Mount Everest). As it turns out, about 63 percent of the atmosphere is below that height.

What are the bends?

Nitrogen under normal atmospheric pressure is nearly insoluble (cannot be dissolved) in blood. Under pressure, the solubility increases. Thus, as a diver goes deeper, the blood holds more and more nitrogen, which dissolves in the blood during gas exchange in the lungs. As the diver ascends, the pressure decreases, and hence, the blood is now supersaturated with nitrogen. The supersaturated nitrogen forms bubbles as it comes out of solution in the blood, or cells. These bubbles collect in veins and arteries. They are very painful, and they can rupture cell walls and block the flow of blood to cells, causing injury or even death.

How is atmospheric pressure measured?

A device to measure gas pressure is called a barometer. There are two major types of barometers, the mercury barometer and the aneroid barometer. Galileo's secretary, Evangelista Torricelli (1606–1647; the unit of pressure called the torr was named after him), developed the mercury barometer in 1643. It consists of a thin, glass tube about 80 centimeters (31 inches) long, which is closed at the top, filled with liquid mercury, and placed upside down in another mercury-filled dish. Depending upon the atmospheric pressure pushing on the mercury in the dish, the level of mercury in the tube will rise or fall because there is no air above it. By measuring the height of the mercury, which would usually be between 737 and 775 millimeters (29 to 30.5 inches) high, the pressure of the atmosphere can be measured.

A common kind of household barometer used today is the aneroid barometer, in which atmospheric pressure bends the elastic top of an extremely low-pressure drum. By measuring the amount that the top bends, a measurement of atmospheric pressure can be determined. The aneroid barometer is often used in airplane altimeters to measure altitude. Since atmospheric pressure decreases as altitude increases, the aneroid barometer is an ideal instrument to use. It is much safer than the mercury barometer because mercury vapor is poisonous, and the mercury must be exposed to the atmosphere for the mercury barometer to work properly.

What happens to a balloon when it is submerged in water?

When an air-filled balloon is placed underwater, the water (which has higher pressure than the air) exerts a force on all sides of the balloon. The pressure from the

A barometer is an instrument used for measuring gas pressures. They are also used to help predict weather, as low-pressure systems tend to bring inclement weather while high pressure systems bring fair skies.

How do divers avoid the bends?

The best way to avoid the bends is to rise to the water surface slowly, allowing the liquid pressure from the water to gradually decrease and prevent any physical damage. Most scuba divers use "nitrox," a gas mixture that contains 35 percent oxygen and 65 percent nitrogen (rather than normal air that contains 20 percent oxygen and 80 percent nitrogen) to reduce, but not eliminate, the possibility of getting the bends.

water causes the air to compress inside the balloon. The deeper the balloon is taken into the water, the greater the pressure, and therefore, the smaller the balloon will become. The pressure from the water will compress the balloon until the air pressure within the balloon can supply an equal amount of force against the water.

Why do closed containers sometimes dent or even collapse on cool days?

Just as a balloon can become smaller when placed under water, a sealed container can change its shape and even collapse under certain atmospheric conditions. For example, a container that stores gasoline for a lawn mower is usually sealed when it's not being used. If the container were sealed on a warm day, when there was little atmospheric pressure, subsequently, on a cool, high-pressure day, the gasoline container could be crushed. A second reason is that gasoline vapor exists in the tank, and at low temperatures, more of the vapor condenses to liquid, reducing the pressure of the gases in the tank.

If atmospheric pressure is so powerful, why don't we get crushed?

Our bodies have air and other gases inside them, and the air inside our bodies is at the same pressure as the air outside our bodies in the atmosphere. Therefore, the pressures are equal, and we can move freely in our atmospheric environment. Most of the rest of our bodies are made up of liquids, like water and blood, and cannot be compressed. Deep-sea divers experience much higher pressures from water as they go deeper and deeper; divers who go to extreme depths and then ascend too rapidly can suffer the bends—a painful and dangerous health problem.

Why do athletes sometimes go to high elevations to train?

Long-distance runners have always trekked to the mountains to train in higher elevations because of the lower atmospheric pressure that the mountains provide. Since the air is not as dense as it is at lower elevations, the lungs need to work harder to supply the body with a sufficient amount of oxygen. Many athletes feel that training in such conditions gets their bodies used to lower amounts of oxygen. Therefore, when running in a competition at lower elevations, they can compete quite well because their bodies are used to working hard to get oxygen.

BUOYANCY

Why do some objects sink and others float?

An object will sink in a fluid if the downward forces on it are larger than the upward forces. There are usually two downward forces in a fluid: the force of the liquid above the object and the object's weight (the force of gravity). The upward force is the force of the liquid below it.

Let's think of a cubic object of height h and area of the top and bottom A. Its volume, V, is then given by $V = hA$. Density is the mass divided by the volume, or $\rho = m/V$. Consider the difference in water pressure between the bottom and top of this cube. Pressure is force divided by area, so, using $P_{bottom} = P_{top} + \rho_{water} gh$, we can write $F_{bottom}/A = F_{top}/A + \rho_{water} gh$, so $F_{bottom} = F_{top} + \rho_{water} ghA$. Now, we recall that hA is the volume of the cube, and $\rho_{water} hA$ is therefore the mass of the water, m_{water}, whose place is taken by the cube of matter.

The net downward force on the object in the water is then the force on the top plus the object's weight less the force on the bottom. That means $F_{net} = F_{top} + m_{object} g - F_{bottom}$. From the results of the paragraph above, $F_{net} = m_{object} g - m_{water} g$. Therefore, if the object's mass is larger than the mass of the water whose place it takes, it sinks. If the mass is smaller, then it will rise. Weight is the mass times the gravitational field strength, g, so the net force is $F_{net} = W_{object} - W_{water}$. The water whose place the object takes is normally referred to as the water "displaced" by the object.

What is the force of buoyancy?

The force of buoyancy, or the buoyant force, is the net upward force of the water. From the equation obtained in the answer to the last question, you can see that when you place an object in water, its weight is reduced by the weight of the water displaced. The reduction in weight, $W_{object} - W_{water}$, is the buoyant force.

Why do rocks sink but wood floats?

Rocks are almost always denser than water. That is, if you compare the mass of a stone to the mass of the same volume of water, the stone's mass will be greater. Therefore, its weight will also be greater, so $W_{stone} - W_{water}$ will be positive, and there will be a downward net force. Therefore, the stone will move downward through the water until it rests on the bottom.

Wood, on the other hand, is almost always less dense than water. If the wood is pushed under water, $W_{wood} - W_{water}$ will be negative. There will be an upward force on the wood, and it will rise. How far will it rise? As some of the wood rises above the level of the water, the volume and the mass, and therefore the weight, of the water displaced will be reduced. When the weight of the wood and the weight of the water displaced are equal, the net force on the wood will be zero, and it will no longer move.

Why is a stone easier to lift when it is in water?

Even though the stone sinks because it is denser than water, there is still a buoyant force on it given by $W_{\text{stone}} - W_{\text{water}}$, so the stone's weight is reduced.

Can things float in a gas?

There is a difference in pressure of any fluid between the bottom and the top of an object in it given by $P_{\text{bottom}} = P_{\text{top}} + \rho g h$. Even though the density of a gas, ρ, is much smaller than that of a liquid, there is still a pressure difference and therefore a buoyancy. There is another difference between a liquid and a gas. A gas can be compressed, so the density of the atmosphere decreases as you rise. When a hot-air balloon is launched, the air within it is heated, and it will rise. As it rises, the density of the air decreases, so the buoyant force decreases. At some height, the weight of the balloon and basket equals the weight of the air displaced, and the balloon stops rising. The operator of the balloon can go higher by making the gas in the balloon hotter and thus less dense.

How do blimps remain at a chosen altitude?

A blimp is a nonrigid airship that floats in the air solely due to the buoyant gas within its giant, balloon-like bag. It typically carries more than 5,000 cubic meters of helium at a density about seven times less than air. An airship floats in the air in the same manner as a ship floating in water. The weight of the airship must equal the buoyant force of the gas inside the bag. To increase the altitude of the blimp, the pilot increases its

Hot-air balloons work by expanding the gas within the balloon so that the air within is at a lower pressure than the surrounding atmosphere.

buoyancy by adding gas from pressurized tanks to the blimp, which expands the flotation bladders, displacing the heavier air and increasing the buoyant force until the reduced weight of the blimp equals the reduced weight of the air. To lower the airship, the buoyancy is decreased by releasing gas from the flotation bladders, which then decrease in size, displacing less of the heavier air.

What are blimps used for?

Since the first airship flight in 1852 by Henry Giffard in France, the dirigible, or airship, was used predominantly for military purposes. In the first and second world wars, airships were used for bombing and surveillance on both sides of the Atlantic. Commercial passenger transportation on airships was conducted for only a few years, while today's blimps are used for advertising and for televising sporting events from high elevations.

What gas is used in blimps?

Hydrogen is the most buoyant gas on Earth, and it is twice as buoyant as helium, the next most buoyant gas. Although hydrogen is more effective in lifting an airship off the ground, hydrogen gas is extremely dangerous because it burns easily. The German airship (or zeppelin) the *Hindenburg*, the world's largest airship at the time, was destroyed on May 6, 1937, in Lakehurst, New Jersey, when it exploded into a huge fireball while attempting to land. Thirty-six people died in the explosion.

After that tragedy, helium became the gas used in airships. In 1937, the United States was about the only source of helium in the world, mostly from one gas well in

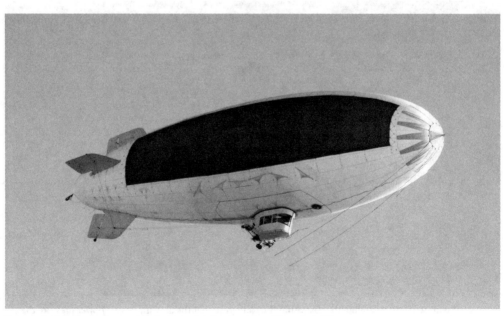

Because helium is a safer gas than hydrogen, it is what is used in today's blimps.

Texas. At that time, the Nazis wanted to buy helium for their zeppelins, but the United States refused to sell it to them as it was considered a strategic resource.

Helium is formed primarily from the radioactive decay of uranium and thorium in rocks deep underground. Helium is used today for a wide variety of purposes from blowing up party balloons to cooling devices to make them superconductive. As a result, the amount of helium available is rapidly decreasing.

If a helium balloon is let go and floats away, what happens to it?

A rubber balloon will expand as its altitude increases. This expansion is caused by the lower atmospheric pressure at higher altitudes. Eventually, the helium's volume increases so much that the rubber balloon breaks, releasing the helium. The broken balloon will fall back to Earth—possibly into the ocean—and the helium will escape upward and eventually leave Earth's atmosphere.

Who was Archimedes, and what discovery about buoyancy did he make?

Archimedes (c. 287–c. 212 B.C.E.) lived in the city of Syracuse on the island of Sicily, then part of Greece. He was charged by the king of Sicily to find out if his crown was made of pure gold or of an alloy of gold and silver. Archimedes had to do this without destroying the crown. One day, when bathing in the public baths, he noticed that when he stepped into the bath, the water rose! He had seen this many times before, but this time he recognized that this common occurrence could help him solve his problem.

A huge steel ship, like an aircraft carrier, can float instead of sink like a stone because it can displace enough water to compensate for its weight and because the air inside of the ship makes its density actually less than the water around it.

How does a hippopotamus sink to the bottom of a riverbed?

A hippopotamus usually spends more than half its day in the water. To eat, hippos—which can reach almost 3 meters (10 feet) in length and can weigh 5 tons—must sink to feed on the vegetation that grows on the bottom of rivers. The hippo, however, has a problem to solve: its low density forces it to float at the surface, and it is not agile enough to dive quickly down to the bottom and come back up again. To reach the bottom, it needs to increase its density so the buoyant force is not strong enough to keep the animal afloat. To do this, the hippo exhales, reducing the air in its body to increase its density.

Legend has it that he was so excited that he ran naked through the streets of Syracuse shouting "Eureka!" ("I have found the answer!").

Archimedes then did experiments in which he hung the crown on one end of a balance and a piece of first gold and then silver on the other end. When the weight of the crown and that of the metal were equal, the balance was horizontal. He then immersed the balance in water. If the crown and the metal had different volumes, the water they displaced would be different, and the balance would tip. He found that the balance tipped when both the silver and gold pieces were on the balance. Archimedes had found that the crown was not pure gold but a mixture of silver and gold. The king had been cheated.

Archimedes had discovered a principle of hydrostatics (liquids at rest) that would one day carry his name: Archimedes's Principle states that an object immersed in a fluid will experience a buoyant force equal to the weight of the displaced fluid.

Why does a small piece of steel sink, while a big ship made of steel can float?

To remain afloat, an object needs to displace an amount of fluid equal to its own weight. If a piece of steel is placed in water, it will sink because it is much denser than water and its volume isn't large enough to displace a weight of water. However, a ship—even one that weighs thousands of tons—can easily stay afloat because it can displace an equal tonnage of water. That is because the ship isn't all steel but contains a large amount of air. Therefore, its density—its mass divided by its volume—can be less than that of water.

What happens to the buoyancy of a ship when passengers or cargo are added?

When cargo and passengers are added to a ship, its weight increases. As the ship's weight increases, it sinks further into the water, displacing a greater weight of water. If it sinks so far that water can spill into the ship, increasing its weight even more, the ship will sink to the bottom of the water.

The amount the ship sinks in the water as a result of cargo and passengers affects its navigation and maneuverability. Large cargo and cruise ships have numbers on the bow of the ship that indicate how far the distance is between the water line and the bottom of the ship. This distance is called the ship's draft. If the ship has a 6-meter (20-foot) draft and the water is only 5.5 meters (18 feet) deep, cargo and passengers must be unloaded to allow the ship to rise.

How much water does a ship need in order to float?

Perhaps surprisingly, not a lot of water is needed to keep a ship afloat—just enough to displace water to equal its weight. If a ship were to enter a canal that was just a bit larger than the size of the ship's hull, it would float as long as there was a small film of water around the entire hull of the ship.

FLUID DYNAMICS

What is fluid dynamics?

Fluid dynamics is the study of fluids in motion. Hydrodynamics is the study of what happens when fluids that behave like water move—for example, in other fluids or through pipes. Hydraulics deals with the use of liquid in motion, usually in a device such as a machine. Oil is the most common liquid used in pumps, lifts, and shock absorbers.

How does a hydraulic lift work?

According to Pascal's Principle, pressure in a liquid is independent of direction. Pressure is force divided by area, so force exerted by a liquid is equal to the pressure times the area. A hydraulic lift has a small piston in a cylinder. If you exert a force on the piston, it will create a pressure in the fluid. The lift also has a second, large-area cylinder and piston. The fluid creates a pressure in this cylinder that exerts a much larger force on the piston because of its larger area. Therefore, the lift has a mechanical advantage. Of course, energy is still conserved, so the small piston must move much farther than the large piston.

A hydraulic lift, such as a car lift used in many automotive repair shops, allows the operator to use very little force to lift an automobile off the ground. It does this by pushing liquid from a small-diameter cylinder and piston through a thin tube that expands into a larger-diameter cylinder and piston, which is located beneath the vehicle to be lifted. Since the liquid cannot be compressed like air, the liquid from the small cylinder is pushed into the large cylinder, forcing the large piston to move upward.

Where are hydraulic lifts often used?

In addition to their valuable use in auto-repair shops, hydraulic lifts are used in elevating crane and backhoe arms, adjusting flaps on airplanes, and applying brakes in automobiles. The noncompressible characteristics of liquids are what make hydraulic devices so useful. Oil is often used rather than water because it does not freeze.

What is pneumatics?

Whereas hydraulics uses liquids to achieve mechanical advantage, pneumatics uses compressed gas. Since gases can be compressed and stored under pressure, releasing compressed air can provide large forces and torque for machines such as pneumatic drills, hammers, wrenches, and jackhammers.

What makes a fluid flow?

As in all of physics, objects move as a result of a net force. In a fluid, the net force comes about because there is a difference in pressure between two points. The fluid flows in the direction of decreasing pressure.

How do fluids move?

When fluids move slowly, they exhibit steady, or laminar, flow. When the speed increases, the flow can become turbulent. In some cases, laminar flow is desired; in others, turbulent flow is better.

This heavy crane boom uses hydraulics to lift the boom.

What is the difference between turbulent and laminar flow?

When a fluid flows in a container, such as a pipe, the film of fluid that touches the container does not move because of friction with the container's surface. But the fluid in the middle of the stream does move. In laminar flow, the transition from not moving to moving at full speed is continuous. Each thin film of water moves slightly faster than the one closer to the surface. In turbulent flow, this transition from not moving to full-speed motion occurs suddenly, and the water moves in tiny circles in this region.

Laminar flow actually has more friction than turbulent flow. A thrown baseball, for example, has more drag in the laminar flow region surrounding it as it moves through the air.

AERODYNAMICS

What is aerodynamics?

Aerodynamics is an aspect of fluid dynamics dealing specifically with the movement of air and other gases. Engineers studying aerodynamics analyze the flow of gases over

> ## Why does a river's current run faster where the river is narrower?
>
> **W**hen water flows down a river, its current is measured by the volume of water that passes by a cross-section of the river divided by the time taken for the water to pass. For example, if the current of a river is 2,000 cubic meters per minute, this means that in every minute, 2,000 cubic meters pass by every part of the river. If the river narrows, the same 2,000 cubic meters of water still must pass in one minute because the water behind it continues to flow downstream. The narrower part of the river has a smaller cross-sectional area, so the speed must increase. This effect is called the principle of continuity.

and through automobiles, airplanes, golf balls, and other objects that move through air. They also study the effect of moving air on buildings, bridges, and other static objects.

What is Bernoulli's Principle?

In 1738, a Swiss physicist and mathematician named Daniel Bernoulli (1700–1782) discovered that when the speed of a moving fluid increases, such as the wind blowing through the corridors of a city, the pressure of that fluid decreases. Bernoulli discovered this while measuring the pressure of water as it flowed through pipes of different diameters. He found that the speed of the water increased as the diameter of the tube decreased (the continuity principle) and that the pressure exerted by the water on the walls of the pipes was less as well. This discovery would prove to be one of the most important discoveries in fluid dynamics.

How does an airplane wing create lift?

The upward force on an airplane wing caused by the air moving past it is called lift. According to Newton's Third Law, if the air exerts an upward force on the wing, then the wing must exert a downward force on the air. There are three causes of this downward force. The first is the tilt, or angle of attack of the wing. The tilt deflects the airflow downward. The second is the Bernoulli Principle. Because of the shape of the wing, the speed of the air on the top surface of the wing is faster than at the lower surface, so the pressure is lower. Therefore, the downward force on the wing is reduced. The third reason is the fact that air "sticks" to the surface of the wing. The air coming off the upper surface is moving downward. This exerts an additional downward force on the air. The downward forces produce the lift needed to keep the plane airborne.

What is aerodynamic drag?

Drag is a force that opposes the motion of an object through a fluid. An object is often said to be "aerodynamic" when its drag forces are kept to a minimum. There are typically two types of drag on an airplane: parasitic and induced.

Parasitic drag is the force when an airplane wing, automobile, or any other object moves through a fluid. The amount of drag depends on the density of the fluid, the square of the speed of the object, the cross-sectional area of the object, and its shape. A large fuselage, like that of a 747, has more drag than a small fighter airplane. A teardrop-shaped object has less drag than a rectangular block. A parachute is designed to have a high drag.

Induced drag is a consequence of the lift generated by the wing. It is a function of the angle of attack of the wing—the lower the angle of attack, the smaller the induced drag. It occurs at the outer edge of the wing, where the downward motion of the air caused by the wing meets the undisturbed air next to it. Induced drag causes vortices, the spiral motion of air that can be extremely dangerous to planes flying behind or below. Induced drag can be reduced by putting small, tilted surfaces on the wing tips.

What are streamlines?

Streamlines are lines that represent the flow of a fluid around an object or through another fluid. Streamlines are often made visible by putting thin films of smoke in the air. They are used in wind tunnel testing of airfoils (wings) and automobiles. A wind tunnel is a closed chamber with vents in the front and rear of the tunnel that allow wind and the streams of smoke that make streamlines visible to pass around an object. If the streamline smoke appears to be flowing in a gentle pattern without breaking up, then the object is considered aerodynamic. If the smoke breaks up upon encountering sections of an object, the flow has become turbulent. If the flow is turbulent, the drag may be increased.

An engineer checks the fan in a wind tunnel. Smoke introduced into wind tunnels makes streamlines visible, which in turn help with the analysis of the efficiency of airfoils and automobiles.

Why do golf balls have dimples?

Golf balls are dimpled so they can stay in the air longer when struck. Without dimples, the flow of air around the ball is laminar, and the ball drags a thin layer of air completely around the golf ball. This causes a wake behind the ball that slows it down. The dimples, first introduced in 1908 by the Spalding Company, break up this air layer, reducing the drag.

Golf balls can also experience lift. When hit with a slight backspin, the air passing over the top of the ball flows in the direction opposite the motion of the ball. This creates low pressure above the ball. On the bottom of the ball, the air motion is in the same direction as the ball, so the pressure is higher. According to Bernoulli's Principle, such a pressure difference provides a lifting force on the ball. This is called the Magnus force, and it gives the ball just a little longer time in flight, which allows it to travel farther.

How does a curveball curve?

A curveball in baseball experiences the Magnus force like a dimpled golf ball. The pitcher can give the ball a backward spin, creating lift, or a sideways spin, causing the ball to curve sideways. When a pitcher throws a knuckleball, the ball spins very slowly. The laces on the ball create small forces that make the ball move erratically through the air.

Why does a discus thrown into the wind go farther than one thrown with the wind?

In most sports, throwing or traveling with the wind at your back (called a tail wind) is a lot easier than working against the wind (called a head wind). In football, teams flip a coin to determine who will be kicking with the wind. In sailing, it is easier and faster to travel perpendicular to or with the wind and takes more skill to travel against the wind. In track, the world record in the 100-meter dash could more easily be broken if running with a strong tail wind. In most sports, having the wind at your back can be a major advantage.

In the field event of discus throwing, however, the advantage comes when there is a head wind. In fact, it has been documented that a discus can travel up to 8 meters (26 feet) farther while experiencing a head wind of only 10 m/s (meters per second). Although the discus still experiences a drag force from the head wind, the lift that the discus gets from pressure differences over and under the disc is substantially more significant than the drag force. Because the discus will remain in the air longer, it will travel farther.

What shape should an object have to be the most aerodynamic?

You might think the narrower and more needle-like an object is, the lower its drag force will be. Although a needle-shaped head cuts easily through the wind, the problem emerges at the tail end, where the wind becomes turbulent and forms small eddy currents that hinder the streamline flow of air. The optimum shape depends on the velocity of the object.

For speeds lower than the speed of sound, the most aerodynamically efficient shape is the teardrop. The teardrop has a rounded, nose that tapers as it moves backward, forming a narrow, yet rounded tail, which gradually brings the air around the object back together instead of creating eddy currents.

At high velocities, such as what a supersonic jet airplane or a bullet may travel, other shapes are better. For turbulent flow, the least drag comes from having a blunt end, which intentionally causes turbulence. The rest of the air then flows smoothly over the region of turbulence behind the object.

How are airplane controls different from car controls?

Cars travel on two-dimensional surfaces and therefore only need two separate controls: the accelerator and brake to control the forward movement and the steering wheel to control side-to-side movements. Airplanes, on the other hand, travel in three-dimensional space. The forward thrust on an airplane is controlled by the throttle, and the "braking" is achieved by closing the throttle and increasing the drag, usually by deploying the plane's flaps. Yaw, which is responsible for the side-to-side movement of a plane, is controlled by the plane's rudder.

To control the pitch, or up-and-down orientation of the nose, the pilot uses elevators or horizontal control surfaces near the plane's rudder. To roll the plane (the rotation of the plane about an axis that goes from the nose to the tail), the pilot uses control surfaces on the back end of the wings called ailerons. The Wright brothers recognized that roll control was crucial to successful flight. They invented a method of warping the wings, creating a primitive but useful aileron.

THE SOUND BARRIER

What is a shock wave?

Just as a boat moving through the water forms a series of V-shaped waves, airplanes create conical (cone-shaped) waves as they fly through the air. The waves that the airplane produces are waves of compressed air.

When an aircraft reaches the speed of sound, the plane's pressure waves that move at the speed of sound overlap each other, creating a shock wave. The shock wave creates one single, loud sonic boom heard by observers on the ground. When the plane travels slower than the speed of sound, the sound waves do not overlap, and instead of hearing a sonic boom, observers simply hear the delayed sound of the plane.

You can think of sound waves as being similar to water waves, emanating from a central source and spreading out in a regular pattern unless they are interfered with.

Why is the speed of sound called Mach 1?

Ernst Walfried Josef Wenzel Mach (1838–1916) was an Austrian physicist and philosopher of science who made fundamental discoveries about shocks and shockwaves. In 1929 the Swiss aeronautical engineer Jakob Ackeret (1898–1981), one of the foremost pioneers of fluid dynamics of his time, proposed that the ratio between the speed of a fluid flow and the speed of sound through that fluid be called the Mach number in honor of Ernst Mach's work. The term stuck and is still used today; Mach 1 is thus the speed of sound, Mach 2 is twice the speed of sound, Mach 3 three times, and so on. Any velocity greater than Mach 1 is referred to as supersonic.

Who was the first pilot to break the sound barrier?

On October 14, 1947, Chuck Yeager (1923–) broke the sound barrier in his Bell X-1 test plane, *Glamorous Glennis*. To reach the sound barrier, the X-1 was carried in the belly of a B-29 bomber to an altitude of 3,660 meters (12,000 feet), where it was dropped. The X-1's rocket engine ignited, and Yeager took the plane to an altitude of 13,100 meters (43,000 feet). At this altitude, Yeager was able to break the sound barrier by traveling 660 miles per hour. The X-1 experienced a turbulent set of compression waves just before he broke past the barrier at Mach 1.05. Yeager kept the plane at this supersonic speed for a few moments before he cut off the rocket engine and headed back toward Earth.

What did pilots and engineers have to consider when breaking the sound barrier?

To reach the sound barrier in an airplane was a major goal for many in the aeronautical field, although the goal carried some uncertainties. Pilots and engineers alike wondered and feared what would happen to a plane's maneuverability when it broke through the shock wave as well as what would happen to the plane's structure itself.

By the end of World War II, fighter planes had powerful engines and experienced pilots. A number of pilots died when their planes broke apart in midair, often when in

Why did Chuck Yeager fly so high to break the sound barrier?

Sound travels approximately 760 miles per hour in the warm, dense air found close to sea level. The cooler and less dense air is, the lower the speed of sound. Since air is less dense at higher elevations, physicists and engineers felt it would be easier to break the sound barrier at those elevations. Knowing the temperature and density of the air at 12,200 meters (40,000 feet) above sea level, scientists determined that the speed of sound would be reduced to only 660 miles per hour. As a bonus, engineers found that not only was the speed of sound slower at such elevations, but when the air has such low density, the parasitic drag (the drag due to friction) is very low as well. Therefore, to break the sound barrier, Yeager traveled as far up as 13,100 meters (43,000 feet) above sea level to both reduce the sound barrier and decrease the parasitic drag.

dives. There were two problems with these aircraft: first, the wings were not swept back, and second, they were driven by propellers. As the shock wave forms near Mach 1, it bends backward from the nose of the plane like a bow wave on a boat. If the shock wave encounters the wings (that is, if the wing extends through the shock front), tremendous forces are placed on the wings. In a supersonic plane, the wing is always designed to be fully behind the shock front because the shock front can tear the wing off the plane. The propeller causes a pulsation in the pressure on the wing: every time one of the blades goes by, it produces a region of slightly higher pressure behind it, followed by a region of low pressure. All of these things came together and helped cause the midair structural failures of the World War II fighter planes.

Why are the angles of airplane wings important for supersonic flight?

When a plane breaks the sound barrier, the shock wave in front of the plane has a difficult time moving out of the plane's way. To break the sound barrier with less difficulty, aeronautical engineers have designed more aerodynamic fuselages and efficient wing designs. As mentioned above, wings on supersonic planes must remain behind the shock front to prevent structural failure and allow the plane to maneuver safely. The swept-back wing design, as found on both military and commercial airplanes, allows the airplane to accelerate easily and rapidly before major pressure builds up around the wings. The delta wings, as found on many jet fighters, are large and extremely thin to keep the wings behind the shock front while increasing lift and reducing drag.

Big problems can also occur when using swept-back wings. As a plane moves faster, the center of lift on the wings can move too far backward, causing unbalanced forces on the plane, which can affect the maneuverability and safety of the plane.

What was the fastest non-rocket aircraft ever built?

The fastest non-rocket (also called "air-breathing") plane ever flown was the SR-71 Blackbird. Its first flight was on December 22, 1964, and it reached a top speed of Mach 3.4 during flight testing. Over several decades, the fleet of SR-71 airplanes flew thousands of missions. The last of the SR-71 Blackbirds were retired in 1999 but not before they set flight-speed records that last to this day, including a top reported speed of 2,242.48 mph (3,608.92 kph) on a Los Angeles-to-Washington, D.C., flight that took only 64 minutes.

What was the X-15?

Just like Chuck Yeager's Bell X-1, which was the first plane to break the sound barrier, the X-15A-2, the fastest aircraft ever built, was dropped from the belly of a bomber plane—in this case, a B-52, one of the largest military airplanes ever built. When released, the X-15A-2's rockets ignited, taking it to a maximum speed of 4,534 miles per hour. That speed, which is equivalent to 7,297 kilometers per hour, is Mach 6. The X-15 series flew 199 flights before being retired in 1969.

What was the best-known commercial supersonic airplane?

From 1976 to 2003, the Concorde—a collaborative project between British and French aerospace companies and airlines—flew commercial flights at speeds exceeding Mach 2. The Concorde was a sleek plane with a delta wing design and a pinpoint nose that tilted down during takeoff and landing, carrying up to 120 passengers at an altitude of up to 18,000 meters (60,000 feet) above sea level. It was a fast but inefficient airliner, and although it was a high-profile business venture, it was never profitable. The plane was retired from service in 2003 but not before it had set a number of commercial aviation records, including the fastest commercial transatlantic flight ever: a Concorde took off from New York and landed in London 2 hours and 53 minutes later.

THERMODYNAMICS

What is thermodynamics?

Thermodynamics is the study of the change of thermal energy in objects and materials that makes them warm and cold, how they interact with each other, and the relationship between energy, heat, and work. It can be a challenging area of physics, in part because most of the vocabulary dates from the time before scientists understood what makes an object hot. Terms like "heat," "heat capacity," and "latent heat" suggest that warm objects contain some material that reacts to temperature. It wasn't until the early 1800s that our present understanding started to develop. Some 200 years later, our common usage of these terms is still based on earlier ideas.

What is thermal energy?

Thermal energy is the random kinetic energy of the moving particles—such as atoms and molecules—that make up matter. Objects expand when heated, so the bonds holding the particles together stretch. That means they have more elastic energy. So, thermal energy is the sum of the kinetic and elastic energy of the atoms and molecules and the bonds that hold them together. It is energy that is inside the object, so it is called a form of internal energy.

THERMAL PHYSICS

Who discovered what makes an object hot?

Benjamin Thompson, Count Rumford (1753–1814), who was born in the Massachusetts Bay Colony but did most of his scientific work in the Kingdom of Bavaria (which is now part of Germany), deserves a great deal of credit in discovering what makes things hot. Before his experiments, most scientists thought that hot objects contained an invisible fluid called caloric. Experiments done before Rumford showed that when you heated an object it didn't gain weight, so caloric must be weightless as well as invisible. This result made many scientists suspicious of the caloric explanation.

In 1789 Rumford drilled holes in bronze cannons through which a cannonball would be shot. He found that both the cannon and the metal chips that resulted from the drilling became hot. He determined the amount of water that could be raised to the boiling point by both the cannon bodies and the chips and showed that the caloric theory did not agree with his results. He finally concluded that in hot objects, the particles that made up the material moved faster than they moved in cold objects. Using our present terminology, they had more kinetic energy. In their motion they vibrate back and forth; they do not move together from one place to another like a thrown ball.

What is temperature?

Temperature is a quantitative measure of how hot an object or substance is. In many substances, temperature is proportional to the thermal energy in the object, but the relationship between temperature and energy depends on many factors.

What is entropy?

Entropy can be thought of as the dispersal of energy through a substance or object. The greater the dispersal or spreading, the larger the entropy. For example, when salt and pepper are in separate piles, the two substances are in distinct locations. If you mix them together, they are no longer in separate regions of pure salt and pure pepper but are dispersed throughout the combined pile. Furthermore, it is not possible for the salt and pepper grains to separate themselves. Thus, the mixing of salt and pepper increases the entropy of the system.

Another way to explain entropy is that it is a measure of how much disorder there is in a system of many particles. In the example above, the system of salt and pepper grains are orderly and well organized when they are in two separate piles but very disorderly when they are mixed together.

The amount of thermal energy in an object or system depends on both its temperature and its entropy. In a glass of ice water, for example, the water and the ice cubes are the same temperature, but water has more entropy per unit mass than ice, so the water contains more thermal energy than the ice.

MEASURING TEMPERATURE

How is temperature measured?

Temperature is measured using a device called a thermometer. There are many different kinds of thermometers, but they all have at least one property that depends on temperature.

For many years, the most common type of thermometer contained a thin column of liquid in a glass tube. When the temperature goes up, the liquid expands, and the height of the column increases. Earlier thermometers used mercury, a metal that is a liquid at room temperature. Because mercury is poisonous, all thermometers sold today contain red-colored alcohol. The glass tube has markings on it from which the temperature can be read.

What kinds of thermometers are there that don't use liquid?

Many thermometers today are electronic. Most of them contain a tiny bead of a semiconducting material whose electrical resistance varies with temperature. The bead is called a thermistor, or thermal resistor. Others contain a tiny semiconducting diode. The voltage across this diode varies with temperature. (Voltage and resistance will be discussed later in this book.) To measure very high temperatures, a wire made of the metal platinum is used because its resistance also varies with temperature, and platinum is not significantly affected by high temperatures.

Another kind of electronic thermometer uses two wires made of different materials that are welded together. Typically, the wires are made of copper and a nickel-containing alloy. This kind of thermometer is called a thermocouple and is often used in gas furnaces or water heaters to make sure that the pilot flame is burning when the gas to the main burners is turned on.

A less accurate but convenient thermometer is a strip of plastic whose color changes with temperature. The plastic contains liquid crystals. The geometric arrangement of molecules in the crystals depends on temperature, and so does their color.

If two different metals are bonded together in a strip, then when the temperature changes, the strip will bend. Because there are two metals, often brass and steel, the device is called a bimetallic strip. This kind of thermometer is often used in household thermostats.

How can the temperature of an environment be controlled?

A device that maintains a constant temperature is called a thermostat. A traditional home thermostat contains a coiled bimetallic strip. When the temperature drops below a set point, the strip trips a switch that turns on the furnace. More modern home thermostats use an electronic thermometer and electronic circuits that turn the furnace and air-conditioning system on and off.

What is a thermograph, and how is it used?

Thermographs, which detect the amount of infrared radiation emanating from objects or regions, use colors to display the temperature on an image. Typically, red indicates the warmest temperatures, while blue indicates cooler temperatures. Thermographs are used throughout science but are well noted for their use in detecting humans in wilderness areas, identifying areas of homes that need more insulation, and in measuring the temperature over regions of Earth.

Who invented the first thermometer?

Galileo Galilei (1564–1642) is credited with developing the first thermometer in 1592. His thermometer was open to the atmosphere, so it measured not temperature alone but rather a combination of temperature and atmospheric pressure.

It was not until 1713 that Daniel Gabriel Fahrenheit (1686–1736) developed the first closed-tube mercury thermometer. The following year, Fahrenheit also defined the temperature scale that bears his name and is still used today.

What is the Fahrenheit temperature scale?

Temperature scales are artificial in the sense that they are related to temperatures important to humans. The German physicist Daniel Gabriel Fahrenheit developed the first well-known temperature scale in 1714. Equipped with the first mercury thermometer, Fahrenheit defined a scale in which the freezing point of water was 32°F and the boiling point was 212°F.

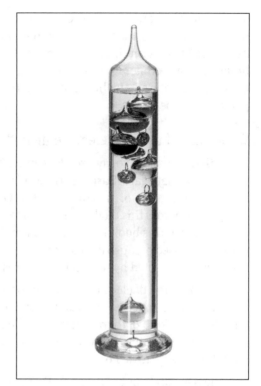

This thermometer is similar to one developed by Galileo. The objects inside the tube have varying densities that rise or fall, depending on the temperature.

Why, on the Fahrenheit temperature scale, does water freeze at 32 degrees?

Daniel Fahrenheit actually did not define 32° as the freezing point of water. Instead, he defined 0° as the freezing point of a particular mixture of water and salt. Since salt lowers the freezing point of water, the freezing point for this mixture was lower than it would have been for plain water. Upon defining the degree intervals between the freezing and boiling points of the water-plus-salt mixture, he found that water by itself freezes at 32°F.

Who developed the Celsius temperature scale?

On the Celsius scale (also called the centigrade scale), the freezing point of water is 0°, and the boiling point is 100°. The Celsius scale is named after a person whose life's work was dedicated to astronomy. Anders Celsius (1701–1744), a Swedish astronomer, spent most of his career studying the heavens. Before developing the Celsius temperature scale in 1742, he published a book in 1733 documenting the details of hundreds of observations he had made of the aurora borealis, or northern lights. Celsius died in 1744 at the young age of 43.

What do Celsius temperatures feel like?

A day with a high temperature of 100 degrees Fahrenheit, or 38 degrees Celsius, is generally considered to be very hot, while 0 degrees Fahrenheit, or –18 degrees Celsius, is considered very cold. Here is a handy table to convert Celsius temperatures to Fahrenheit:

Temperature	Celsius	Fahrenheit
Very hot day	40	104
Hot day	30	86
Room temperature	20	68
Sweater weather	10	50
Water freezes	0	32
Cold day	–10	14
Very cold day	–20	–4

How do astronomers measure the temperature of the sun?

There are two big challenges to measuring the temperature of the sun. First, the sun is so hot that it would immediately vaporize just about any thermometer. Second, the sun is so far away that bringing a thermometer there would be very impractical. The solution is to measure the electromagnetic radiation emanating from the sun's surface.

How does that work exactly? Think about a hot iron: you can feel the energy radiating from it without touching it. That radiation is in the form of infrared waves leaving the iron. When an object gets extremely hot, it glows red, and when it gets even hotter, it can take on a whitish glow. The temperature of a very hot object can be measured by the amount of radiation flowing from it as well as by the light it emits.

Scientists measure the temperature of the sun and other stars by analyzing their color and brightness. From such measurements, astronomers have determined that the surface of the sun is approximately 6,000 degrees C (11,000 degrees F).

If a mercury thermometer breaks, is it dangerous to touch the spilled mercury?

Mercury is a very dangerous metal that can cause great damage, especially to the kidneys, brain, and nervous system. Mercury from a broken thermometer should not be touched but instead scooped up with equipment and disposed of as a hazardous substance. Although mercury poisoning will probably not occur unless larger doses than what's in a thermometer are ingested or breathed after the mercury vaporizes, proper precautions should always be used when handling mercury. Note that mercury can be found not only in thermometers but also in barometers, which measure atmospheric pressure.

What are the surface temperatures of the planets in our solar system?

Just like temperatures on Earth, temperatures on other planets can vary. For the planets that have atmospheres (mixtures of gases surrounding the surface of a planet), the average temperature stays relatively constant because the atmosphere acts as a type of insulator. These planets have only small variations in the temperature when a section of the planet faces away from the sun. Mercury, on the other hand, with no atmosphere and an elliptical orbit, has very large differences. In the table below, the temperatures of Mercury, Venus, Earth, and Mars are taken on the planetary surface, while those of Jupiter, Saturn, Uranus, and Neptune are taken at the tops of the clouds, there being no solid surface on these planets.

Planet	Temperature Range (°C)
Mercury	−184 to +420
Venus	+440 to +480
Earth	−55 to +55
Mars	−152 to +20
Jupiter	−163 to −123
Saturn	−178
Uranus	−215
Neptune	−217

Why is the temperature on Venus so high?

Venus is so hot because it exhibits a runaway greenhouse effect. As its name implies, the greenhouse effect occurs on planets with atmospheres, and it causes a planet's surface to be warmer than it would be without that atmosphere. In a greenhouse on Earth, transparent glass walls, doors, and roofs let visible sunlight in, which then strikes the objects inside the greenhouse and converts into heat. The heat tries to escape as invisible infrared radiation, but the glass blocks the infrared light. Heat builds up inside the greenhouse, and its temperature is much warmer than the air outside. When the greenhouse effect happens on a planet, gases in that planet's atmosphere prevent heat from leaving the planet's surface as quickly as it would without those gases present. Carbon dioxide and water vapor are common gases that trap heat very effectively, so planets with thick atmospheres containing large amounts of these gases can get much warmer than they otherwise would be.

Scientists at NASA believe that Venus might have once had oceans and could have looked something like the photo above, but a runaway greenhouse effect made the planet uninhabitable.

Is there a highest temperature that can be reached?

There is no maximum temperature set by the laws of physics. The highest temperatures achieved on Earth to date have been at the centers of nuclear explosions, where the temperature can reach hundreds of millions of degrees. Out in the universe, temperatures can get even higher; in the centers of very massive stars and near the boundaries of supermassive black holes, temperatures can reach many billions of degrees.

On Venus, the greenhouse effect "ran away." The heat trapped by the Venusian atmosphere caused the surface temperature to get so high that the rocky crust began to release greenhouse gases like carbon dioxide. The atmospheric insulation consequently became even thicker, which caused more heat to be trapped, which caused the temperature to rise higher still, which caused even more greenhouse gases to be released. After finally reaching thermal equilibrium, Venus is now the inferno we see today.

ABSOLUTE TEMPERATURE

What is the lowest possible temperature?

The lowest possible temperature is called absolute zero. It is the temperature at which molecular motion is at a minimum and cannot be further reduced. While absolute zero can never be reached in a physical system that contains matter (see the Third Law of Thermodynamics later in this chapter), the present record low temperature achieved in the laboratory is only a few billionths of a degree.

What is the Kelvin scale?

The Kelvin temperature scale, developed in 1848 by the British scientist William Thompson, Lord Kelvin (1824–1907), is widely used by scientists throughout the world. Absolute zero is the temperature at which thermal energy is at a minimum. Each division in the Kelvin scale, called a kelvin (K), is equal to a degree on the Celsius scale, but the difference is where zero is. On the Celsius scale, 0° is the freezing point of water, while in the Kelvin scale, the zero point is at absolute zero. Therefore, 0 K is equal to –273.15°C, and 0°C is equal to 273.15 K. (The degree symbol is not used with Kelvin units.) The Kelvin scale is used for very low or very high temperatures when water is not involved.

How do the Fahrenheit, Celsius, and Kelvin temperature scales compare?

The following table provides some examples for the three commonly used temperature scales.

Temperature	Fahrenheit	Celsius	Kelvin
Absolute zero	−459	−273	0
Water freezes at Earth's surface	32	0	273
Normal human body temperature	98.6	37	310
Water boils at Earth's surface	212	100	373

How do you convert from one temperature scale to another?

You can use this table below to convert between Fahrenheit, Celsius, and Kelvin to the nearest degree.

From	To	Formula
Fahrenheit	Celsius	$°F = (9/5)(°C) + 32$
Celsius	Fahrenheit	$°C = (°F − 32) (5/9)$
Kelvin	Celsius	$K = °C − 273$
Celsius	Kelvin	$°C = K + 273$

STATES OF MATTER

What are states of matter?

Matter can be in different states, or phases, depending on the temperature, pressure, and entropy of the matter at the time. The three standard states of matter are solid, liquid, and gas. When gas is electrically charged, that is sometimes considered to be a fourth state of matter known as plasma.

What is the difference between solid, liquid, gas, and plasma?

In the solid phase, a material's atoms or molecules are held in rigid positions by the chemical bonds between them. They can vibrate but cannot change positions. In the liquid phase, molecules or small groups of molecules can move easily past one another. In the gas phase, the atoms and molecules have almost no forces between them, so they are free to move independently and can be much less dense than solids or liquids. To become plasma, one or more electrons must be removed from the atoms or molecules in a gas.

Where can we find plasma in everyday life?

Plasmas can be found in fluorescent lights, some television displays, and so-called neon signs. Lightning is a channel of plasma that runs briefly through the atmosphere. At very high altitudes, Earth's atmosphere is plasma. Out in the universe beyond Earth, the sun and the stars are mostly plasma, and much of the "vacuum" of outer space is actually very, very sparse plasma.

This plasma lamp—an apparatus you often see at science fairs and novelty shops—emits streams of plasma, electrically charged particles that are found in everything from stars to television displays and fluorescent lights.

What is unusual about the phases of water?

The chemical formula for the water molecule is H_2O. Solid H_2O is simply called ice, liquid H_2O is simply called water, and gaseous H_2O is simply called vapor or steam. In most liquids, the spacing between the molecules is slightly larger than in solids, giving them a lower density than their solid versions. The spaces are actually larger in ice than in water, however, meaning that ice has a lower density than water. This unusual property of H_2O means that ice floats in water—a very important phenomenon that helps make life on Earth possible.

What makes matter change from one state to another?

Although the space between molecules typically increases between the solid, liquid, and gas phases, the space between molecules does not actually determine the state of matter of a substance. Rather, it is the amount of thermal energy in a substance. When that amount changes, the matter can change from one state to another if a phase boundary is crossed. The phase boundaries are determined by the temperature, pressure, and entropy of the matter. For example, the phase boundary between water and ice at the atmospheric pressure at sea level on Earth occurs at a temperature of 0 degrees C (32 degrees F). Water at that temperature, however, has more entropy than ice at that same temperature, so to melt ice into water, extra energy must be added, even though the temperature stays the same.

Can matter change state from solid to gas without becoming liquid?

Yes, depending on the substance, the temperature, and the pressure. On Earth at sea level, for example, H_2O goes from solid (ice) to liquid (water) to gas (steam) as the temperature increases. Under those same conditions, however, CO_2 (carbon dioxide) goes from solid (dry ice) to gas without becoming liquid first. This process is called sublimation.

What determines the amount of energy required to increase the temperature of a substance?

Every substance has its own specific heat capacity, or the amount of heat needed to increase one kilogram of the substance one degree Celsius. For example, the specific heat capacity of water is 4,186 joules (4.186 kJ). The following table lists the specific heat capacities of some common solids, liquids, and gases.

Substance	Specific Heat Capacity (J/kg °C)
Aluminum	897
Copper	387
Iron	445
Lead	129
Gold	129
Silver	235
Mercury	140
Wood	1,700
Glass	837
Water	4,186
Ice	2,090
Steam	2,010
Nitrogen	1,040
Oxygen	912
Carbon Dioxide	833
Ammonia	2,190

How is specific heat capacity used?

Specific heat capacity is the energy per kilogram per change in temperature. So, to find the amount of energy needed to heat something, you need to know the material, its mass, and the change in temperature desired. For example, if you want to increase the temperature of 2 liters (2 kilograms) of water from room temperature (20°C) to the boiling point (100°C), you need to multiply the specific heat capacity of water (4,186 J/kg°C) by 2 kg and by 80°C to obtain 669,760 J, or about 670 kJ.

Suppose the water is placed in an aluminum pan with a mass of 300 g (0.3 kg). How much extra energy is needed to heat the pan? The answer is found by multiplying the specific heat capacity of aluminum (897 J/kg°C) by the mass (0.3 kg) and by the temperature change (80°C) to obtain 144,000 J or 144 kJ. So, the total energy needed to heat both the water and the pan is 814 kJ.

How is the high specific heat capacity of water useful?

Note in the previous example how much more energy is needed to heat the water than the pan. That is because water has a very high heat capacity compared to most common substances. That makes water a good material to use, for example, for cooling an automobile engine by circulating the water through the engine, where the water is heated, and then through the radiator, where flowing air can cool the water. Globally, water can store a great deal of heat from the atmosphere, which helps to keep the temperatures near ocean shores and throughout the world more stable.

How much energy is needed to change ice to water to steam?

The amount of energy needed to change phase is called latent heat. The latent heat involved in the transition from solid to liquid is called the latent heat of fusion, while the energy involved in the transition from liquid to gas is called the latent heat of vaporization. For water, the latent heat of fusion is 334 kJ/kg. The latent heat of vaporization is much higher: 2,265 kJ/kg.

Energy must be added to go from ice to water and water to steam, but if steam condenses to water, it releases 2,265 kJ for each kilogram of steam condensed. That's the reason that steam burns are so dangerous. Almost all of that energy is transferred to your skin. If water freezes, it releases 334 kJ for each kilogram of water frozen. In a freezer, that amount of energy must be removed by the freezing mechanism.

Why do water droplets sometimes accumulate on the outside of glasses and soda cans?

The water does not seep through the container but comes from the air surrounding it. Water vapor is the gaseous form of water that is in air below the boiling point of water. As discussed above, it takes a larger amount of energy to vaporize water, so the molecules of water in the air have more thermal energy than do the molecules in the colder glass. So, when the water molecules strike the glass, they transfer much of their thermal energy to the glass. The colder water molecules join together to form water droplets on the glass. The process is called condensation. Condensation also occurs on windowpanes when the outside is cold and the interior air is warm and humid.

Those water droplets on your soda can are the result of water vapor condensation occurring on the cold can's surface.

How do liquids evaporate?

A substance does not have to boil to change from liquid to gas state. The boiling point is where the pressure of the water vapor equals the atmospheric pressure. At all temperatures, there is a great variation in the kinetic energy of the molecules that make up the substance. When the molecules with high energy reach the surface, they have the possibility of escaping into the air. The process of changing from liquid to gas at a temperature below the boiling point is called evaporation.

When the molecules with higher energy leave the liquid, the remaining molecules have a lower average energy and thus a lower temperature. It is cooled by evaporation.

How can evaporation be a cooling process?

Evaporation leaves a cooler liquid behind, so the material on which the liquid rests is cooled. Evaporation is an effective way to cool our bodies. Perspiration leaves a coating of water on our skin that evaporates, cooling the skin. Alcohol evaporates more quickly than water and has an even greater cooling effect. For that reason, parents are advised to rub alcohol on the skin of babies who are running dangerously high temperatures. It can reduce the fever and keep the baby out of danger.

HEAT

What is heat?

Heat is the energy transfer from warmer objects to cooler objects. Thermal energy can be transferred in three general ways: conduction, convection, and radiation.

What is conduction?

If a warm object is in contact with a cooler one, the method of heat transfer is called conduction (see Figure 13). The faster molecules in the hotter object strike the slower ones in the cooler one, transferring their thermal energy. The average speed, and thus kinetic energy, of the molecules in the warmer object are reduced, while those in the cooler object are increased.

What is convection?

If a warm object is in contact with air or water, it heats the fluid. Hotter fluid has a lower density, so it rises. Its place is taken by colder fluid, creating moving currents of the fluid, called convection currents. Convection is a very efficient way of transferring energy from hot to cold objects.

What is radiation?

Thermal energy can also be transferred even if the warmer object isn't in contact with any other object. The vibrations of the molecules create electromagnetic waves that carry energy. These waves are called radiation. The warmer the object, the more energy

How do clouds form?

As warm air rises into the atmosphere through convection currents, the air expands as it experiences less atmospheric pressure. During the expansion, the warm water vapor quickly cools and condenses, forming water droplets in the air. When the droplets begin to accumulate, they attach themselves to particles in the air (like dust or smoke) and form clouds.

in the waves. More energy goes from the warmer to the cooler object than in the reverse direction, so radiation cools the warmer object and warms the cooler one.

How do different substances conduct heat?

When part of a substance is heated, its thermal energy is increased. The fast-moving atoms or molecules strike the slower-moving, cooler atoms. They begin to move faster and so gain thermal energy. Their temperature increases. The ease of conduction depends on the material. Most metals are good conductors—even a small difference in temperature produces heat flow. Other materials are poor conductors. There can be large temperature differences without significant heat flow. In that case, one part of the substance is hot while another part is cold.

Heat conductivity is higher in metals that have freely moving electrons. So, copper, silver, gold, and aluminum are good conductors. Stainless steel is a poor conductor.

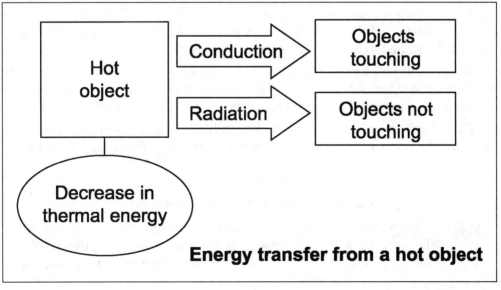

Figure 13.

In a nonmetal, conductivity depends on the substance's ability to transfer vibrations of the atoms. The conductivity of ice, concrete, stone, glass, wood, and rubber is less than 1/100 that of metals. Conductivity depends on the material, its thickness, the area it covers, and the temperature difference.

Light gases have better conductivity than heavier gases. For example, the heavy gas argon is used to fill the space between dual-pane windows because of its lower heat conductivity.

How does convection create wind?

During a day at the shore, the sun warms the ground and the water. The ground has a lower specific heat, so its temperature increases more than the temperature of the water. The ground heats the air above it, which rises in convection currents, and cooler air from over the ocean flows toward the shore to "fill in the gap" left by the rising warm air. This flow of cooler air from the ocean toward the shore creates what is known as a sea breeze.

In the evening, when the sun dips below the horizon, the ground cools faster than the water. Therefore, the air over the ocean is warmer than the air over the shore, and the reverse takes place. The warmer ocean air rises, while a breeze flows from the shore to the water.

Convection can cause breezes and winds not just at the shore but anywhere when there is a difference in temperature and pressure in the air from one place to another. Similarly, winds can blow either warm air or cool air depending on the atmospheric conditions.

How does clothing keep us warm?

Our skin is cooled primarily by convection currents in the air. Clothing, especially wool, traps air in small pockets, which reduces or eliminates convection. There can still be conductive cooling through the cloth, but cloth is a good insulator.

Home insulation, such as styrofoam panels, fiberglass, or blown-in cellulose, acts like clothing in reducing heat transmission by convection. Home insulation is rated by its R value of resistance to energy transfer. R is the inverse of conduction: $R=1/U$.

Following the same principles, air pockets within snow and ice function as excellent insulators. Many small mammals build snow dens to keep themselves warm, thereby

How can ice help keep plants warm?

Many farmers protect their crops during subzero temperatures by spraying water on the crops. When the water freezes, it covers the crops with a thin layer of ice. The plants are then insulated by the poor conductive properties of the ice.

taking advantage of the insulating properties of the snow. People in the Arctic build igloos that also keep their occupants warm by reducing loss by convection.

Why does a tile floor feel cold, but a rug on that floor feels warm?

Heat flow occurs when there is a difference in temperature between two objects. Heat only moves from warmer to cooler objects. The larger the temperature difference is between two objects, the greater the amount of heat flow there is.

A tile floor feels cooler than a carpeted floor because tile is a better conductor of heat. Both materials are the same temperature; however, when your warm foot makes contact with the tile, it heats the top layer, and the heat flows quickly through the tile. That makes the tile feel cool to your foot. The carpet is a poorer conductor, which means much less heat flows from your foot to the top layer of carpet and on to the rest of the carpet. As a result, the carpet feels warmer to your foot. This is an example of how heat flows by conduction.

What are some common methods of heating a home?

Homes using forced-air heat have a furnace that heats the air in it and a fan that blows the hot air into heating ducts that allow the hot air into the rooms. The hot air rises and forces the colder air out of the room through return ducts, the entrances to which are usually near the floor. Hot-water heat uses pipes to carry hot water. The pipes run through accordion-like metal fins that promote the convection of air beyond the hot-water pipe. This warmed air then circulates through the room. Electrical baseboard heat works in the same way. Electric resistance wiring in the floor or ceiling can warm the air in contact with these surfaces, again creating convection currents. Convection is the movement of thermal energy via a fluid (liquid or gas).

Why is it warmer to wear dark-colored clothes than light-colored ones?

Dark-colored objects absorb more electromagnetic radiation than light-colored clothes. The absorption of this radiated energy heats the dark material, increasing its temperature. This, in turn, increases the temperature of the air between the clothing and the person. So, wearing white clothing, which absorbs much less radiative energy from the sun, is typically cooler in hot climates, especially when one is standing in the sunlight.

Winter whites might look pretty, but if you wish to stay warmer on cold days, wear dark colors like black or navy blue or dark brown. You'll absorb more heat that way.

What is a calorie?

Heat, like work, is a process that transfers energy. The joule, named after the British scientist James Prescott Joule, is the international standard unit for measuring energy. A calorie is a unit of energy often used to measure heat flow. One calorie is the amount of energy needed to increase one gram of water by one degree Celsius. One calorie is equivalent to 4.186 joules.

What is a BTU?

Another unit used to measure heat flow is the British thermal unit, or BTU. One BTU is the amount of energy needed to increase the temperature of one pound of water by one degree Fahrenheit, and it is equivalent to 252 calories. Although the BTU is a nonmetric unit, it is still commonly used in the United States to measure energy production and usage. For example, the total amount of energy produced in all of the power plants in the United States in the year 2018 was about 100 quadrillion BTUs.

LAWS OF THERMODYNAMICS

What are the laws of thermodynamics?

As scientists learned about thermodynamics, they observed and developed four laws of nature that govern the physics of heat, work, and energy. Amusingly, they are numbered not 1, 2, 3, and 4 but rather 0, 1, 2, and 3.

What is the Zeroth Law of Thermodynamics?

The Zeroth Law is so named because it wasn't added as a law until after laws one through three were developed, but it was obvious to physicists in retrospect that it should precede the other three in order. It is based on thermal equilibrium between two bodies. As has been stated, if two objects have different temperatures, heat will flow from the hotter to the colder. If there is no temperature difference, there is no net heat flow: they will be in thermal equilibrium. The Zeroth Law states that if objects A and B are in equilibrium and B and C are in equilibrium, then A and C are also in equilibrium.

As an example, suppose object B is a thermometer. You put it in contact with object A. Heat flows until they are at the same temperature. You then move the temperature to object C. If the thermometer shows no change, then B and C are in equilibrium, and we can conclude that objects A and C are at the same temperature.

What is the First Law of Thermodynamics?

The First Law of Thermodynamics is a restatement of the conservation of energy. It says that the energy loss must equal the energy gain of a system. It relates the heat input and output, the work done on the system and the work the system does, and the change in internal energy, or temperature change.

For example, the cylinder and moving piston of an automobile engine is heated by the burning of gasoline in the cylinder. The piston moves out, doing work, and heat is transferred to the coolant because of the higher temperature of the cylinder and piston. As long as the temperature of the cylinder and piston does not change, the heat input equals the work output plus the heat output. Energy isn't gained or lost—it just changes form.

Another way to state the First Law is that net heat input equals net work plus change in thermal energy (Figure 14). Note that heat can be either positive (heat input) or negative (heat output). Work can also be positive (work done by the system) or negative (work done on the system). Thermal energy can also go up or down.

What is the Second Law of Thermodynamics?

One simple way to describe the Second Law is that it is impossible to convert heat completely into work in a cyclic heat engine; there is always some heat output. The Second

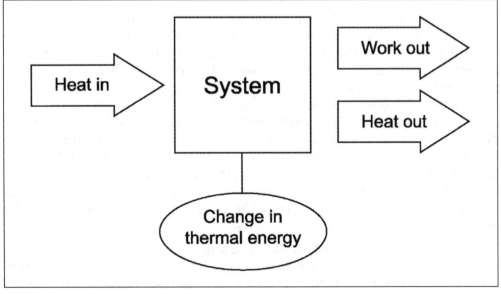

Figure 14.

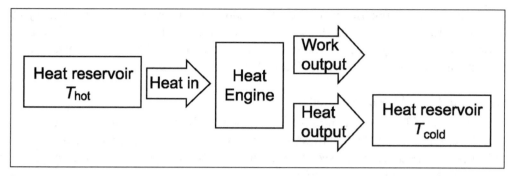

Figure 15.

Law can be shown in Figure 15, where the input heat is taken from a "reservoir" that maintains a constant high temperature, and the output goes to a second reservoir at a constant low temperature.

Note that the First Law would allow a heat engine to have no heat output; the heat input would equal the work done.

How was the Second Law of Thermodynamics developed?

In the early 1800s, many scientists and engineers worked to improve the efficiency of steam engines. The French military engineer and physicist Sadi Carnot (1796–1832) tried to answer two questions: Is there any limit to the amount of work available from a heat source, and can you increase the efficiency by replacing steam with another fluid or gas? Carnot wrote a book in 1824 called *Reflections on the Motive Power of Fire* that was aimed at a popular audience, using a minimum of mathematics.

The most important part of the book was the presentation of an idealized engine. This engine could be used to understand the ideas that can be applied to all heat engines. A heat engine is a device that converts heat into work in a cyclical process. That is, the engine periodically returns to its starting point. A steam engine and a gasoline- or diesel-power automobile engine are all heat engines where the pistons return to their starting positions. A rocket is not a cyclic heat engine. A diagram of Carnot's simplified model allowed him to give answers to his two questions. He found that the efficiency—that is, the work output divided by the heat input—depended only on the temperatures at which heat enters and leaves the system, or efficiency = $(T_{hot} - T_{cold})/T_{hot}$. It doesn't depend on the fluid or gas used in the engine. Real engines would have lower efficiencies, but no engine could have a higher efficiency. For example, a steam turbine gets its heat from high-temperature steam and puts its output heat into a much colder location like a lake or river.

The Second Law of Thermodynamics, as explained by Carnot, means that there is no such thing as a perpetual motion machine. Although such a machine could obey the Law of Conservation of Energy, the heat it produces means that the machine must eventually stop.

What happened to Sadi Carnot and his ideas about engines?

Sadly, Sadi Carnot died in a cholera epidemic at age 36. Because people were concerned about the transmission of this deadly disease, all his papers and books were buried with him after his death. Thus, only a few of his works have survived.

Despite the limitations of Carnot's model, his work inspired Rudolf Diesel (1858–1913) to design the theoretical engine named after him to achieve higher efficiencies than steam engines could at the time. Carnot's book was therefore very important to the design of practical engines, and its principles are still taught today.

How efficient are vehicles and electric generators today?

In the United States, about 75 percent of energy used for transportation is wasted as heat and exhaust. Electrical generators are somewhat better but still only 31 percent efficient—they waste about 69 percent of the energy. Engineers are working on both improving efficiency and in making use of the "rejected" energy. For example, the warm water that carries away the waste heat in an electrical generating plant could be used to heat homes close to the plant.

How do refrigerators and air conditioners work?

When a liquid evaporates into a gas, it is cooled. Heat flows into the system (see Figure 16, next page). The opposite process, the condensation of a gas into a liquid, results in

This turbine plant in California uses natural gas to heat water into steam that then produces energy in the turbines that generate electricity. Not all the energy from the natural gas can be converted to electricity, and lost energy escapes as heat.

131

an increase in thermal energy and an output of heat. A refrigerator circulates a refrigerant, a liquid that evaporates at a low temperature, through tubes. The gas is then compressed by an electrically driven compressor. The pressure and temperature of the gas increases. Coils of the tubing outside the refrigerator cool the liquid and heat the air around them. As it cools, the refrigerant condenses back to a liquid that goes through a tiny hole, called the expansion valve. The pressure drops, evaporating the liquid and making the gas cold. The tubing containing the cold gas is inside the refrigerator, making it cold and cooling the food inside it.

An air conditioner works in a similar way. The evaporator is in the unit inside the house, and the compressor is outside. Because of the work put into the compressor, heat is removed from the air inside the house and transferred to the outside air. Figure 16 shows work and heat flows in a refrigerator or air conditioner.

What are the environmental impacts of refrigerants?

The first home refrigerators used ammonia as a refrigerant, but ammonia is toxic. In the 1930s, a chemical called Freon was developed by the DuPont Company of Wilmington, Delaware. Freon is a chlorofluorocarbon (CFC). If Freon escapes, it carries chlorine atoms to the upper atmosphere. There, ultraviolet radiation from the sun separates one of the chlorine atoms from the CFC. That atom converts ozone to oxygen, contributing to the destruction of the ozone layer, an essential barrier against harmful ultraviolet sunlight.

Freon's destructive nature was known since the 1970s, but it was not until the early 1990s that legislation was implemented banning the use of Freon in new air conditioners and refrigerators. In 2002 there was an estimated six million tons of Freon in existing products. Unfortunately, when the chlorine destroys an ozone molecule, the chlorine is not destroyed but continues to live for a while, destroying more ozone. In fact, more Freon is still headed toward the upper limits of the atmosphere, for it can take several years for Freon to reach such elevations.

After laws were passed banning Freon, DuPont and other corporations developed alternatives to Freon that replace the chlorine atoms with hydrogen atoms. These substances do not harm the ozone layer and are used today in refrigerators, air conditioners, and aerosol cans.

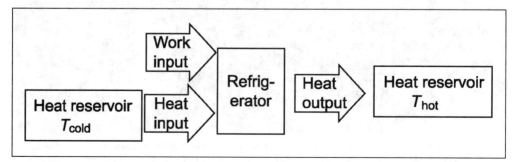

Figure 16.

> ### What is one way to remember what the laws of thermodynamics are?
>
> It is possible to state the First, Second, and Third Laws of Thermodynamics metaphorically with a pithy statement that compares heat, energy, and engines to gambling in a casino: "You can't win, you can't break even, and you can't get out of the game."

What was the social importance of using less-harmful refrigerants?

Freon and other chlorofluorocarbons (CFCs) posed a global environmental health threat in the twentieth century that was caused by humans. When governments worldwide worked together to require the use of new, less-harmful refrigerants, the threat was greatly reduced. By the early twenty-first century, the loss of stratospheric ozone had been brought under control. The successful restoration of the ozone layer serves as an important example of how the world's nations can work together to solve large, global environmental issues.

What is the Third Law of Thermodynamics?

The Third Law of Thermodynamics states that absolute zero, the lowest possible temperature, can never be reached. The entropy of a system is zero at absolute zero. A procedure can remove a portion of the entropy but not all of it. Thus, it would take an infinite number of repetitions of the procedure to reach absolute zero. It has been possible to achieve temperatures as low as a few billionths of a kelvin, but it has never been and never will be possible to reach absolute zero.

What does entropy have to do with the laws of thermodynamics?

Entropy is the energy of disorder. As was mentioned early in this chapter, it is not possible to make something disorderly—like a mixed pile of salt and pepper—into something orderly—like a pile of only salt and another pile of only pepper—without adding more energy than was originally there due to the entropy. In other words, the entropy of a closed (isolated) system can never decrease. This is another way to state the Second Law of Thermodynamics.

Who helped develop the concept of entropy?

The German physicist Rudolf Clausius (1822–1888) was not satisfied with Sadi Carnot's use of the term "waste heat" in Carnot's development of the Second Law of Thermodynamics. Clausius eventually developed another version of the Second Law that involves the concept of entropy. He defined entropy as the dispersal of energy: the greater the dispersal or spreading, the larger the entropy.

133

What does entropy have to do with life and the universe?

The total amount of entropy in the universe is always increasing. In a living organism like a person, however, disorderly matter (like carbon, iron, and oxygen atoms) is continually built into orderly systems (like DNA, cells, and organs). Life is thus a special example of entropy decreasing in small, limited parts of the universe while it is increasing everywhere overall.

If we take this idea further, life may be so fundamental to the way the universe works that the presence of life defines the universe itself. In one model called the biocentric universe, abstract ideas like time and space come about because living things observe them and construct them into a shared existence. In a way, this is like an extension of the old riddle: if a tree falls in the woods and nobody hears it, does it make a sound? If the decrease of entropy in living systems is what has led to consciousness, intelligence, and thought, perhaps it is also connected with the order and structure of the cosmos as a whole. Biocentrism is an intriguing scientific philosophy that has yet to be experimentally confirmed.

After Clausius had given Carnot's work a firmer foundation, later work by the Austrian Ludwig Boltzmann, the American Josiah Willard Gibbs, and the Englishman James Clerk Maxwell developed the statistical basis for entropy that is used today.

What do entropy and thermodynamics have to do with time?

Suppose you place ice in water. Ice and water are separate. The water has higher thermal energy and the ice has lower, so the system has low entropy. When the ice melts, the two can no longer be separated—the process is now irreversible. The thermal energy is dispersed throughout the system, so the entropy has increased. But wait, you might say: the ice water (the system) has cooled the air around it (the environment), decreasing its entropy because hot gas has greater entropy than cold gas. Calculations, however, show that the increase in the ice-and-water mixture is greater than the decrease in the air. This example demonstrates how one way to state the Second Law of Thermodynamics is that the entropy of the system and the environment can never decrease.

This continual increase of entropy in the universe suggests a direction of time, sometimes called the "thermodynamic arrow of time." The "forward" direction of time is the one in which entropy increases.

WAVES

What is a wave?

A wave is a traveling disturbance that moves energy from one location to another without transferring matter. Oscillations in a medium or material create mechanical waves that propagate away from the location of the oscillation. For example, a pebble dropped into a pool of water creates vertical oscillations in the water, while the wave propagates outward horizontally along the surface of the water.

What are the two main forms of waves?

Transverse waves and longitudinal waves are the two major forms of waves.

In a transverse wave, the wave itself and its energy move away from the source perpendicular to the direction of the oscillations. A transverse wave can be created, for example, by repeatedly shaking a string or rope up and down. The string moves up (the highest point is called the "crest") and down (the lowest point is called the "trough") while the wave travels.

The oscillations in longitudinal waves move in the same direction that the wave is moving. A longitudinal wave can be created, for example, by repeatedly pushing and pulling on a long, soft spring.

Can the two main forms of waves be combined?

Yes, they often do combine. Ripples in a pond, for example, are a combination of transverse and longitudinal waves that can move outward in circles. As in the case of transverse and longitudinal waves alone, energy is transferred, but matter is not moved.

What determines the velocity of a wave?

The velocity of a wave depends upon the material or medium in which it is traveling. Typically, the stronger the coupling between the atoms or molecules that make up the medium and the less massive they are, the faster the wave will travel. All waves of the same type (transverse or longitudinal) travel at the same speed through the same medium. For example, a sound wave in air at 0°C will travel at 331 meters per second regardless of the sound's frequency or amplitude.

The velocity of water waves depends both on the properties of the water and on the frequency of the wave.

Electromagnetic waves are special transverse waves that can travel either through empty space or through material. Their velocity depends on the electric and magnetic properties of space or the material but not on the frequency or amplitude of the wave.

What happens when a longitudinal wave travels through a medium?

The medium (for example, air or water) in longitudinal waves alternately pushes close together (this is called compression) and separates (this is called rarefaction). One very good example of longitudinal waves in a medium are sound waves, which are a series of back-and-forth, longitudinal oscillations of atoms or molecules that form alternate regions of high and low pressure in a medium such as air.

What are amplitude, period, frequency, and wavelength?

These are terms used to describe the properties of a wave. A wave's amplitude is the distance from the midpoint of a wave to the point of maximum displacement (a crest in a transverse wave or the compression of a longitudinal wave). The period is the amount of time it takes a wave to complete a full cycle (from crest to trough back to crest, or compression to rarefaction back to compression). The frequency is the number of cycles of the wave that occur in one second. The wavelength of a wave is the distance from crest to crest or from compression to compression.

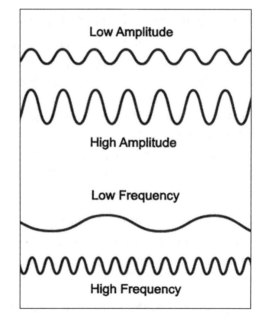

What is the relationship between a wave's period and frequency?

The period and the frequency of a wave are always reciprocal quantities. For example, if the period of a wave is 0.001 second, then the frequency is 1,000 cycles per second or 1,000 hertz (Hz); and if the period of a wave is 1,000 seconds, then the frequency is 0.001 cycle per second or 0.001 Hz. Mathematically, if a wave's frequency is given by f, and the period of a wave is given by T, then $f = 1/T$ and $T = 1/f$.

Amplitude describes the height of a wave; frequency describes the number of waves within a certain period of time.

What is the relationship between a wave's velocity, frequency, and wavelength?

Suppose you shake a rope up and down at a constant rate. The rate or frequency is the number of times your hand is at the top of its motion per second. As the waves move along the rope, the distance between the crests of the rope—that is, the wavelength—will remain the same. The wavelength depends both on the frequency of oscillation and on the velocity of the wave along the rope. The relationship is velocity = frequency times wavelength ($v = f\lambda$) or wavelength = velocity/frequency ($\lambda = v/f$).

Note that if the frequency of a wave increases, the wavelength decreases; the velocity of that wave, however, is a property of the medium (rope) and doesn't change. In other words, the frequency and wavelength of a wave are inversely proportional to each other.

On what does the amplitude of a wave depend?

The amplitude depends on the amount of energy that the source of the wave puts into the wave. For example, the amplitude of a transverse wave in a rope depends on how hard you shake it. For a sound wave, it depends on how much compression the loudspeaker or musical instrument creates. The amplitude of a wave does not depend on its frequency, wavelength, or velocity.

Does the amplitude of a wave depend on its distance from the source?

The energy carried in a wave depends on the wave's amplitude and its velocity. Waves can be put into two general categories: those that spread, like water waves on a pond, sound waves, or electromagnetic waves; and those that are confined to a narrow region, like waves on a rope or electrical oscillations on a wire. A water wave spreads on the surface of a pond, lake, wide river, or ocean. As it spreads, its energy is spread over an increasingly large area, so the energy transmitted to a particular location is reduced when the source is farther away. Therefore, the amplitude of the water wave is also reduced in proportion to the distance from the source. Sound and electromagnetic waves usually spread in two dimensions. Again, as they spread, the energy carried is also spread, so as the distance from the source is increased, the amplitude is decreased, but this time as the square of the distance.

On a rope or wire, the wave doesn't spread. A different mechanism often reduces the amplitude. In a rope, there is friction between the fibers, which changes some of its kinetic energy to thermal energy. If a signal is sent through a wire as an oscillating voltage, the resistance of the wire will convert some of the electric energy to thermal energy, reducing the voltage and thus the amplitude of the wave. This loss can be reversed by placing amplifiers along the wire, which can put more energy into the wave and increase its amplitude.

WATER WAVES

What type of wave is a water wave?

Ocean or water waves look like transverse waves, but they are actually a combination of both transverse and longitudinal waves. The water molecules in a water wave vibrate up and down in tiny, circular paths. The circular path of the water wave creates an undulating appearance in the wave.

How fast does wind have to blow to produce different kinds of water waves?

Wind rubbing against the water surface is a major cause of waves. Since the water cannot keep up with the wind velocity, the water rises and then falls, creating the familiar wavelike motion. Depending upon the wind velocity and the distance the wind has been able to travel over the water, different-sized waves are generated.

Wind speeds over bodies of water are often measured in units of knots, or nautical miles per hour. This table shows the general forms of water waves that are created by winds.

Type of wave	Wind Velocity	Effects
Capillary waves	< 3 knots	Tiny ripples—the longer they are generated, the larger their amplitude
Choppy waves	3–12 knots	Combined capillary waves that have traveled and formed larger waves
Whitecaps	11–15 knots	Amplitude of wave must be more than 1/7 the wavelength in order to break into a whitecap
Ocean swells	No specific speed	Form over long distances from a combination of different waves

How do speeds of ocean waves vary?

The speed of an ocean wave depends on the distance between two successive crests—its wavelength. The longer the wavelength, the faster the wave travels. A small surface wave, such as a ripple created by the wind, travels quite slowly because it has such a short wavelength. Swells—the larger waves created by constant winds—have longer wavelengths and travel at higher velocities. The energy that the wave carries depends on the square of the height of the wave, which explains why high waves can cause so much damage to shorelines.

Why do ocean waves break as they approach the beach?

Water waves rarely break, or form whitecaps, when they come in contact with a cliff or mountainside shoreline. They usually only break as they approach a gradual decrease in depth such as a beach. A shoreline with a gradual decrease in depth will produce more spectacular whitecaps than a wave that encounters a steep decrease in depth.

Waves break as they approach a shoreline because the lower part of the wave moves more slowly than the top part of the wave due to increasing friction with the shallower ocean bottom.

The reason waves break is the result of the way the wave velocity depends on the depth of the water. Consider a water wave with a large amplitude. As the wave moves toward the beach, at first, it travels at a constant velocity. As the ocean depth begins to decrease, the bottom of the wave gradually encounters more and more friction with the beach, causing the lower part of the wave to travel more slowly than the upper part. As the lower part slows down, the crest, moving faster, moves over the trough. When there is not enough water to support the crest, the wave breaks or forms a whitecap.

How is surfing like downhill snowboarding?

Surfing is done on water, while snowboarding is done on snow. Still, they are similar in that in both cases, the athlete and the board travel down a hill. For snowboarding, the hill is a mountain covered with snow, while in surfing, the hill is the rising water of a breaking ocean wave. An ideal surfing wave has a large amplitude as it reaches an extremely gradual decrease in ocean depth. While the surfer moves down the wave, the water on the front edge of a crest continually rises underneath the surfer, allowing the surfer to ride down the wave without actually moving downward.

What kinds of beaches are best for surfing?

The best surfing beaches are located along the edges of oceans when wind conditions have produced waves with large wavelengths. Another requirement of a good surfing beach is a gradual decrease in water depth.

139

Some of the best surfing is done in Hawaii—at Waikiki Beach on Oahu in the summer and the north shores of Oahu and Kauai in the winter. In the continental United States, the best surfing is in southern California. The U.S. shore of the Pacific Ocean, famous for its long wavelengths and gradually decreasing-depth beaches, has some of the best surfing in the world.

What is a tsunami, or tidal wave?

A tsunami, or tidal wave, is not caused by windy conditions or tides but by major underwater disturbances such as earthquakes and volcanic eruptions. The seismic disturbances create huge upward forces on the water, the opposite of dropping rocks into water. A tsunami is a series of several waves with a period of more than 30 minutes between each crest. The ocean first recedes from the beach; then water rushes inland at a very high speed.

Large tsunamis can be extremely destructive upon reaching the shoreline due to their amplitudes.

What disaster was caused by a tsunami in Japan on March 11, 2011?

On March 11, 2011, a huge undersea earthquake off the coast of Tohoku, Japan, created a tsunami that reached a height of 40 meters (130 feet). The tsunami traveled more than 10 kilometers (6 miles) inland in some places, destroying or damaging more than

A scene of the aftermath of the 2004 tsunami shows the devastation of Banda Aceh, Indonesia. Almost a quarter of a million people died because of the natural disaster.

one million buildings. Even though the Japanese people had substantial disaster preparations in place, more than 15,000 people were killed, some in evacuation shelters that were washed away.

To add to this terrible disaster, the earthquake and tsunami led to a series of events that caused the Fukushima Daiichi nuclear power plant to melt down and explode, spreading radioactive contamination in the region. To this day, there are still places as far away from the former power plant as 30 kilometers (18 miles) where no one is allowed to live.

What was the deadliest tsunami in modern history?

The most deadly tsunami ever recorded occurred December 26, 2004, in the Indian Ocean. The earthquake that caused it was off the west coast of Sumatra, Indonesia. That quake released an amount of energy equivalent to 550 million times the energy released in the Hiroshima nuclear bomb. One part of the ocean bottom was lifted by 4 to 5 meters (13 to 16 feet) and moved horizontally 10 meters (33 feet). The tsunami, traveling at a speed of 500 to 1,000 kilometers per hour, had a low amplitude (about 60 centimeters, or 24 inches) in mid-ocean, but when it crashed into the coasts from Thailand to India and as far as South Africa, it had an amplitude as high as 24 meters (79 feet). More than 200,000 people were killed and more than a million made homeless.

ELECTROMAGNETIC WAVES

What is an electromagnetic wave?

An electromagnetic wave is a special kind of wave that consists of two transverse waves: one an oscillating electric field, the other a corresponding magnetic field perpendicular (at right angles) to the electric field. Unlike other waves, electromagnetic waves do not need a medium to travel through; they can travel through a vacuum, including the vacuum of space.

What is the relationship between electromagnetic waves and light?

Electromagnetic waves are light waves at all colors and wavelengths, visible or invisible. Scientists, in fact, refer to electromagnetic waves as "light waves" and electromagnetic radiation, whether or not humans can see it, as simply "light." Whether or not we can detect a particular electromagnetic wave depends on the wave's frequency or wavelength. In a vacuum, electromagnetic waves travel at the speed of light (299,792,458 meters per second, or 186,282.4 miles per second). That's because they *are* light.

What is the electromagnetic spectrum?

The electromagnetic spectrum is the wide range of electromagnetic (EM) waves from low to high frequency. The spectrum ranges all the way from low-frequency radio waves to microwaves, infrared radiation, visible light, ultraviolet radiation, X-rays, and very-high-frequency gamma rays. In the middle of the electromagnetic spectrum is a small region containing the frequencies of visible light, which follow the colors of a rainbow in order of low to high frequency: red, orange, yellow, green, blue, indigo, and violet.

How can an electromagnetic wave be created and detected?

Electromagnetic waves can be created by accelerating electrons. The electrons create an oscillating electric field. This field in turn creates an oscillating magnetic field, which creates another oscillating electric field, and so on. The energy carried by the waves radiates into the area around the moving charges. When it strikes a material whose electrons can move freely, it causes these particles to oscillate.

Who predicted the existence of electromagnetic waves?

In 1861 James Clerk Maxwell (1831–1879) demonstrated the mathematical relationship between oscillating electric and magnetic fields. In his *Treatise on Electricity and Magnetism*, written in 1873, Maxwell described the nature of electric and magnetic fields using four differential equations, known to physicists today as "Maxwell's equations." Putting the four equations together, Maxwell predicted the existence of the electromagnetic wave.

Who confirmed that electromagnetic waves exist?

Heinrich Hertz (1857–1894), a German physicist, was the first person to demonstrate that electromagnetic waves existed. He designed a transmitter and receiver that produced waves with a 4-meter wavelength. He used standing waves to measure their wavelength. He showed that they could be reflected, refracted, and polarized and could produce interference. It was Hertz's breakthroughs in electromagnetic waves that paved the way for the development of radio. In 1930 Hertz was honored by having the unit of frequency, which is cycles per second, named after him as the hertz (Hz).

How did communication with electromagnetic waves start?

An electromagnetic wave with no changes in amplitude or frequency carries no additional information; it cannot be used for communication. The first method of using these waves to communicate was to switch them on and off in regular patterns. Letters were represented by a combination of long and short pulses using what is called Morse code, named after the American Samuel F. B. Morse (1791–1872), who developed the code to transmit information over wires—the telegraph.

Later, in 1876, the American Alexander Graham Bell (1847–1922) invented the telephone, which allowed communication over wires by voice rather than just dot-and-dash pulses.

Who invented communication with radio waves?

In 1895 Guglielmo Marconi (1874–1937), a 20-year-old Italian inventor, created a device that transmitted and received elec-

Samuel F. B. Morse was famous for inventing the Morse code, that allowed people to first transmit messages over telegraph wires.

tromagnetic waves over a distance of about one kilometer (3,280 feet). Later improvements to his antenna and the development of a crude amplifier enabled him to receive a British patent for his wireless telegraph. In 1897 he transmitted signals to ships 29 kilometers (18 miles) from shore, and in 1901 he was able to send wireless messages across the Atlantic Ocean. As a result of Marconi's work on radio transmitters and receivers, he was the cowinner of the 1909 Nobel Prize in Physics. Over the next decade, transmitters and receivers were improved enough that they could be installed in oceangoing ships.

What was the early history of wireless radio communications?

For voice communications across long distances, such as with telephones, the voices had to be amplified to be heard. In 1906 the American inventor Lee De Forest (1873–1961) invented a vacuum-tube amplifier he called the Audion. It took until 1915 for a radio receiver to be sold using Audions. In 1916, De Forest developed an Audion-based transmitter that allowed dance music to be transmitted 40 miles. A number of other experimental stations demonstrated the transmission of music by "wireless." Eventually, people started calling wireless transmission "radio." A large number of amateur radio operators, nicknamed "hams," made significant advances in radio broadcasting around this time.

When the United States entered World War I in 1917, all radio stations not owned by the government were shut down, and it became illegal for people to listen to any radio transmission. During that war, radio was used to communicate between ships and land and from ship to ship.

In this 1901 photo, Guglielmo Marconi demonstrates his radio transmitter and receiver. The invention earned him the 1909 Nobel Prize in Physics.

How did commercial radio stations develop?

After World War I, amateurs (nonmilitary users of wireless) in the United States were restricted to using only one wavelength of radio waves: 200 meters, which had a frequency of 1,500 hertz. Radio waves with wavelengths shorter than that were thought to be useless for government use.

Technological innovation quickly showed that this assumption was incorrect. Within a year, one amateur wireless user was able to send signals 3,000 miles. Companies began to use radio for specialized needs like to send time information to jewelers to allow them to set their clocks. From 1919 through 1921, radio was mostly used to transmit musical concerts. The first transmission of a football game occurred in November 1919. In 1921 transatlantic voice transmissions were made.

By 1922 newspapers had developed radio stations transmitting news, weather reports, crop reports, and lectures. Large companies, such as General Electric, Westinghouse, AT&T, and RCA, began to be involved in developing commercial broadcasting. From 1922 to 1923, as the number of stations grew without regulation, chaos reigned. In 1928 the government announced new assignments in the frequency band 550 to 1,600 kHz—the AM radio band that we still use today. Many more assignments have been added for commercial use since World War II, including the FM radio band and cellular telephone bands.

How does an antenna transmit and receive signals?

An antenna for radio and television signals can used to transmit and receive electromagnetic radio waves. Oscillating voltages produced by a radio or television transmitter cause the electrons in a metal wire or rod—that is, the transmitting antenna—to oscillate, creating an oscillating electric field that in turn creates an oscillating magnetic field that creates another oscillating electric field. The combined electric and magnetic wave moves away from the antenna at the speed of light.

A receiving antenna is a metal rod, wire, or loop. When an electromagnetic wave strikes the antenna, it causes the electrons in the metal to oscillate at the same frequency as that of the wave. The oscillating electrons produce a voltage in the receiver that is eventually decoded and results in the sounds or pictures produced by a radio or television.

What frequencies of electromagnetic waves are used today for communication?

The table below shows the regions of the electromagnetic spectrum and their uses. The unit for frequency is hertz (Hz). Hertz is the name for the number of oscillations or cycles per second of a wave. The letters kHz mean one thousand hertz, MHz a million hertz, and GHz a thousand million (or a billion) hertz.

Frequency Range	Wavelength Range	Name & Abbreviation	Use
Less than 30 kHz	> 10 km	Extremely low frequency (ELF)	Submarine communication
30 kHz to 300 kHz	10 km to 1 km	Low frequency (LF)	Maritime mobile, navigational
300 kHz to 3 Mhz	1 km to 100 m	Medium frequency (MF)	Radio broadcasts, land and maritime mobile radio
3 Mhz to 300 Mhz	10 m to 1 m	Very high frequency (VHF)	Television broadcasts, maritime and aeronautical mobile, amateur radio, meteorological communication
300 Mhz to 3 Ghz	1 m to 10 cm	Ultrahigh frequency (UHF)	Television, radar, cell phones, military, amateur radio
3 Ghz to 30 Ghz	10 cm to 1 cm	Superhigh frequency (SHF)	Radar, space and satellite microwave communication, wireless home telephones, wireless computer networks
30 Ghz to 300 Ghz	1 cm to 1 mm	Extremely high frequency (EHF)	Radio astronomy, radar

How does an antenna's size and shape affect how it works?

The length of an antenna determines the frequency that it best receives. The most efficient antennas have a length equal to half the wavelength of the wave it is receiving. This allows the induced electrical current in the receiving antenna to resonate at that particular frequency. If the antenna is a simple rod, it is most sensitive when its length is one-quarter the wavelength.

A loop- or coil-shaped antenna is typically used for the low-frequency, long-wavelength signals in the AM radio band. That's because a half-wavelength straight wire antenna would be longer than a football field! Shorter wires or rods can be used and are more efficient if coils of wire are used to "load" the antenna.

FM radio and television antennae used in most homes and buildings are designed to receive a broad range of frequencies but with less sensitivity. Antennas for the ultra-high frequencies used in digital, high-definition televisions are much shorter and can easily be mounted outside on rooftops or on top of television sets.

PUTTING INFORMATION ON ELECTROMAGNETIC WAVES

What's the difference between analog and digital signals?

All transmitters create what is called a carrier wave at a specific frequency. For licensed commercial broadcasters, the Federal Communications Commission (FCC) assigns the frequency. The carrier wave transmits no information. For information to be carried, some property of the carrier wave must be changed, or encoded, in such a way that a receiver can reconstruct, or decode, the information being sent. The two basic ways of encoding and decoding are analog, which uses continuous electrical signal, and digital, which breaks the signal into many small pieces. The earliest methods of radio broadcasting used analog signals from a source like a microphone.

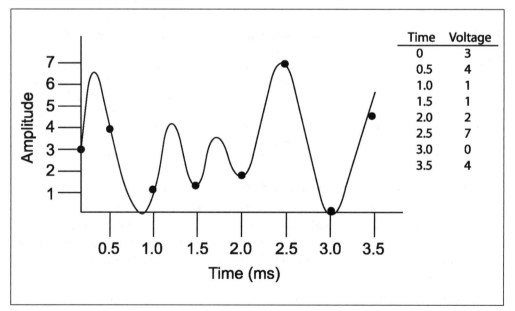

Time	Voltage
0	3
0.5	4
1.0	1
1.5	1
2.0	2
2.5	7
3.0	0
3.5	4

Figure 17.

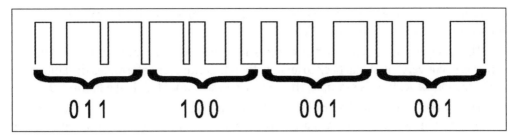

Figure 18.

How does analog-to-digital conversion work?

Sound striking a microphone produces a varying voltage output. That is, the output of a microphone has the same shape of the amplitude of the sound wave that strikes it. Figure 17 represents an analog signal from a microphone.

To create a digital signal, the analog output of the microphone is converted into a series of numbers. For example, consider Figure 17 again. Suppose the signal is "sampled" two thousand times each second. That is, the voltage is recorded every 0.5 millisecond. The dots show the results of the sampling. Voltage can be only in whole numbers. The series of numbers representing the waveform would then be those shown in the table to the right of the smooth graph.

Next, the voltages are converted into binary digits, a series of 0s and 1s. The table on the right shows the conversion to binary for numbers 0 through 7. The binary digits reading left to right represent the numbers 4, 2, and 1. Thus, 5 = 4 + 1, or 101.

Now the binary numbers are converted into a series of voltages to be sent to the transmitter. One method is called Pulse Width Modulation, or PWM. A narrow pulse

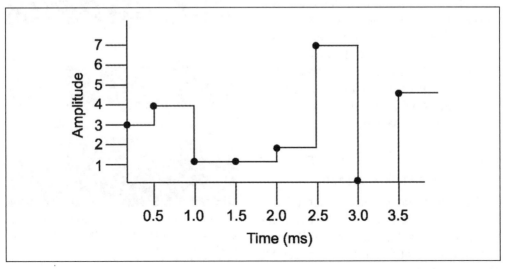

Figure 19.

represents a 0, and a wider pulse represents 1. For example, the first four samples would produce the wave train shown in Figure 18.

The receiver is instructed that three bits should be converted into a number. If the receiver is to convert the signal back to analog, then that number determines the amplitude of the analog wave. In this example, the converted wave would look like Figure 19.

This doesn't look like the original analog signal, does it? Our example had two significant problems. First, the "wiggles" in the wave are missing. Second, with only 8 choices of voltages, the vertical resolution is too small. To make a more accurate conversion, you need to sample the wave more often—at least once per millisecond—and to allow many more than 8 voltage choices. To obtain the quality of sound in a typical compact disc or streaming video, the sampling interval is at most 22 μs (22 microseconds), and there are at least 4,096 voltage choices.

What do AM and FM on a radio mean?

From its beginnings, commercial radio stations used analog broadcasting methods. AM (amplitude modulation) and FM (frequency modulation) are methods of transmitting information on a radio. In each case, a property of the "carrier" wave is changed or modulated. These variations are what hold the information that the carrier wave transmits.

AM was the first method invented, in part because it is simpler to transmit and receive AM radio waves. The American electrical engineer Edwin Howard Armstrong (1890–1954) demonstrated transmission and reception of FM in 1935. Listeners were amazed at the way that FM eliminated "static" because noise, like that due to thunderstorms, changes the amplitude but not the frequency of signals. The FM receiver's output does not depend on the amplitude of the signal and is therefore unaffected by interference with amplitude.

Due to opposition from the radio networks and receiver manufacturers, commercial FM broadcasts were delayed and did not become widely available until after World War II. Today, most of the music broadcasts on commercial radio use the FM channels, while most news and talk radio channels broadcast on AM.

Radios once only had AM (Amplitude Modulated) frequencies available; then came FM (Frequency Modulated) in 1939. Today, there is talk among radio professionals that AM stations might become a thing of the past.

Where are AM and FM broadcasts on the electromagnetic spectrum?

In the United States, commercial AM stations broadcast at frequencies between 550 kHz and 1600 kHz. Commercial FM stations broadcast at frequencies between 88 MHz and 108 MHz.

Within each band, the individual stations need to be spaced apart so their signals don't interfere with one another. In the AM band, the frequencies of radio stations are spaced at least 10 kHz apart. At the very high frequencies used by FM stations, more bandwidth is available, and stations are spaced by 200 kHz.

Why are AM and FM broadcast frequencies spaced the way they are?

Suppose the analog signal is a 440 Hz tone, the A above middle C on a piano. A typical AM transmitter would have a carrier wave frequency of 1 megahertz (MHz). The result is a signal with three different frequencies: 1 MHz, 1 MHz + 440 Hz, and 1 MHz − 440 Hz. Thus, the total signal requires a set of frequencies 880 Hz wide. Most human speech sounds have a maximum frequency of 4000 Hz, or 4 kHz. Sounds with a frequency of 4 kHz require a range of frequencies, called the bandwidth, 8 kHz wide, just about the spacing between adjacent stations.

The ear can detect frequencies as high as 15 kHz, so for hi-fidelity transmission of sounds (for example, music broadcasts), the bandwidth of radio broadcasts should ideally be at least 30 kHz. FM stations are able to do that more easily than AM stations because of their much higher frequencies.

What other radio frequencies are used and how?

The electromagnetic spectrum has many other users for communication purposes. Police and firefighters usually use AM, while aircraft, where noise reduction is important, use FM. Until June 12, 2009, television stations in the United States used AM for the picture and FM for the sound; today, picture and sound are both transmitted with digital signals. Other users include the armed forces, marine ship-to-shore services, weather broadcasts, commercial mobile phones, the citizens' band (CB) radio, and amateur wireless "ham" radio operators. They all share the HF, VHF, UHF, and SHF bands.

How is stereo sound transmitted?

Stereo sound means that two separate sounds, the left (*L*) and right (*R*), are produced from a pair of speakers. Because stereo broadcasts must also be usable by receivers that

---

What are sidebands?

Modulating a carrier wave produces additional frequencies above and below the frequency of the carrier. These frequencies are called sidebands. There is identical information in the two sidebands, so many radio services filter out one of the two sidebands, resulting in a single-sideband broadcast, or SSB. SSB is an alternative method of producing an analog broadcast signal, and it can work with either AM or FM radios. It has half the bandwidth of a double-sideband broadcast, so more radios can use the same part of the spectrum.

cannot reproduce stereo, the signal that all receivers can detect consists of the sum of the left and right channels $(R + L)$. In an FM radio signal, The difference of the two channels $(R - L)$ is broadcast 38 kHz above the $R + L$ signals. Non-stereo ("mono") receivers cannot detect these very high frequencies. A stereo receiver can, however; and from the $R + L$ and $R - L$ signals, it produces separate R and L signals to be amplified and sent to the corresponding speaker, recreating the stereo output sound.

FM stations actually have enough bandwidth to accommodate a third signal. This signal can be sold to users to provide background music for stores or elevators. In addition, educational broadcasts can use the third channel to deliver lessons to schools or broadcast in a second language.

How far away can FM and AM stations be received?

All electromagnetic waves travel in a straight line while in a uniform medium such as the lower atmosphere. Therefore, most radio waves only have what is called a line-of-sight range. That means that if a mountain range or the curvature of Earth were in the way of the radio signal, the receiver would be out of range and would not receive the signal. This is why most broadcasting antennae are placed on tall buildings or mountains; this helps increase the line-of-sight range.

Why can AM radio signals sometimes be received from very far away?

Waves with frequencies below about 30 MHz are sometimes able to reflect off the charged particles in Earth's ionosphere; this is referred to as "skip." Instead of passing through the ionosphere and going into interplanetary space as higher-frequency electromagnetic waves do, the lower frequencies on the AM band can be reflected back toward Earth to increase their range dramatically. After sunset, the ionosphere's altitude permits stations to be heard hundreds or even thousands of kilometers from a transmitting tower.

A handful of AM stations are "clear channel" stations. That is, there is only one station in the continental United States broadcasting on that frequency. Those stations can be heard across almost the entire country without interference from other stations. Broadcast FM stations, with frequencies of 88–108 megahertz, penetrate the ionosphere and usually can be heard at most only 80–160 kilometers from the transmitter.

How do cellular phones work?

Cell phones use the UHF part of the electromagnetic spectrum, 800 to 900 MHz, 1,700 to 1,800 MHz, and 2,100 to 2,200 MHz. The service area for a cell phone provider is divided up into hexagonally shaped cells, each one served by base stations with antenna towers at three corners of the cell. The stations can both receive and transmit information to cell phones. The stations are connected to a network that uses fiber-optic cables.

When a cell phone is turned on, it searches for available services according to a list stored in the phone. It selects the correct frequencies to transmit and receive data, and then it sends its serial and phone numbers to the system, registering itself in that cell.

The network makes sure that the phone number is part of its system and that there is money in the account to pay for a call.

After registration is complete and a call is made to that phone, the network can direct the call to the correct cell. The cell phone always searches for the strongest signal from a tower. If the phone moves during the conversation, the signal strengths will change, and the phone will "hand off" the call to a different base station.

Transmitting and receiving messages in the electromagnetic spectrum, cell phones automatically select and correct appropriate frequencies, send serial numbers to servers, and register themselves in the cell in which they are located at the time.

Are cellular phones analog or digital?

Cell phones digitize the voice signals they receive and transmit. Special circuits, called digital signal processors, then compress the voice signals and insert codes that can detect errors in transmission. Compression is achieved by sending only the changes in the digital signals, not what stays the same.

Cell phone systems also send many different conversations at the same time. One method, called TDMA (time division multiple access), splits up three compressed calls and sends them together. CDMA (code division multiple access) uses TDMA to pack three calls together and then puts six more calls at two other frequencies. Each of the nine (or more) calls is assigned a unique code so that it can be directed to the correct recipient. Spread CDMA systems use a wide band of frequencies that permit even more simultaneous calls.

What are LTE and UTMG?

The two main systems used in the first decade of the twenty-first century to manage simultaneous cell phone calls were CDMA (code division multiple access) and GSM (global system for mobiles). In 2010, cellular service providers worldwide agreed to a new LTE (long-term evolution) standard for all new cell phones. LTE is more efficient than CDMA or GSM in its use of electromagnetic spectrum bandwidth, so many more cell phone numbers and calls can be supported with LTE. As technology keeps changing and phones become more advanced, CDMA and GSM networks will be needed less and less; by the year 2021, only LTE and more advanced networks will be in use in what is now called the universal mobile telecommunications system (UTMG).

What are 2G, 3G, 4G, and 5G wireless networks?

The "G" stands for generation. Early cell phones that could only make analog voice phone calls are called first-generation devices, or 1G. Networks that came afterward used digital methods, allowing many more simultaneous users; these were called second generation, or 2G. The increased capacity of 2G systems allowed e-mails and text messages to be exchanged between cell phones. In the year 2000, standards were released for 3G (third-generation) networks so that these devices could download television-like video and exchange video images easily.

By 2009, smartphones were in common use, with many features that were formerly available only on computers, GPS locators, video cameras, and telephones. The first 4G (fourth-generation) broadband cellular networks began operation

Fifth-generation (5G) wireless networks are expected to become fully operational in the 2020s and be 10 to 100 times faster and more efficient than 4G.

soon after that. The current 4G networks are typically five to ten times faster than 3G and have data transmission rates of 100 million to 1 billion bits per second. In the 2020s, networks that can be called 5G (or fifth generation) will be able to provide service for at least one million wireless devices per square kilometer—ten times more than 4G can currently support.

MICROWAVES

What is a microwave?

"Microwave" is the term generally used to describe electromagnetic waves with a wavelength between a few hundred microns (1,000 microns = 1 millimeter) and a few centimeters. Their frequency is about 3 GHz, and they can be used as carrier waves for wireless communication. Many common molecules can be excited to resonance and get hot when struck by microwaves of specific frequencies. Interestingly, the cosmic background radiation of the universe left over from the Big Bang consists mostly of microwaves.

How are microwaves used for communication?

Microwaves are used in homes equipped with wireless internet devices, wireless telephones, Bluetooth devices, and satellite radio and television. In addition, industries, governments, and the military use microwaves for communicating from one installation to another.

Microwaves can be transmitted in two different ways. The first is the line-of-sight approach, where the microwave transmitter is pointed at a microwave receiver (these

> ## Can a microwave oven be used to dry things?
>
> Water molecules are warmed and eventually boil off into steam by the microwaves in an oven. Thus, anything that is wet could be dried in a microwave oven. However, there is a very important safety consideration that must be made before placing the object inside the oven: the object being dried must not contain a great deal of water itself. Microwaves are very useful for drying wet books, papers, or magazines but must never be used to dry things like plants or small animals. Living things would be killed by the resonance and heating of water molecules inside their bodies.

must be no more than 30 kilometers apart). The second method of transmission is to send the signal up to a satellite that receives it and retransmits it back down to a receiving dish.

How are microwaves used in the kitchen?

In addition to having a great range of frequencies to transmit information, microwaves are generated in microwave ovens around the world. A device called a magnetron is used to produce streams of waves with a frequency of 2.4 GHz and scatter them throughout the oven. These particular microwaves excite water and fat molecules and cause them to rotate and vibrate much faster, increasing their thermal energy and heating the foods that contain them. Different kinds of molecules absorb the microwave energy at different rates, so some foods are heated more than others by the same amount of microwaves. Microwave-safe containers are made of materials that do not absorb the microwaves and thus remain cooler to the touch.

Why is there a grating on the door of a microwave oven?

People using microwave ovens want to see the food cooking inside the oven but not be bombarded by potentially harmful microwaves. To prevent the escape of microwaves through the plastic or glass door, a grating consisting of small holes is used to reflect the microwaves back into the oven. The microwaves, which have a wavelength of about 12 centimeters (5 inches), are too big to pass through the holes, but visible light, whose wavelength is smaller than the opening, can easily pass though the grating. Although the grating protects people from the microwaves, some microwaves can still leak out through the door seal, however, if it's not cleaned occasionally.

Why shouldn't metal objects be placed in microwave ovens?

Manufacturers caution consumers about placing metal containers and aluminum foil inside microwave ovens for two very good reasons. The first reason is that metal and aluminum may impede cooking. Microwaves warm food by transmitting energy to water and fat molecules within the food. If food is placed under aluminum foil or in a metal container, the microwaves will be reflected from the metal and won't be able to reach the water molecules and cook the food.

The second reason for not placing metal objects inside a microwave oven is for the safety of the microwave oven itself. Metal acts as a mirror to microwaves. If too much metal is placed in the oven, the microwaves will bounce around the oven in waves that can damage the magnetron that produces the microwaves. If a piece of metal in the oven is the correct size, it can even act as a microwave receiver, and the induced voltages can produce sparks that can start a fire.

THE PRINCIPLE OF SUPERPOSITION

What is the principle of superposition?

When two waves overlap, they don't crash and destroy one another; instead, they pass through each other without interaction. The graphs in Figure 20 show two waves approaching, overlapping, and moving away. They continue to move at the same velocity throughout. The arrows show their motions. The dotted drawings show the individual waves, while the solid drawings show the results.

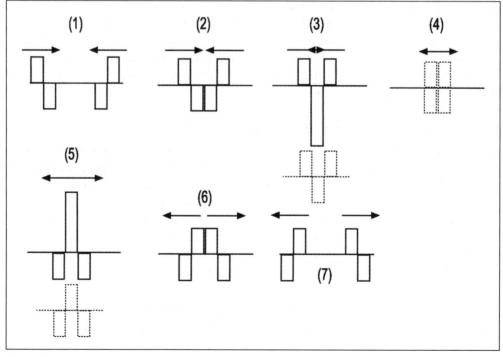

Figure 20.

The amplitudes of the two waves add together, producing a larger wave if they are both positive or both negative. They produce a smaller wave if one is positive and the other negative. In fact, as shown in stage (4), it is possible for two such waves to produce no amplitude at all. The large amplitudes are called constructive interference. The reduced amplitudes are called destructive interference.

What are standing waves?

The example used above shows what happens when two single waves going in opposite directions meet. A continuous wave is a set of single waves, one after another. You can produce such a wave by shaking one end of a rope up and down at a constant frequency. Now if the other end of the rope is tied to something that doesn't move, the wave will be reflected back toward you. If you shake the rope at the correct frequencies, the two waves will overlap each other and will seem to stand still, producing a standing wave.

Two distinct regions on a standing wave can be seen. At certain points, the rope won't be moving. That point is called a node. The point where the motion of the rope is largest is called an antinode.

What do nodes and antinodes in a standing wave represent?

Nodes are locations of destructive interference where the two waves moving in opposite directions have opposite amplitudes; the crest of one wave and the trough of the other are at the same location. Antinodes are at locations of constructive interference where the two waves have both positive or both negative amplitudes; that is, both are either crests or troughs.

The frequencies that produce standing waves depend on the length of the rope and the velocity of the wave on the rope. The lowest frequency will have nodes at the two ends and an antinode in the center. The next higher frequency will have a node in the center and ends and two antinodes. At each higher frequency, the number of nodes and antinodes increases by one.

How are standing waves generated in musical instruments?

Many musical instruments depend on standing waves to produce their sound. Standing waves are created on the strings of a guitar, piano, or violin and in the air columns of a trumpet, flute, or organ pipe. The string is made to oscillate either by plucking it (pulling it aside and then letting go) or by bowing it, where the

Many musical instruments, such as clarinets and violins, depend on standing waves to produce sound.

stickiness of the rosin and horsehair on the bow also pulls the string aside and then releases it. In a piano, a felt-covered hammer strikes the string, starting it vibrating. In a trumpet or other brass instrument, the player's vibrating lips create the traveling sound wave that is reflected when it reaches the open end of the instrument. In a flute or organ pipe that mimics a flute, the player blows air over a hole. The moving air interacting with the hole produces a periodic change in the pressure of the air inside the tube, which creates the traveling sound wave. In a clarinet or saxophone, the player blows through a narrow gap between the mouthpiece and a flexible piece of bamboo called the reed. Oboes and bassoons have two reeds separated by a thin gap. The stream of air causes the reed to vibrate, periodically stopping the airflow and causing the sound wave.

What happens when you change the standing wave in a musical instrument?

To change the pitch produced by an instrument, the standing wave inside the instrument must be altered. Changing the length of a wind instrument, or the tension and length of the strings for a string instrument, changes the frequency of the standing wave that is produced, which changes the pitch of the note or sound. Pressing a key on a trumpet inserts an additional length of tubing into the instrument. On a flute, clarinet, or saxophone, a hole on the instrument is covered or uncovered, changing the effective length of the instrument. On a piano, each note uses a string of a particular length. On a violin or guitar, the player's finger is used to change the length of a string. Thicker strings have lower pitches than thinner strings of the same length. Increasing the tension on a string increases the frequency of its standing waves and thus raises the pitch.

RESONANCE AND IMPEDANCE

What is resonance?

All objects that can vibrate have a natural frequency of oscillation. If you hold one end of a ruler on a desk, push down on the other end, and suddenly release it, the ruler will vibrate; the natural frequency depends on the material and its size and shape.

Resonance occurs when an external oscillating force is exerted on an object that can vibrate. When the frequency of the external force equals the natural frequency, then the amplitude of the oscillation reaches a maximum. This condition is called resonance. Only a very small external force is needed to create a large oscillation amplitude.

You can explore resonance with a massive object like a yo-yo or heavy metal washer on a string. Hold the top of the string still and pull the object to one side and watch it oscillate at its natural frequency. Then shake the top of the string at the same frequency and watch the amplitude of the oscillation increase. You will have found the resonance frequency. If you raise or lower the shaking frequency away from the resonance frequency, you'll find that the amplitude of the oscillation gets smaller.

Where does resonance occur on the playground?

Children can discover resonance early in life when playing on a swing set. They use their arms and legs to pump themselves back and forth on the swing. They eventually learn that if they pull back on the chains of the swing every time they are at the largest backward displacement, they will eventually achieve the largest swinging amplitude. If, however, they pull back at other times, or if someone pushes on the swing at the wrong times, the amplitude will be decreased because the external frequency is not at the natural frequency.

Can resonance cause a crystal glass to break?

Many years ago, the great singer Ella Fitzgerald (1917–1996) was shown to perform a physics experiment in an advertisement for Memorex audiotape cassettes. The company claimed that the famous singer could create a pure tone at just the right

Sound can make glass break when amplified because the sound waves can cause the glass to oscillate and bend. When enough kinetic energy is used, the sound waves can distort the glass to the point where it cracks or shatters.

157

frequency to cause a crystal wineglass to break and that the Memorex tapes recorded and played back sounds so accurately that the glass would break even when the recorded voice of Ms. Fitzgerald was played. The advertising slogan read, "Is it live, or is it Memorex?"

Although it may be hard to think that glass is something that can bend, if you tap the rim of a thin wineglass, you can hear it "ping." The shape of the rim of the glass oscillates. When an amplified sound wave pushes on the glass, it distorts its shape. Some of the kinetic energy in the sound wave is transferred into the kinetic energy of the oscillating glass. When the frequency of the sound wave matches the natural frequency of the glass, the amplitude of oscillation may indeed be large enough to shatter the glass.

Crystal wineglasses might shatter when resonating, but can they make music?

If the energy of a resonant standing wave is large enough, a crystal wineglass might shatter, but when the amplitude is smaller, the wineglass can produce a sound instead. You can try rubbing your finger around the moist lip of a wineglass; if you do it just right, the glass seems to ring or sing or hum. The tone is created because the rubbing of the wet finger on the glass causes it to have a standing wave pattern. The resonating glass generates enough energy to vibrate the surrounding air and create a steady humming sound.

In 1761, Benjamin Franklin (1706–1790) invented a musical instrument he called an armonica. (Today, it is known as a glass harmonica.) Wineglasses of various sizes were fastened to a rotating shaft. Musicians would rub their fingers on the appropriate glasses to make music. More than 100 classical music composers, including Wolfgang Amadeus Mozart, wrote music for the glass harmonica.

What is impedance matching?

Impedance is the opposition to wave motion exerted by a medium. When a wave travels from one medium into another, the impedance changes, causing some of the energy of the wave to be reflected back into the original medium. Therefore, not all of the wave's energy travels into the new medium. An impedance-matching device between the two media allows for a smooth transition in impedance and reduces reflections.

What are transformers?

An impedance-matching device is called a transformer. Instead of an abrupt change between the two media, a transformer provides a smoother, gradual transition from the old to the new medium. Depending upon the wave and the medium, different transformers, such as quarter-wavelength and tapered transformers, can be used to help minimize reflection.

An example of a tapered transformer can be found in soundproof rooms or sound studios. Any sound that is produced is supposed to be absorbed by the impedance-matching material on the walls. Special foam, tapered in a V-like shape, is used as a transformer to gradually absorb all the sound into the walls. The gradual changeover from the air medium to the wall medium prevents sound from reflecting back into the air.

An example of a quarter-wavelength transformer can be found on many camera lenses and eyeglasses. The quarter-wavelength-thick coating on a lens is used to reduce reflections off the lens surface, allowing more light to enter into the lens.

Electrical transformers are used to match impedances by changing the varying voltages and currents in an electronic circuit. Modern electronic circuits make relatively light use of transformers because of their weight and size.

SOUND

What is the source of sound?

Sound waves are created by mechanical vibrations or oscillations that force the surrounding medium to vibrate. A tuning fork is an excellent example of a vibrating sound source. When struck by a rubber mallet, the tines of the tuning fork vibrate, causing the air molecules around them to move back and forth at the same frequency, creating areas of compression (where the molecules are close together and air pressure is slightly increased) and rarefaction (where the molecules are spread apart and thus the air pressure is reduced). These pressure changes then move away from the tines, creating the longitudinal sound wave.

What type of a wave is sound?

A wave that consists of compressions and rarefactions—such as a sound wave—is called a longitudinal wave. The medium— the material through which the wave is traveling—does not get transferred from sender to receiver; the molecules only vibrate back and forth about a fixed position. The wave does, however, carry energy from its source to the receiver.

Who determined that sound needs to travel through a medium?

In the 1660s, English scientist Robert Boyle (1627–1691) proved that sound waves need to travel through a medium in order to transmit sound. Boyle placed a bell inside a vacuum and showed that as the air was evacuated from the chamber, the sound of the bell became softer and softer until there was no sound.

Considered one of the founders of modern chemistry, Robert Boyle is best known for Boyle's Law, which describes the relationship between volume and pressure of a gas in a closed system, but he also proved that sound requires a medium to travel through.

What did Isaac Newton add to the knowledge of sound media?

Although he is mainly known for his work in mathematics, classical mechanics, and the principles of geometric optics, Isaac Newton (1642–1727) did make several important discoveries in the field of sound. His major contribution was his work on sound wave propagation. He showed that the velocity of sound through any medium depended upon the characteristics of that particular medium. Specifically, Newton demonstrated that the elasticity and the density of the medium determined how fast a sound wave would travel. Because Newton worked before the field of thermal physics was developed, his theory has errors, but they are not significant to the calculation of sound speed.

What is the speed of sound?

The speed of a sound wave (also called an acoustic wave) depends on, among other things, the pressure and temperature of the medium through which the wave travels. At sea level on a typical spring day here on Earth, sound travels through the air at approximately 340 meters per second, or 760 miles per hour.

The speed of sound, by the way, is much slower than the speed of light, which travels just under 300 million meters per second, or 186,282 miles per hour. The difference in the speed of sound compared to the speed of light can be observed at a baseball game, when spectators sitting in the outfield bleachers see the batter hit a ball before they hear the crack of the bat.

What is the speed of sound through a medium other than air?

This table illustrates some examples of the speed of sound in different media.

Medium	Speed (meters per second)
Air (0°C)	331
Air (20°C)	343
Air (100°C)	366
Helium (0°C)	965
Mercury (liquid)	1,452
Water (20°C)	1,482
Lead (solid)	1,960
Wood (oak)	3,850
Iron	5,000
Copper	5,010
Glass	5,640
Steel	5,960

On what does the speed of sound depend?

A simple model that explains the main factors affecting the speed of sound is a collection of balls (molecules) connected to each other by springs (bonds between molecules). Vibration from one ball will be transferred by the springs to neighboring balls in succession throughout the collection. The stiffer the springs and the lighter the balls, the faster the vibrations will be transferred. The springs are a model of the bulk elasticity (how the volume changes when the pressure on it changes) of the material, and the balls and their spacing are models of the density of the material. In general, the speed is slowest in gases, fastest in solids. Even though liquids and solids are about 1,000 times denser than gases, the greater elasticity of liquids and solids more than compensates for the larger density. In gases, the speed depends on the kind of molecule and temperature. For air at sea level, which has roughly the same mix of molecules everywhere on Earth, the speed depends only on temperature.

What is the sound barrier?

The sound barrier is the speed that an object must travel to exceed the speed of sound. The speed of sound is often used as a reference with which to measure the velocity of an aircraft. The speed of sound, 331 meters per second at 0°C, is called Mach 1. Twice the speed of sound is Mach 2, three times the speed of sound is Mach 3, and so on.

What is a sonic boom?

A sonic boom occurs when an object travels faster than the speed of sound. The boom itself is caused by an object, such as a supersonic airplane, traveling faster than the sound waves themselves can travel. The sound waves pile up on one another, creating a shock wave that travels through the atmosphere, resulting in a "boom" when it strikes a person's ears. A sonic boom is not a momentary event that occurs as the plane breaks the sound barrier; rather, it is a continuous sound caused by a plane as it travels at such a speed, but the shock wave travels with the plane, so each person hears it only when the airplane passes their location.

All objects that exceed the speed of sound create sonic booms. For example, missiles and bullets, which travel faster than the sound barrier, produce sonic booms as they move through the atmosphere. The shock wave created by an F-15 fighter plane is even visible.

Does sound travel faster on a hot or cold day?

Air molecules move faster in hot and humid environments due to their increased thermal energy. Since sound relies on molecules bumping into one another to create compressions and rarefactions, increased speed of molecules makes the sound waves move faster. The speed of

Supersonic planes, such as this F-15 fighter jet, can go beyond Mach 1. When they do, they generate a "sonic boom" caused by sound waves piling up and creating a shock wave.

sound increases by 0.6 meters per second for every degree Celsius increase in temperature. Water vapor in the air also increases the speed of sound because water molecules are lighter than oxygen and nitrogen molecules. Therefore, sound travels faster on hot, humid days than on cool, dry days.

The following formula gives the speed of sound in air: Speed of sound = $(0.6$ m/s$)T$ + $(331$ m/s$)$, where temperature, T, is measured in degrees Celsius, and the speed of sound is measured in meters per second.

How far away was that bolt of lightning I just saw?

When a lightning bolt goes from cloud to cloud or from cloud to ground, it suddenly heats the air through which it passes. This sudden increase in temperature causes thunder to occur at the same time as the lightning. Although it takes virtually no time to see the lightning, depending on the observer's distance from the lightning, it can take quite a while to hear the thunder.

The speed of sound at room temperature is about 1,100 feet per second. A mile is 5,280 feet, so it takes about 5 seconds for sound to travel one mile. So, to determine how far away lightning has struck, when you see the lightning, count the number of seconds it takes before hearing thunder. Divide the number of seconds between the lightning and thunder by five to determine how many miles away the thunder and lightning occurred. For example, if you see a flash of lightning and approximately 10 seconds later you hear the thunder, divide the 10 seconds by the 5 seconds per mile to find that the lightning occurred 2 miles away.

HEARING

How does a person hear sound?

The ear is the organ used to detect sound in humans and some animals. The ear consists of three major sections: the outer ear, middle ear, and inner ear. The outer ear, the external section of the ear, consists of a cartilage flap called the pinna. The pinna's size and shape form a transformer to match the impedance of the sound wave in air to that at the end of the ear canal by gradually funneling the wave's sound energy into the ear. To hear more sound, people can increase the size of the pinna by cupping their hand around the back of the pinna—in effect, increasing its size and funneling capabilities.

What happens after sound enters the ear?

Once the sound has entered the ear canal, it moves toward the eardrum, where the longitudinal waves cause the eardrum to move in and out depending on the frequency and amplitude of the wave. The middle ear includes the eardrum; the hammer, anvil, and stirrup, the three smallest bones in the human body; and the oval window on the inner ear. The eardrum is 17 times larger than the oval window, and this difference in area makes the inner ear act like a hydraulic machine, increasing the changes in pressure on the eardrum to that on the oval window at least 17-fold. The three bones link the eardrum to the oval window. They act like a lever system to further increase pressure on the oval window. The mechanical advantage of the lever system varies with frequency, peaking at about 5 in the frequency range of 1 to 2 kilohertz. Thus the middle ear is like a complex machine that can amplify the pressure on the eardrum by a factor approaching 100 as it transfers the energy of the sound wave to the inner ear. Muscles connected to the eardrum and the three bones can react to very loud sounds and reduce the sensitivity of the ear, thus protecting it from damage.

How does the human inner ear work?

The inner ear is a series of tubes and passageways in the bony skull. It consists of the cochlea, a spiral-shaped tube that changes the longitudinal sound wave into an electrical signal on the nerves connecting the ear to the brain. The inner ear also has three semicircular canals that sense the body's motions and give rise to a sense of balance. The oval window is at one end of the cochlea. Sound waves transmitted through the middle ear to the oval window cause a traveling wave in the fluid of the cochlea. This wave in one of the three tubes within the cochlea, the Organ of Corti, causes hair cells called cilia to tilt back and forth. The tilting causes chemicals to pass through channels in the nerve, creating electrical impulses that travel along the auditory nerve to the brain for analysis. The further the hair cells are from the oval window, the lower the sound frequency to which they are sensitive. Thus, different nerves are excited by different frequencies, allowing the ear to distinguish the sounds' frequencies.

What are the frequency limits of the human ear?

The ear allows humans to hear frequency ranges between about 20 hertz and 20,000 hertz, but it is most sensitive to frequencies between 200 hertz and 2,000 hertz.

The lower and upper fringes of this bandwidth can be difficult to hear, but many people—especially younger people—can hear these frequencies quite well. As people age, their sensitivity to high frequencies diminishes. Damage to the hair cells caused by exposure to loud sounds also reduces the ear's sensitivity to high frequencies.

What are the bandwidths of hearing for different animals?

The following table gives the frequencies of the bandwidths of some animals.

Animal	Lowest Frequency	Highest Frequency
Human	20 Hz	20,000 Hz
Dog	20 Hz	40,000 Hz
Cat	80 Hz	60,000 Hz
Bat	10 Hz	110,000 Hz
Dolphin	110 Hz	130,000 Hz

Dolphins can detect frequencies of between 110 and 130,000 Hz. Their hearing is over six times more sensitive than that of a human being.

Why does my voice sound different on a recording?

How you hear yourself is unique to you. When you speak, you hear yourself through sound waves propagating through your body in addition to the waves propagating through the air. To make a sound from your throat, you vibrate your vocal cords, which in turn vibrate the surrounding tissue. This tissue includes muscle, bone, and cartilage. Waves travel through these media at varying speeds and create slightly different sounds when they are transmitted through the skull to the inner ear. Your voice sounds different to you when listening to a recording because you're hearing it without all those extra paths the sounds travel through your skull.

ULTRASONICS AND INFRASONICS

What are ultrasonic sounds?

Ultrasonic sounds are those frequencies above human hearing. Frequencies above 20,000 Hz are almost never audible to people, but they are easily heard by those animals whose hearing is sensitive to such frequencies. For example, dolphins use ultrasonic frequencies to communicate, and bats use ultrasonic sounds as a tool for navigation and hunting.

What is ultrasound?

Ultrasound is a method of looking inside a person's body to examine tissue-based and liquid-based organs and systems without physically entering the body. Ultrasound systems direct high-frequency sound (usually between 5 and 7 MHz) into particular regions of the body and measure the time it takes for the sound wave to reflect back to the machine. By analyzing the pattern of reflections received, a computer can create a visual representation of the interior of the body.

What are some medical uses of ultrasound?

Ultrasound is sometimes used instead of X-rays because it does not use ionizing radiation and thus is safer for the person being examined. Obstetricians use ultrasound to examine the progress of a fetus or problems that it might be experiencing. Ultrasound is also used to observe different fluid-like organs and systems in the body such as the nervous, circulatory, urinary, and reproductive systems. An ultrasound is often the first picture or movie taken of an embryo or fetus that is developing in a mother's uterus.

Ultrasound can also be used to pulverize kidney "stones." In this application, very intense, tightly focused beams of high-frequency sound are directed at the stone. The stones are shattered, and the small pieces can be passed through the urinary tract with little or no pain.

What is sonar?

Sonar, an acronym for "SOund NAvigation Ranging," is a method of using sound waves to determine the distance an object is from a transmitter of sound. The sonar contains a transducer that converts an electrical impulse to sound when it transmits and a sound wave to an electrical impulse when it receives. Sound waves, usually brief pulses of ultrasound, are emitted from the transducer, reflected off an object, and reflected back to the transducer. Electronic circuits measure the length of time it took for the sound waves' round trip, and they use the speed of sound to calculate the distance the object is from the transducer.

What is sonar used for?

Sonar is used predominantly as a navigational tool. Machines such as depth finders on boats, distance meters used in real estate and construction, and motion detectors for security devices all employ sonar. Some animals also use sonar; they do not need to use machines but have evolved biological systems that produce, receive, and interpret sound as well as or even better than a mechanical sonar system could. Dolphins and bats, among other animals, use sonar for navigation, hunting, and communicating with one another.

What is infrasound?

Whereas ultrasonic sounds are frequencies above the human bandwidth of hearing, infrasonic (or "subsonic") sounds are those frequencies below the human bandwidth of hearing, or 20 hertz. Infrasounds as low in frequency as 0.001 hertz are produced by a variety of natural sources like earthquakes and volcanoes, as well as man-made structures and nuclear explosions. Elephants are known to make sounds as low as 12 hertz and can communicate in this way over large distances.

165

Tornadoes generate subsonic sounds that cannot be heard by humans but can be detected by instruments that can then be used to warn people up to a hundred miles away of approaching danger.

How can infrasound be used to provide early storm warnings?

Using sensors that detect infrasound, scientists discovered quite by accident that the spinning core or vortex of a tornado produces sounds that are a few hertz below the human bandwidth of hearing. The tornado, much like an organ pipe, produces low frequencies when the vortices are large and higher frequencies when the vortices are small. Since the infrasound waves from tornadoes can be detected from up to 100 miles away, they can help increase the warning time for tornado strikes.

INTENSITY OF SOUND

What is sound intensity?

Sound intensity is the power in the sound wave divided by the area it covers. Power is energy per unit time, so the intensity is the wave energy passing through a surface point divided by the time taken. The energy transferred by the wave is proportional to the square of the amplitude of the wave. In the case of sound, the amplitude measures how much greater the peak pressure in compression is than the average pressure.

Why does sound diminish as you move farther from its source?

Sound waves do not travel in a narrow path but spread out into the surrounding medium as spherical waves. As long as the energy in the sound doesn't change, the power per unit area decreases as the area increases. In some cases, there is some transfer of the sound's energy into thermal energy of the medium. Thus, there can be some loss of sound energy with distance, but this loss is usually small.

How quickly does sound intensity diminish as you move away?

While the total energy in the sound wave remains the same as the sound spreads over a sphere, the intensity decreases as the area of the sphere increases. The area of a sphere is proportional to the square of its radius, so the intensity is inversely proportional to the square of the distance. That is, sound intensity diminishes according to the inverse square law. For example, if a person stands 1 meter (3.3 feet) away from a source, the sound intensity might be an arbitrary unit of 1. If that same person moved so she was 2 meters (6.6 feet) away from the source, the intensity would be 1 over the square of the distance, or one-fourth the intensity. Again, if the listener moved 3 meters (9.8 feet) away, the intensity would be one-ninth the intensity it was at the 1-meter mark.

When doesn't sound intensity diminish following the inverse square law?

Sound doesn't always spread uniformly, and it can be focused or directed so that its intensity doesn't diminish as it would in open air. You may have experienced this if you shout while you walk through a tunnel. You hear echoes, and a person at another part of the tunnel will hear you much more clearly than he or she would in an open area.

Whispering chambers are often found in science museums. They are rooms with walls in an elliptical shape. If you stand at one focus of the ellipse and another person stands at the other focus, you can hear each other speak even if you whisper while other people in the chamber are talking loudly.

How can sound travel through the ocean without quickly spreading out?

Sound can be transported through the ocean with less spreading using an acoustic waveguide, similar to the way light can travel in an optical fiber (see the chapter "Light and Optics"). The ocean water can separate into layers at different temperatures. The speed of sound depends on water temperature, and if there is a layer of cold water where sound speed is lower under a layer of warmer water where the speed is higher, then some of the sound moving through the cold layer will be reflected at the boundary back into the cold layer, keeping it from spreading. The name of this channel is the "SOund Frequency And Ranging" (SOFAR) channel. It is 100 to 200 meters below the surface in the cold waters off Alaska but 750 to 1,000 meters deep in the warmer Hawaiian waters.

What is a decibel?

A decibel (dB) is the internationally adopted unit for the relative intensity of sound. The sound intensity of 0 decibel is the threshold of human hearing, 10^{-16} watts per square centimeter. This corresponds to a pressure of 2×10^{-5} newtons/m^2 or two ten-billionths of atmospheric pressure. The ear is extremely sensitive! The decibel scale is a logarithmic scale, meaning for every 10 decibels, the intensity is increased by a factor of 10. For example, a change from 30 decibels to 40 decibels means the sound will be 10 times more intense. A change from 30 decibels to 50 decibels would mean the new sound would be 100 times more intense.

How many decibels are some typical sound environments?

The following table shows a typical sound environment, how many times louder those levels are than the threshold of human hearing, and the relative intensity of that sound compared to the threshold of hearing.

Sound	How many times more intense	Relative intensity (dB)
Threshold of human hearing	1	0
Quiet breathing	10	10
Leaves rustling in the wind	10^2	20
Close whisper	10^3	30
Library	10^4	40
Normal talking	10^5	50
Vacuum cleaner	10^6	60
Cars driving by	10^7	70
Motorcycle	10^8	80
Factory	10^9	90
Lawn mower	10^{10}	100
Rock concert	10^{11}	110
Threshold of pain	10^{12}	120
Jet engine	10^{13}	130
Rocket launch	10^{14}	140
Hearing loss threshold	10^{15}	150

Why do people hear ringing in their ears after a rock concert?

After leaving very loud rock concerts, many concertgoers often complain of ringing in their ears. The ringing sound results from the damage to the cilia in their inner ears by the sustained intense sounds. Usually, the ringing has stopped the day after the concert, but permanent damage has already been done; those cilia cells will not recover. The effects of such hearing loss may take years and repeated loud sounds to become noticeable; over time, however, they can be devastating and even lead to deafness.

What are good ways to protect my hearing at a rock concert?

The first protection against damage to the cilia in your inner ears is to increase the distance from your ears to the speakers. In plain English, the farther away you are from the speakers, the lower the intensity of the sound. By simply doubling the distance, the intensity becomes one-fourth of what it was originally. That can work in open areas where the sound can spread, but in arenas or halls, the sound can reflect off the ceilings and walls and does not decrease with distance.

Loud music from live concerts can cause hearing loss over time by damaging the cilia within the ear. Standing or sitting farther back from speakers can help prevent harm, while musicians themselves often wear earplugs.

Another method of protecting your ears is to dampen the sound waves as they enter the ear. Many musicians, both rock and classical, to avoid gradual hearing loss, use earplugs to decrease the amplitude of the wave entering the ear. The fluid in the cochlea transfers less energy to the cilia than if the listener were wearing no hearing protection at all.

What are some governmental standards for hearing protection?

Federal regulations mandating the use of hearing protection in the workplace state that if an employee works for eight hours in an environment in which the average noise level is above 90 decibels, the employer is required to provide free hearing protection to those employees. One example is in the landscaping business. Since the average decibel level for a lawn mower is about 100 decibels, free earplugs or earmuffs must be provided to any employees who work for eight or more hours per day.

What level of sound will make my ears hurt?

The threshold of pain for humans depends on the person in question, but it typically ranges between 120 and 130 decibels. Such pain can be experienced at extremely loud rock concerts and next to jet engines and jackhammers, for example.

What's the difference between loudness and sound intensity?

Sound intensity is a physical property that depends on energy. Loudness describes how a listener responds to sounds, and it is different for each person because different people's ears don't respond equally to all frequencies. A sound with an intensity of 60 decibels will sound louder at some frequencies than others. Loudness also depends on the type of tone such as whether it is a very pure tone, a more complex tone, or noise.

The ear is most sensitive to frequencies between 1 kilohertz and 3 kilohertz, and its sensitivity is much less for both low (20–100 hertz) and high (10–20 kilohertz) frequencies. As people get older, their ears respond less to all frequencies but especially frequencies above 5 kilohertz.

Loudness typically doubles for each 10-dB increase in sound intensity. It is measured in sones. Normal talking, which is between 40 and 60 dB, has a loudness of 1 to 4 sones. Hearing damage from sustained sound, at about 90 decibels, is 32 sones.

ACOUSTICS

What is acoustics?

Acoustics is the branch of physics that deals with the science of sound. Although sound has been studied since Galileo (1564–1642) made some predictions in the early 1600s, the ability to study sound has grown tremendously in the past few decades with the advent of electronic sound generators and measuring devices. In addition to the study of sound itself, the field of acoustics has several applied subfields, including speech and hearing (audiology), architectural acoustics, and musical acoustics.

What is architectural acoustics?

Theaters, concert halls, churches, classrooms, and soundproof rooms have to be designed and constructed so that their acoustic properties match their uses. To achieve these goals in architecture, an acoustical physicist or engineer must consider not only the size and shape of the hall but also the acoustic properties of materials used on the floors, ceilings, and walls of the room as well as chairs and other objects in the room. Even the audience and air humidity affect the acoustic properties of the room.

Generally speaking, hard surfaces such as concrete, plaster, wood, and tile reflect sound, while soft materials like carpet, heavy drapes, and plush upholstered chairs absorb sound. Room shapes and sizes that create standing waves should be avoided because sound at frequencies at which the standing waves occur will not be uniform. They will be loud in some places and soft in others.

What is reverberation time?

The reverberation time for a sound is the time it takes for the echoing sound to diminish by a factor of 60 decibels—that is, to 1/1,000,000th of its original intensity. The longer the reverberation time, the more echoing is heard because the sound has reflected off walls and other surfaces. The shorter the reverberation time, the less echo is heard.

In some spaces, it is important that the speaker, singer, or musician or group be heard clearly in all parts of the room. A good concert hall has intimacy; music should sound as if it were being played in a small hall. It should have liveliness and warmth, created by a fullness of bass tones. The sound should be clear, and the audience should be able to locate the source of the sound. The sound should be uniform over the entire hall, and sounds from the stage should be blended by the time they are heard by the audience. Finally, the hall should be free of noise from within (such as air-handling systems) and from outside sources.

What role does reverberation play in acoustics?

Reverberation time greatly influences the quality of sound heard in a concert hall or theater. If the reverberation time for middle and high notes is too short, the sound will diminish almost instantaneously, and the room will sound "dry." A "full" bass tone requires a longer reverberation time for low notes. If the reverberation time is too long, however, as it is in many gymnasiums, the echoing effects will interfere with the new sounds, making music sound "mushy" and words of a speaker difficult to understand.

Acoustical engineers typically design concert halls to achieve reverberation times of between one and two seconds. Rooms designed for speech, like lecture halls or classrooms, should have reverberation times of less than one second; movie theaters a little more than one second; and rooms designed for organ music as much as two seconds. Some large cathedrals have extremely long reverberation times of 10 seconds or more.

What determines the amount of sound reflection or absorption?

A good absorber of sound matches the impedance of sound waves between air and the new medium, while a poor absorber has very different impedance and reflects the sound back into the environment. Typical absorbers have an open structure with holes of various sizes into which the sound waves can penetrate and transfer their energy to thermal energy of the material.

What materials absorb sound well?

Different materials will absorb certain frequencies of sound better than other frequencies; generally, though, some of the best absorbers of sound are soft things. Materials such as felt, carpeting, drapes, foam, and cork are good at matching the impedance of a sound wave and reflecting back very little sound. Materials such as concrete, brick, ceramic tile, and metals, by contrast, are effective reflecting materials of sound. That is why gymnasiums (with hardwood floors, concrete walls, and metal ceilings) have relatively

171

long reverberation times, while concert halls furnished with upholstered seats, carpeted floors, and long drapes have relatively short reverberation times. People are also effective sound absorbers, so a full concert hall has different acoustic properties than an empty one.

What was the first concert hall designed by an acoustical physicist?

Boston Symphony Hall in Boston, Massachusetts, was the first concert hall designed specifically to enhance the sound of an orchestra playing to an audience in the hall. Built in 1900, it developed an excellent reputation for sound quality, mostly due to the choices of sound-absorbing materials and the strategic placement of reflecting material. The strategy was to use sound-reflecting materials to create strong initial reflections while using sound-absorbing materials to absorb most of the sound's energy that would ordinarily reflect off the high ceiling and side walls in the rear of the hall.

Who was the acoustical physicist that designed Boston Symphony Hall?

The American physicist Wallace Clement Sabine (1868–1919) went to Harvard University to undertake graduate studies in physics, and he wound up joining the faculty after obtaining his A.M. degree. In 1895 the university opened the Fogg Art Museum, but its lecture hall had terrible acoustics; Sabine was asked to fix it, and in the process, he founded the field of ar-

Although built in 1900, Boston Symphony Hall is still considered one of the best-designed concert halls in the world when it comes to acoustics.

What went wrong with the acoustics at the Lincoln Center concert hall?

The architects made Philharmonic Hall at Lincoln Center in New York much wider than Boston's Symphony Hall, which was famous for its excellent acoustics. This wider hall did not have enough initial reflections and sounded dry. It was equipped with "clouds," reflectors hung from the ceiling, but reflections from them were too delayed to be effective. Another problem was that performers on the stage couldn't hear each other well.

In 1973 Mr. Avery Fischer contributed more than $10 million to support a complete reconstruction of the hall, and the hall was named after him. (Later, it was renamed David Geffen Hall.) The changes improved the acoustics, but the large stage reduced the loudness of bass sounds, and the initial reflections were still too strong. Curved surfaces on the stage made of extremely tight-grain maple wood have improved the bass problem, and reflectors consisting of 30,000 dowel rods were installed on the side walls.

chitectural acoustics. Sabine later served as the dean of the Harvard Graduate School of Applied Science. Today, the unit of sound absorption, the sabin, is named in his honor.

Sabine discovered the mathematical relationship between sound intensity, absorption, and reverberation time. He discovered that having strong reflections immediately after a sound was produced would enhance the acoustics of a space, but if the sound were reflected midway between the source and the listener, it would detract from the acoustics because the sound's travel time would be significantly different. Boston Symphony Hall, finished in 1900, was the first building specifically designed to have excellent acoustics. It is still considered today as having some of the finest acoustics in the world.

What is the best shape for a concert hall?

Building a concert hall is a mixture of science, engineering, art, and, perhaps surprisingly, even politics. Politics is important because concert halls can be expensive to build, and the people who provide the funds have goals for the use of the hall.

Consider, for example, the history of Philharmonic Hall at Lincoln Center in New York City. It was originally designed to be similar to the Boston Symphony Hall—long and narrow. However, a campaign led by one of the major newspapers in the city argued that the hall should seat more than 2,400 people, so the architects made the hall larger. But when it was completed in 1962, critics were very unhappy with the sound.

How can the acoustics of a space be changed electronically?

With the exception of many classical music concerts and operas, most musical performances today use electronic amplification. Loudspeakers can be located throughout the performance space, and their response to different frequencies can be adjusted to compensate for any acoustical shortcomings in the architecture. The electrical signal to speakers far from the stage can be delayed so that sound from all speakers arrives at the same time. Furthermore, these changes can be adjusted so that the hall has the best acoustics for any type of use.

MUSIC

What is the difference between a sound's frequency and its pitch?

Frequency, like sound intensity, is a physical property. Pitch, like loudness, is a description of how the ear and brain interpret the sound. Pitch is primarily dependent on frequency, but it also depends somewhat on loudness, timbre, and envelope, which will be discussed below. Humans hear pitch in terms of the ratio of two or more tones. The ear perceives two notes to be equally spaced if they are related by a multiplicative factor. For example, the frequency of corresponding notes of adjacent octaves differ by a factor of 2. Notes in common chords are related by ratios of 3:2, 4:3, 5:4, and so on. In the same way, the perceived difference in pitch between 100 hertz and 150 hertz is the same as between 1,000 hertz and 1,500 hertz.

The Maggio Musicale Fiorentino Orchestra's string section, conducted here by Zubin Mehta, would be hard to hear were it not for the beauty of oscillating strings producing sounds over the instruments' bridges and wooden bodies. The shape and size of the instruments have a great effect on the sounds produced.

Do musical tones consist of one frequency?

A pure tone has one frequency. A musical tone, however, has many frequencies. To understand why, consider a stringed instrument like a guitar, piano, or violin. The string can oscillate in response to being plucked, hit, or bowed. Standing waves will be formed as they would be if you shook a rope back and forth. By shaking it at different frequencies, you can make it oscillate in several different modes. The lowest frequency results in nodes only at the ends. Twice this frequency produces nodes at the ends plus one in the middle. Three times the lowest frequency give nodes at the ends plus two nodes at 1/3 and 2/3 its length. If you pluck, hit, or bow the string of a stringed instrument, you cause it to vibrate in many of those nodes at the same time, depending on the location you plucked it.

What are fundamental frequencies and harmonics?

In a musical tone, the lowest frequency of oscillation is called the fundamental frequency. Plucking the string 1/4 the distance from one end results in oscillations at 2, 3, 4, 6, and 7 times the fundamental frequency. The higher frequencies are called harmonics. For example, if the fundamental frequency were middle C, 256 hertz, then the second harmonic would have a frequency of 512 hertz, the third 728 hertz, the fourth 1,024 hertz, and so on.

How do stringed instruments produce sounds?

The sound made by a vibrating string is by itself very weak. On acoustic stringed instruments, the strings pass over a wooden bridge that transmits the oscillations to the top plate of the body of the instrument. Low-frequency sounds also excite oscillations in the air and in the bottom plate of the guitar's body. Sounds from the oscillations of the air pass through the sound holes in the top plate into the surrounding air. The amplitude of the fundamental frequency is the largest. The relative amplitudes of the higher harmonics depend not only on the string but also on the shape and size of the body of the instrument.

The relative intensities of the higher harmonics depend on the instrument. The spectrum of sound produced by the instrument is characterized by these relative amplitudes. The spectrum is also called the quality of sound, or the timbre. The sound quality also depends on how the sound starts and stops.

How do wind instruments produce sounds?

In wind instruments like flutes, clarinets, and saxophones, the column of air is the oscillating object. The musician must create the oscillation. Perhaps you have blown over the top of a soda bottle and created a tone. When you blow, some of the air goes into the bottle. That increased air pressure is reflected off the bottom of the bottle and returns to the top, where it deflects the blown air upward. This process repeats, resulting in a tone whose frequency depends on the length of the bottle. The energy in your breath is converted into the energy of oscillation of the air in the bottle.

A flute works in a similar way, where the player blows over a hole in the side of the flute. The other end of the flute is open. The sound wave is reflected because the impedance in the tube is different from that of the room air. The spectrum of a flute contains all harmonics. If the player blows harder and changes the location of her top lip, she can make the flute play one or two octaves higher. In other words, the fundamental frequency of the flute is increased by a factor of two or four. The frequency of a flute or piccolo can be changed by opening holes along the side of the instrument. This shortens the length of the oscillating air column, increasing the frequency.

In a saxophone or clarinet, the vibrations are caused by a thin piece of wood called the reed. The player blows through a gap between the reed and the instrument's mouthpiece. The pulse of air is reflected off the end of the instrument and returns to the reed, pushing it open to admit another pulse of air. Double-reed instruments like the oboe and bassoon work in the same way. The clarinet is shaped like a cylinder. Its spectrum consists of only the odd harmonics: 1, 3, 5, 7, and so on. A clarinet can be played in a higher register by opening a small hole near the mouthpiece that strongly reduces the amplitude of the fundamental tone. The new pitch is an octave and a fourth higher than the lower register. Saxophones are not shaped like cylinders but like cones. As a result, all harmonics are included in its spectrum, and opening the register key raises the instrument's pitch by one octave.

How do brass instruments produce sounds?

In a bugle, trumpet, trombone, French horn, or other brass instrument, the oscillations of the air within the instrument are caused by the player's lips. The lips act as a valve, releasing pulses of air into the instrument, which causes the oscillations in the air column. In a brass instrument, the fundamental tone is absent. By adjusting the tightness of the lips, the player can cause the instrument to play at the 2nd, 3rd, or 4th harmonic, and so on. The valves on a brass instrument add small lengths of tubing, lowering the pitch. In a trombone, the length of the tube can be varied continuously, allowing any frequency to be played. The spectrum of a brass instrument depends strongly on its pitch and loudness. The louder it is played, the more energy there is in the higher harmonics.

What is an overtone?

Many instruments, especially bells, oscillate in modes that may be, but are not always, harmonics. These higher modes are called overtones. Remember that a harmonic is a mode of vibration that is a whole-number multiple of the fundamental mode; the first harmonic is the fundamental frequency, the second harmonic is twice its frequency, and so on. Overtones include harmonics, but harmonics do not include overtones. (By the way, the first overtone is not the fundamental; what musicians call "the first overtone" is the second harmonic.)

How is a human voice similar to a wind instrument?

The vibrating sources of the human voice are the vocal cords in the throat. Their vibration, at a relatively low frequency of 125 hertz, creates oscillations in the air that fill the throat and mouth. By varying the size and shape of your mouth and position of the tongue, you can change the frequency of the sound as well as the relative amplitude of the harmonics. Try it yourself! Sing a constant pitch while you vocalize the vowel sounds—"a," "e," "i," "o," and "u." Note how you change the shape of your mouth and position of the tongue when you go from one vowel sound to another.

How does the spectrum of a sound relate to its waveform?

Jean-Baptiste Joseph Fourier (1768–1830), a French mathematician and physicist, made discoveries in a number of fields, including the greenhouse effect of Earth's atmosphere. He developed mathematical tools known as Fourier series and Fourier transforms that are used in a wide variety of applications. The Fourier Theorem states that any repetitive waveform can be constructed from a series of waves of specific frequencies. The reverse is also true—so if you add together waves of frequencies f, $2f$, $3f$, and so on, specifying the amplitudes of each wave, you can construct any complex waveform. Using Fourier analysis, you can then record the waveform of a musical instrument and determine the amplitudes of the harmonics of which it is made—in other words, its spectrum.

Today, Fourier analysis is very easy to do. Most computers either have built-in microphones or can use an external microphone. Free software can be downloaded from the web that will display the spectrum of a sound.

What is a difference tone?

Difference tones are frequencies that are produced as a result of two different frequencies mixing with each other. This mixing can only occur if a device is nonlinear—that is, if the output frequency is not a multiple of the input.

You can create difference tones with two slide whistles. Place both in your mouth and blow hard. Adjust their lengths so that the two tones are the same pitch. Then adjust one of the whistles while holding the other steady. As you move its pitch away from the other, you will hear a low-pitched sound whose pitch will increase as the two whistles' pitches get further apart. For example, if the high-frequency sound were 812 hertz, while the lower frequency was 756 hertz, the difference tone from the interfering sound waves would be 144 hertz. The nonlinear device in this case is your ear!

Why does a singing duet sometimes sound like a trio?

When two people sing loudly at slightly different pitches, the frequencies can mix, causing a difference tone. You may also experience this phenomenon with fire whistles blaring through a town or the beeps and ringtones emitted by an alarm clock, radio, or phone. In fact, many difference tones are intentionally created to enhance a particular sound.

How does a synthesizer work?

A synthesizer is an electronic device that generates, alters, and combines a variety of waveforms to produce complex sounds. Often, a piano-type keyboard allows the musician to select the notes to be constructed. Synthesizers may use electronic circuits to create the tones or use software that controls a circuit that converts a digital number to a voltage. Some synthesizers use a computer rather than a keyboard to select the notes. The computer can then control electronic circuits through a musical instrument digital interface (MIDI). The most common method of creating the synthesized sounds is to use a frequency-modulated synthesis, which creates higher harmonics that match those of a musical instrument being imitated or creates an entirely new musical sound.

Different instruments have characteristic attacks, sustain times, and decays. Attack describes how fast the amplitude rises from zero to its full value. Sustain times describe how long the tone amplitude remains the same, and decay describes how the amplitude decreases at the end of the played note. Synthesizers can create hundreds of different sounds, typically called voices.

Some synthesizers like this one use keyboards to select the sound to be generated, while others use computers. The ability of a synthesizer to imitate various instruments is accomplished using frequency-modulated synthesis to create "voices."

NOISE POLLUTION

Why is noise pollution dangerous?

In the past, noise pollution was thought to create health effects only if the intensity was large enough to cause hearing damage. Studies over the past several decades, however, have found that long-term exposure to noise can cause potentially severe health problems—in addition to hearing loss—especially for young children. Constant levels of noise (even at low levels) can be enough to cause stress, which can lead to high blood pressure, insomnia, and psychiatric problems and can impact memory and thinking skills in children. In a German study, scientists found that children living near the Munich Airport had higher levels of stress, which impaired their ability to learn, while children living further from the airport did not seem to experience the same problem.

Does that motorcycle have to be that loud?

Product designers and manufacturers try to achieve just the right sound or noise for their products. Whether it's a vacuum cleaner, lawn mower, motorcycle, or any other product, the sound it makes needs to be quiet enough so as not to cause stress yet also needs to be loud enough for the product's owner to feel the product is powerful. Muffler technology can be used, for example, to reduce a motorcycle's noise greatly, but many manufacturers are afraid that motorcycle enthusiasts will not purchase a bike if its sound is not "powerful" enough. On the other hand, some people prefer products to be as silent as possible. Electric and hybrid cars are sometimes so quiet that people with limited sight cannot easily tell when such a car is approaching. For safety considerations, engineers may need to develop noise-generating devices for those kinds of vehicles.

What can be done to reduce noise pollution?

Since noise creates stress and can lead to other health problems, industries and governments around the world are working to reduce noise levels, especially around populated regions. One method of reducing noise pollution around airports has been to reroute airline traffic so that it passes over less-populated areas. Sound barriers have been installed along many highways to absorb or reflect sound away from houses built alongside the roads. In countries such as Austria and Belgium, roadways are being constructed with a material called whisper concrete that their inventors claim reduces noise by five decibels. Finally, Swedish engineers have developed a road surface made of pulverized rubber that can reduce the noise level by as much as 10 decibels.

What limits have been established to reduce exposure to noise pollution?

The World Health Organization has recommended that noise during sleep be limited to a level of 35 decibels, and governments are beginning to place restrictions on noise levels in both residential and business environments. In the Netherlands, for example, regulations specify that new homes may not be built in areas of high noise levels—those that exceed an average of 50 decibels. In the United States, employers must provide hearing protection for those who endure noise levels of 90 decibels for eight or more hours per day.

What is active noise cancellation?

ANC (active noise cancellation) creates a waveform that is the opposite of the noise so that the noise is cancelled. An ANC headset consists of one or two microphones that detect the noise, electronic circuits that invert the waveform, and a headphone driver. Low-frequency noise is more successfully cancelled than noise at higher frequencies, so passive noise reduction is used to reduce high-frequency noise. As a result, low-frequency noises found in helicopter, jet-engine, and muffler noise can be controlled using ANC, while the high, squealing sounds of jet-engine noise are more difficult to cancel out.

What is psychoacoustics?

When the physics of acoustics is connected with human psychology, you get psychoacoustics—the study of how the mind reacts to different sounds. This field of study is especially important to consumer product manufacturing because a consumer associates particular sounds with certain products or sensations. For example, people associate low-frequency rumbling sounds with power and torque, while higher-frequency sounds often represent high speeds and out-of-control occurrences. Psychoacoustics can play a major role in the development and commercial success of many products, including cars, appliances, and even clocks and watches.

THE DOPPLER EFFECT

What is the Doppler effect?

The Doppler effect is the change in frequency of a wave that results from an object's changing position relative to an observer. A well-known example of the Doppler effect is when an ambulance zooms by you and makes a "whee-yow" sound. The high-pitched "whee" is caused by sound waves that are bunched together because the ambulance is moving in the same direction as the emitting sound waves. The bunching together of sound waves creates an increase in the frequency and results in a higher-pitched sound. The lower-pitched "yow" sound occurs when the vehicle moves away from the propagation of the sound wave. Since the ambulance moves away from the sound wave, the spacing between successive waves becomes greater. This decrease in the frequency of the sound wave results in a lower pitch.

Who is the Doppler effect named after?

Johann Christian Doppler (1803–1853), the Austrian mathematician for whom the Doppler effect is named, proposed in 1842 that the color of double stars rotating about one another would depend on whether the star was approaching or receding from Earth. The effect was too small to be measured using the technology of the time. But in 1845

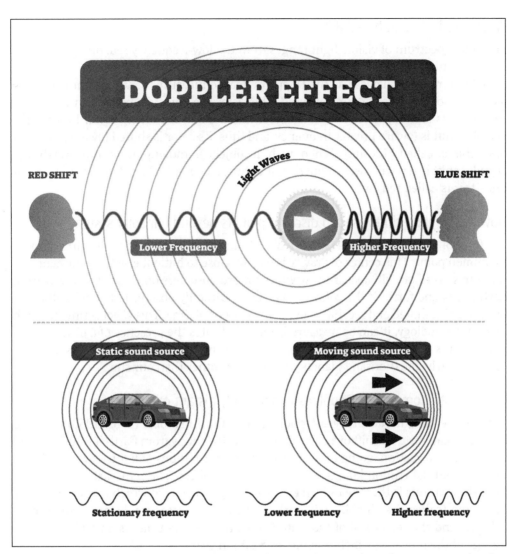

When an object approaching your position creates a sound, the sound waves in front of it are compressed, making for a higher pitch. Then, as it moves away from you, the waves are spaced farther apart, making the sound lower pitched.

Christophorus Henricus Diedericus Buys Ballot (1817–1890) set up an experiment using two sets of trumpeters. One set remained at rest, while the other was on an open railway car traveling at the then-fantastic speed of 40 miles per hour. Although both sets of trumpeters played the same note, the change in tone was clearly heard, confirming the Doppler effect for sound.

Doppler later extended his theory to the case when both the sound source and observer were moving. Eventually, the French physicist Hippolyte Fizeau (1819–1896) extended Doppler's theory to electromagnetic waves—that is, light—as well.

181

What is redshift and blueshift?

The color spectrum of visible light ranges from the low-frequency red, orange, and yellow to the higher-frequency green, blue, indigo, and violet. Astronomers observing the planets, stars, and galaxies use the Doppler effect to measure the velocity at which objects are moving, rotating, or revolving. The faster the object is moving, the more the frequency is shifted. When a yellow object like the sun is moving toward Earth, the Doppler shift is toward greater frequency, and thus, its color is shifted toward bluer colors—this is called blueshift. When such an object is moving away from Earth, the Doppler shift is toward lower frequency, and thus, its color is shifted toward redder colors—this is called redshift.

How are Doppler shifts used to discover planets beyond our solar system?

The force of gravity between a star and a planet causes the planet and star to circle around a common point. This center of gravity is usually close to but not exactly at the center of the star, so the star will appear to "wobble" toward observers on Earth, then away from Earth, back and forth. If the color of that star is carefully observed, its light will be alternately blueshifted and redshifted over long periods of time. This effect is tiny, but with modern technology, it can be measured, and it indicates the presence of a planet affecting the star's motion. Using this technique over the past 25 years, astronomers have discovered nearly one thousand planets orbiting stars other than our sun.

How do astronomers use Doppler shifts to study distant galaxies?

Early in the twentieth century, astronomers like Edwin Hubble (1889–1953) observed that the greater the redshift of a galaxy was, the farther away from Earth it was. This held true in every direction, going outward from Earth for billions of light-years (one light-year is about 9.4 trillion kilometers or 5.9 trillion miles). The remarkable conclusion from these results is that the universe is expanding. The rate at which that expansion is occurring (called the Hubble constant) can be used to measure distances to faraway galaxies, and the reciprocal of that rate (called the Hubble time) is a way to estimate the age of the universe, which is about 13.8 billion years.

RADAR

What is radar?

Radar is is an acronym for "RAdio Detection And Ranging." A radar installation emits electromagnetic waves and detects the waves reflected from an object. It measures the time for the "echo" to return to find the distance of the object. The radar dish is constantly rotating, permitting it to find the direction of the object as well. Radar is used for many different purposes today but was first used in World War II to detect the approach of enemy aircraft.

Who developed radar?

Radar was developed independently in many countries in the 1930s. But in 1935, Robert Watson-Watt (1892–1973), a Scottish physicist, was the leader of a group that created the first radar defense system for the British military. Although a large number of nations, including the United States, Canada, Britain, France, Germany, the Soviet Union, and Japan, worked to develop radar systems during the 1930s, the British system of ground-based radar stations was the first to use radar effectively in warfare. By the early 1940s, radar systems were made small enough to be installed in aircraft so they could engage other aircraft in fights at night.

Radar guns used by police to monitor traffic speeds work by taking advantage of the Doppler effect.

How do police radar guns work to catch speeding cars?

The police use the Doppler effect when checking for speeding vehicles. A radar gun sends out radar waves at a particular frequency. As the radar wave hits a vehicle, the wave reflects back toward the radar gun at a different frequency. The frequency of the reflected wave depends upon the direction and speed of the vehicle: the faster the speed, the greater the frequency change. The radar gun determines the speed of the vehicle by measuring the difference between the emitted frequency and the reflected frequency and computing the speed from that measurement.

Ironically, the Scottish inventor of military radar was the deserving victim of his own radar technology years after World War II. According to Canadian police, Robert Watson-Watt (1892–1973) had been speeding on a stretch of Canadian road, and he was detected by a police radar gun. Watson-Watt willingly paid the fine and drove away.

How do military aircraft today mislead enemy radar?

A French fighter plane developed in the 1990s used a technology called active noise cancellation to help the jet evade radar detection. It received an incoming radio wave and sent out the direct opposite pattern of that wave—in this case, a radar wave half a wavelength out of phase with the incoming radar. When the two waves interfered with one another, they experienced destructive interference, cancelling out the signal. With a return signal, the radar system couldn't detect the plane.

What are stealth aircraft?

Stealth aircraft are planes that are able to avoid radar detection. The materials on the plane's surfaces and their peculiar shapes and angles deflect radar waves away from the plane, and in some instances, the plane's outer fuselage can absorb the radar waves without reflecting it back to the enemy radar transmitter. (Refer to the Fluids chapter for more information about aerodynamics and aviation.)

What is NEXRAD Doppler radar?

NEXRAD, or next-generation weather radar, was an important technological breakthrough for weather forecasting that is still heavily used today. NEXRAD relies on the Doppler effect to calculate the position and the velocity of precipitation. The spherical NEXRAD radar tower emits radar waves 360 degrees around and calculates the frequency shift of the reflected radar waves off rain, sleet, and snow. The NEXRAD computers then translate the information and represent the possible weather problems on a color-coded map for analysis. The maps are readily available in real time over the internet.

The goal and main function of NEXRAD precision radar is to save American money and lives by predicting threatening weather problems and warning the public before tragedy strikes. Meteorologists estimate that this tool for weather forecasting has saved many lives and many millions of dollars of property damage through its early warning systems. One of the most impressive advancements has been in pinpointing tornadoes and hurricanes far more accurately than what was possible before NEXRAD. Each NEXRAD station scans a radius of 125 miles with excellent accuracy and up to 200 miles with somewhat less accuracy.

What is TDWR?

The Terminal Doppler Weather Radar system, or TDWR, is a Doppler weather radar system operated by the Federal Aviation Administration. It was established after NEXRAD first began and is installed at 45 airports. Its range is about half that of NEXRAD, and it can't see well through heavy rain. On the other hand, TDWR uses radar waves with a wavelength of 5 centimeters (2 inches) rather than the 10 centimeters (4 inches) used in standard weather radar. As a result, it can resolve objects with twice as much detail, per-

How is radar used to study other worlds?

In radar astronomy, electromagnetic waves are aimed at planets and moons, either from Earth or from a space probe. By analyzing the reflected signals, the position, velocity, and shape of objects in our solar system can be determined. In the early 1960s, radar was used to determine the exact distance between Earth and Venus and between Earth and Jupiter. Later, radar was installed on the space probe *Magellan* to map the surface of Venus. Radar astronomy has been extremely useful in determining distances to objects in our solar system and in looking onto and under the surfaces of planets and moons to find out the distribution of ice, liquid water, and other substances on those celestial bodies.

mitting it to detect wind shear, microbursts (small, very powerful thunderstorms), and other weather conditions that are particularly dangerous to airplanes and their passengers. As with NEXRAD, TDWR radar images are available to the public over the internet.

RADIO ASTRONOMY

How is radio astronomy different from radar astronomy?

Radar astronomy measures the reflections of transmitted radio waves to determine an object's size, position, velocity, and surface characteristics. Radio astronomy is like optical astronomy but uses the VHF, UHF, and microwave portions of the electromagnetic spectrum rather than the infrared and visible portions. Radio waves penetrate the dust that hides the centers of galaxies and obscures regions where stars are forming. They can also detect hydrogen gas, which constitutes much of the known mass of the universe.

How do radio telescopes work?

A radio telescope looks a lot like a receiver for a radar system or a satellite television service. The basic radio telescope is an antenna with a parabolic dish that focuses incoming radio waves to a single point. Sensitive electronic detectors and systems then collect, decode, and interpret the radio waves to get a picture of the target object or measure its brightness at different parts of the electromagnetic spectrum. Radio telescopes are typ-

Highly sensitive telescope arrays like this one are used by radio astronomers to detect signals created by atoms, molecules, and ions from space.

ically able to detect electromagnetic waves with wavelengths between 0.1 millimeter and 1 kilometer.

What do radio astronomers hear?

Radio astronomers detect what sounds like noise but is actually signals from atoms, molecules, and ions in stars, galaxies, and particles in interstellar and intergalactic space. If radio telescopes are precisely aimed at rapidly rotating neutron stars, the devices can pick up a regular thumping noise as radio emissions from these "pulsars" swing toward Earth's direction as they rotate. Their rotation can be very fast: the Crab Nebula pulsar, for example, spins around 33 times each second! The pulsing noise heard from it with radio telescopes sounds like a rolling snare drum.

What are the largest radio telescopes in the world?

The two largest single-dish radio telescopes in the world are so big that they have been built into natural valleys in mountain ranges.

The largest single radio telescope in the Western Hemisphere is the Arecibo Telescope, 305 meters (1,000 feet) in diameter, located in Puerto Rico. The reflecting dish, made of 40,000 perforated aluminum panels, is located in a mountain valley. The actual antennas are in a 900-ton platform, suspended on cables 137 meters (450 feet) above the dish. The antennas, sensitive to frequencies from 50 MHz to 10 GHz, detect signals collected by and reflected from the dish, and they can be moved to determine the direction from which the signals are coming. Arecibo can also conduct radar astronomy using a 1-megawatt radar transmitter to send signals toward other planets; the telescope then listens for the faint echoes that come back.

In 2016, the FAST radio telescope in Guizhou, China, began operations. It is a single dish that is 500 meters (1,600 feet) across, and it is located in the Dawodang basin—a natural depression in the mountain landscape.

How are radio telescope arrays used?

Arrays of radio telescopes can be carefully arranged across wide areas to make observations of the same object at the same time. The electromagnetic wave information collected by these telescopes can then be combined using computers. Constructive and destructive interference between the signals allows astronomers using the array to make very accurate measurements of the location and size of the object. This technique is known as interferometry.

One of the world's largest radio telescope arrays is the Karl G. Jansky Very Large Array (JVLA). Located on a high plateau near Socorro, New Mexico, the JVLA has 27 separate telescopes working as one unit. Each telescope is 25 meters (82 feet) in diameter and weighs more than 200 tons. The telescopes can be arranged in various shapes up to 36 km (22 miles) across. Another large radio telescope array is the Atacama Large Millimeter/Submillimeter Array (ALMA), designed to observe shorter-wave radio frequencies than the JVLA. Located on a high plateau in the Atacama Desert in northern Chile, each of ALMA's 66 telescopes is a high-precision antenna up to 12 meters (40 feet) in diameter, and they can be positioned from 0.15 to 16 km (10 miles) apart.

What is VLBI?

Larger still than radio arrays like JVLA, systems of radio telescopes around the world can be linked together to perform Very Long Baseline Interferometry (VLBI) astronomical observations. The different telescopes of a VLBI array can be situated thousands of kilometers apart or even halfway around the world. Electric cables can't be used to combine the signals of telescopes so far apart, so each telescope needs a very precise clock to determine the arrival time of the signals they receive from their astronomical targets. Hydrogen masers in outer space, whose frequency is stable to one part in a million billion, are often used. Again, although the signal strength is small, the telescope's wide spacing increases the effectiveness of using interference to measure very fine details of the shape, size, and location of stars and galaxies.

In 2019 a VLBI project called the Event Horizon Telescope obtained a spectacular image of a supermassive black hole and its surroundings. The black hole, located at the center of a galaxy called Messier 87, is 53 million light-years away from Earth and is 6.5 billion times the mass of our sun.

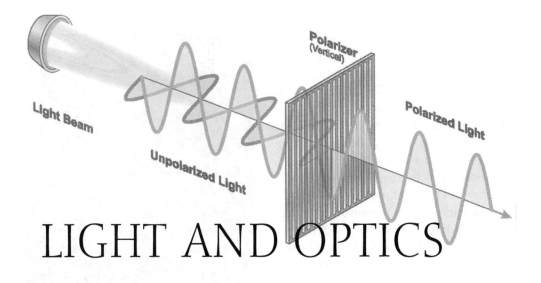

Light Beam

Unpolarized Light

Polarizer
(Vertical)

Polarized Light

LIGHT AND OPTICS

What is optics?

Optics is an area of study within physics that deals with the properties and applications of light. Optics can be used not only with visible light but with the entire electromagnetic spectrum from gamma rays to X-rays, ultraviolet light, visible light, infrared light, microwaves, and radio waves.

What were some early ideas about light?

Does light travel at a finite speed or at infinite speed? Is light emitted by the eye, or does it travel to the eye? These questions were debated for centuries. In ancient Greece, Aristotle (384–322 B.C.E.) argued that light is not a movement. Hero of Alexandria (10–70 C.E.) said it moves at infinite speed because you can see the stars and sun immediately after you open your eyes. Empedocles (490–430 B.C.E.) said it was something in motion, so it must move at a finite speed. Euclid (325–265 B.C.E.) and Ptolemy (90–168 C.E.) said that if we are to see something, light must be emitted by the eye.

Much later, in 1021 C.E., the Arab scientist Alhazen (Ibn al-Haitham, c. 965–c. 1030) did experiments that led him to support the argument that light moves from an object into the eye, and therefore, it must travel at a finite speed. At around same time, the Iranian polymath Abu Rayhan al-Biruni (973–c. 1050) noted that the speed of light is much faster than the speed of sound. The Turkic astronomer Taqi al-Din (1521–1585) also argued that the speed of light was finite and that its slower speed in denser objects explained refraction. He also developed a theory of color and correctly explained reflection.

What ideas about light were considered starting in the seventeenth century?

The science of light underwent great advances in the seventeenth century. The German astronomer Johannes Kepler (1571–1630) and the French philosopher, mathematician, and physicist René Descartes (1596–1650) argued that if the speed of light was not infinite, then the sun, moon, and Earth wouldn't be in alignment in a lunar eclipse. Despite this misconception, Kepler, in his 1604 book *The Optical Part of Astronomy*, essentially invented the field of optics. He described the inverse-square law, the workings of a pinhole camera, and reflection by flat and concave mirrors. He also recognized the influence of the atmosphere on both eclipses and the apparent locations of stars. Willebrord Snellius (1580–1626) discovered the law of refraction (Snell's Law) in 1621. Descartes used Snell's Law to explain the formation of rainbows shortly thereafter.

Also a mathematician, astronomer, and inventor, Dutch scientist Christiaan Huygens made major contributions to physics, especially in optics and mechanics.

The Dutch scientist Christiaan Huygens (1629–1695) wrote important books on optics and proposed the idea that light was a wave. Soon afterward, Isaac Newton's (1642–1727) celebrated experiments using a prism to separate white light into its colors led to Newton's "Theory of Colors," a paper published in 1672. He recognized that telescope lenses would cause colored images and invented the reflecting telescope using a concave mirror that would not have this fault. Newton believed that light was made up of very lightweight particles, or corpuscles.

What is the modern conception of light waves?

The theoretical and experimental work of French physicists Augustin-Jean Fresnel (1788–1827) and Siméon Poisson (1781–1840) firmly established the wave theory of light in 1815 and 1818. Later, the work of James Clerk Maxwell (1831–1879) and Heinrich Hertz (1857–1894) showed that visible light is an electromagnetic wave. Then, amazingly, at the beginning of the twentieth century, German physicists Max Planck (1858–1947) and Albert Einstein (1879–1955) showed that light is also a massless particle.

On the one hand, light has all the properties of a transverse wave. That is, it can transfer energy and momentum. It obeys the principle of superposition and can be diffracted and interfere with itself. On the other hand, it also has the properties of a massless particle. While in a medium, it moves in a straight line at a constant velocity. It can transfer energy and momentum. A full description of the true dual nature of light as both wave and particle is one of the fundamental concepts of quantum physics.

How is visible light defined?

The limits of human vision define the lower and upper boundaries of the electromagnetic waves we call visible light. Wavelength, rather than frequency, is commonly used when describing light because until the past three decades, only wavelength measurements were possible; light frequencies were too high to measure directly.

The lowest frequency of visible light is about 4×10^{14} Hz, which has a wavelength of 700 nanometers (700 $\times$ 10^{-9} m). To human eyes, it is a deep shade of red. The highest frequency of visible light boundary is about 7.9×10^{14} Hz, a wavelength of 400 nanometers. To our eyes, that is a purple or violet color. Light with wavelengths just longer than red light is called infrared; light with wavelengths just shorter than violet light is called ultraviolet.

How is incandescent light emitted?

You have often seen the light emitted by hot objects, whether it is the dull, red glow of the heating coil on an electric range; the orange glow of the element in an electric oven; or the bright yellow-white of the glowing filament of an incandescent lamp. Even the yellow glow of a fire comes from light emitted by hot carbon particles. This kind of light is called thermal radiation or black-body radiation. Energy, often from stored chemical energy, is converted into thermal energy. That energy is transferred to the surroundings by radiation, including light. Unfortunately, producing light in this way is very inefficient because about 97 percent of the released energy is in the form of infrared radiation that warms the environment rather than visible light that can be seen. Because of their large energy use, many countries have begun restricting the use of incandescent lamps.

How is fluorescent light emitted?

Light can be emitted by gases and solids by a process called fluorescence. Neon signs are one example of a gas that glows because electrical energy is converted into light energy. High-intensity lamps use either sodium or mercury vapor to produce intense light. Fluorescent tubes and compact fluorescent lights (CFLs) use electrical energy to excite mercury atoms. The ultraviolet emitted by these atoms causes compounds deposited on the inside surfaces of the lamps to glow. The colors can be chosen to emulate incandescent lamps or daylight. CFLs are many times more efficient than incandescent light bulbs and can convert up to 15 percent of the electrical energy they use into light.

How do lasers and LEDs emit light?

Lasers are used today in a wide variety of ways, including in pointers, CD and DVD players, and supermarket barcode scanners. They typically consist of a small crystal composed of a mixture of elements like gallium, arsenic, and aluminum. The lasers produce intense light of a single color that is emitted as a narrow ray. Lasers can shine a tight beam of light a long distance, and, as a result, they can be dangerous, causing eye damage or blinding people if they are not carefully used.

LED lights use a technology called a light-emitting diode to shine bright light with high efficiency and much less heat than incandescent lights. They use the same kind of materials used in lasers but emit light in all directions rather than in a focused beam. They have been used for decades in ways such as on/off indicators in appliances, traffic lights, car taillights, and television remote controls. In recent years, LED light bulbs have become much more common in homes and businesses; they are costly to produce but are much more energy efficient and longer lasting than incandescent light bulbs.

How is light detected?

Light carries energy, so a light detector must convert light energy to another form of energy. In most cases, light is converted into electrical energy. In a human eye, light strikes a molecule called an opsin. The absorption of light changes the shape of the molecule, which results in an electrical signal sent on the optic nerve.

In most digital light detectors used today, such as in the digital cameras on your cell phone, when light is absorbed in a semiconductor, one or more electrons are released. The charge they carry produces a voltage that is then converted into a digital signal. In old-fashioned photographic film, molecules consisting of silver and chlorine, bromine, or iodine are used. When light strikes these molecules, it transfers its energy to electrons. The molecules are broken apart, and a tiny crystal of metallic silver remains.

THE SPEED OF LIGHT

What were the first attempts to measure the speed of light on Earth?

In 1638, the Italian physicist Galileo Galilei (1564–1642) proposed a method of measuring the speed of light. Galileo would have one lamp, and an assistant a great distance away would have a second lamp. The assistant was to uncover his lamp immediately when he saw Galileo uncover his own lamp. The speed could then be determined by measuring the time it would take the light to travel from Galileo to the assistant and back again. Galileo claimed to have done the experiment several years before 1638, but there was no record of his results. In 1667 the academy of sciences in Florence, Italy, carried it out between two observers a mile apart. They reported there was no measurable delay, showing that the speed of light must be extremely rapid.

What were the first laboratory measurements of the speed of light?

The first measurement of the speed of light in a laboratory was by Hippolyte Armand Fizeau (1819–1896) in 1849. He used a beam of light that passed through the gaps between teeth of a rapidly rotating wheel, was reflected from a mirror eight kilometers away, and returned to the wheel. The speed of the wheel was increased until the returning light passed through the next gap and could be seen. The speed was calculated to be 315,000 kilometers per second. Léon Foucault (1819–1868) improved on this a year later by using a rotating mirror in place of the wheel and found the speed to be

> ## How did early astronomers try to measure the speed of light?
>
> In 1676 the Danish astronomer Ole Rømer (1644–1710) measured the orbital period of Jupiter's innermost moon, Io. He found the period was shorter when Earth was approaching Jupiter than when it was moving away from it. He concluded that light travels at a finite speed and estimated that it would take light 22 minutes to travel the diameter of Earth's orbit. The Dutch astronomer Christiaan Huygens (1629–1695) combined this estimate with an estimate of the diameter of Earth's orbit. He concluded that the speed of light is 220,000 kilometers per second.
>
> In 1725 the English astronomer James Bradley (1693–1762) noted that the location of a star changed with the seasons. He proposed that the shift was due to the addition of the speed of light and the speed of Earth in its orbit, an effect called the aberration of starlight. Bradley observed the shift in several stars and determined that light traveled 10,210 times faster than Earth moves in its orbit. The modern result is 10,066 times faster; Bradley was off by less than 2 percent.

298,000 kilometers per second—off from the correct value by less than 1 percent. He also used this technique to determine that light travels slower in water than in air.

How was the speed of light measured to an accuracy of 0.1 percent?

The American physicist Albert Michelson (1852–1931) greatly improved Foucault's measurement in 1879, using an eight-sided rotating mirror and a plane mirror located on Mount San Antonio, 35 kilometers (22 miles) away from the source on Mount Wilson in California. By measuring the speed of the rotating mirror and the distance between the mirrors, Michelson made the most accurate measurement of the speed of light to that date: 299,910 kilometers per second.

After James Clerk Maxwell's (1831–1879) theory of electromagnetism was understood, it became possible to calculate the speed of light indirectly from the relationships between an electric charge and its electric field and between an electric current and the magnetic field it produces. In 1907 Edward Bennett Rosa (1873–1921) and Noah Ernest Dorsey (1873–1959) calculated the speed of light to be 299,788 kilometers per second in this way. Both this measurement and Michelson's measurement were accurate to better than 0.1 percent.

How did the speed of light get measured to its modern value?

In 1926, Albert Michelson (1952–1931) made a new interferometric measurement of the speed of light that obtained 0.001 percent accuracy. A couple of decades later, research on microwaves used in radar during World War II led to a new method of measuring the speed of light. By 1950 Louis Essen (1908–1997) and A. C. Gordon-Smith reported a result of 299,792 kilometers per second using a cavity resonance experimental

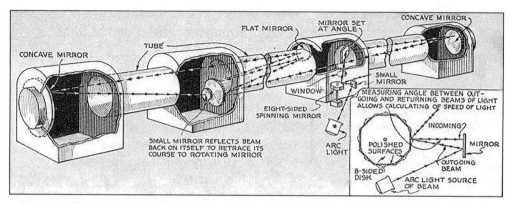

A diagram of Albert Michelson's device for measuring the speed of light. The equipment included a mile-long tube about a meter wide; two large, concave mirrors; and an eight-sided spinning mirror that rotated 512 times per second.

technique, which was even more precise than Michelson's result. In 1958, K. D. Froome (1921–) used a radio interferometer—improving on Michelson's technique—to obtain a measurement of the speed of light with accuracy 10 times greater still. By the 1970s, scientists at the U.S. National Bureau of Standards (known today as the National Institute of Standards and Technology) in Boulder, Colorado, succeeded in directly measuring simultaneously the wavelength and the frequency of an infrared laser. From these two measurements, they were able to calculate the speed of light to be 299,792.4562 kilometers per second with an uncertainty of only 0.0011 meters per second—that is, 0.0000003 percent accuracy!

These new measurements were so accurate that physicists decided to reconsider the definition of the length standard. At that time, the meter was defined as the length of a specific number of wavelengths of a specific color of light emitted by the element krypton-86. That definition was deemed to be less accurate than the accuracy to which the speed of light had been measured. Accordingly, in 1983 the Conférence Générale des Poids et Mesures (General Conference on Weights and Measures) decided to fix the speed of light in a vacuum at 299,792.458 kilometers per second and define the meter as the distance light travels in 1/299792458 of a second.

What is a light-year?

A year is a unit of time, but a light-year is a unit of distance. Specifically, it is the distance that light travels in one year through a vacuum. Since light travels at 299,792.458 km/s and a year contains about 31,557,000 seconds, a light-year is about 9,460,500,000,000 (9.46 trillion) kilometers, or about 5.88 trillion miles.

How long does it take light to travel certain distances?

The following table lists some light travel times to certain distances. (Some of these destinations would require a light beam to travel in a circular or curved path.)

Distance	Time
1 foot	1 nanosecond
1 mile	5.3 microseconds
From New York City to Los Angeles	0.016 seconds
Around Earth's equator	0.133 seconds
From the moon to Earth	1.29 seconds
From the sun to Earth	8 minutes
From Alpha Centauri to Earth	4.4 years
From the Andromeda Galaxy to Earth	2,200,000 years

How does the speed of a light beam depend on the medium it's in?

Waves travel at different speeds when traveling in different materials. In the vacuum of space, light travels at 299,792 km/s. When light encounters a denser medium, like Earth's atmosphere, it slows down ever so slightly to 298,895 km/s. Upon striking water, it slows down rather dramatically to 225,408 km/s, three quarters of its original speed. Finally, when light passes through the dense medium of glass, it slows to only 194,670 km/s. The ratio of the speed of light in a vacuum to that in a medium is called the refractive index, represented by the symbol n. The refractive index is extremely important in the use of lenses such as eyeglasses and microscopes.

How is the brightness of light measured?

There are two general ways to measure the intensity of light. The first is a physical system that measures energy or power transferred. The second system measures the effect of light on the surface it strikes—in other words, how brightly we would see or measure the light.

Can a light beam be frozen in time?

Lene Hau

Remarkably, it is possible to stop a beam of light—that is, freeze it at a certain location with zero speed and then let it go on its way after holding it there for a period of time. In 2001 the Danish-American physicist Lene Vestergaard Hau used a Bose-Einstein condensate (BEC)—a special state of supercold matter—to stop a beam of light. A few years later, she was able to stop a light beam in a BEC, transfer the beam to a different BEC, and then send it forward again. In 2013 a team of German physicists used a technique called electromagnetically induced transparency to trap a beam of light in a crystal for a full minute before releasing it. These experiments may someday lead to a wide variety of applications, including light-based data storage and transfer.

You are familiar with the watt. It's the rate at which energy is transferred. The equivalent unit for light, the luminous power, is the lumen. If you look at the box a lamp comes in, you'll find both the electric power it dissipates in watts (W) and the luminous power in lumens (lm). For example, a 25 W clear bulb emits 200 lm. A 100 W lamp emits 1,720 lm. A 60 W halogen lamp emits 1,080 lm. Compact fluorescent lamps produce more light for the same power: a 25 W lamp emits 1,600 lm.

The intensity of light depends on the degree to which the lamp spreads or focuses the luminous power. If the light goes into all directions, it won't be as intense as it would be from a reflector spot lamp that reflects light to form a narrow beam. The unit in which light intensity is measured is the candela (cd). If a 100 W, 1,720 lm lamp could spread its light into all directions, it would have an intensity of 137 cd. But if the same lamp were a spotlight, concentrating all the light into a 30 degree angle, then the intensity would be about 640 cd.

What is illuminance?

Illuminance is a measure of the total amount of light striking an object divided by the area of the object. For example, if a light source with an intensity of one candela shone on an object one foot away, its illuminance would be one foot-candle. Another way to state it would be that if the flux of one lumen shines evenly on a surface with an area of one square foot, its illuminance would be one foot-candle.

As the distance increases between a lamp and the surface it illuminates, the illuminance provided by the lamp decreases according to the inverse square law. Suppose you have a light luminous power of 1,000 lm 1 meter (3.2 feet) away from the surface. If you now moved the lamp to 2 meters (6.5 feet) away, the illuminance would be 1,000 lm divided by the distance squared, or $1/(2^2)$—that is, one-fourth of the illumination at 1 meter. If you moved the lamp to 3 meters (9.8 feet) away, the illuminance would decrease to one-ninth of the original amount.

What is candlepower?

Candlepower is an old unit of illuminance that is equivalent to 0.981 candela. It received its name from the amount of light produced by a particular kind of candle weighing one-sixth of a pound (76 grams) that burned at 120 grains (7.8 grams) per hour. Today, many people simply use the term candlepower to mean candela because they are almost the same.

POLARIZATION OF LIGHT

What is polarized light?

Light is an electromagnetic wave that consists of an oscillating electric field and an oscillating magnetic field. The two fields are perpendicular to each other. In most cases, the direction of the electric field has no preferred direction because different parts of the source produce light with electric fields in different directions. Such light is unpolarized. Some sources, however, like laser pointers, emit light whose electric field is always in one direction. Such light is called polarized.

How does light get polarized?

Light can be polarized in two ways. If a shiny surface such as water or an automobile window reflects light, it can be partially polarized. Or, you can use a polarizing filter that passes only the light that has the electric field in one orientation. A typical polarizing filter has long, thin molecules, all oriented in the same direction, embedded in plastic. The molecules absorb light that has an electric field parallel to the long dimension of the molecules. Therefore, the light passing through the filter has an electric field perpendicular to the long dimension of the molecule.

When looking through a pair of polarized sunglasses, why do car windows sometimes look spotty?

The windows in cars are typically built of a kind of safety glass that has a thin layer of plastic between two outer layers of glass. The spots seen on these rear windows when wearing polarizing sunglasses are the stress marks of the plastic layer between the two layers of glass. The spots, created during the manufacturing of the safety glass, act like polarizing filters and block some of the light, creating small, dark, circular regions in the otherwise transparent glass.

Why are polarized sunglasses useful?

Polarized glasses are useful when driving, sailing, skiing, or in any situation where unwanted glare is present. Glare is caused by light reflecting off a surface such as water, a

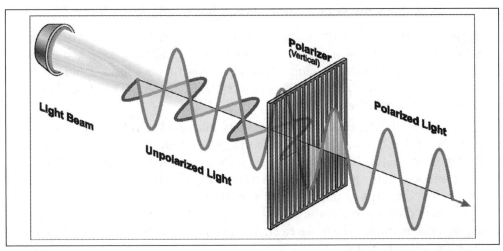

A polarizing filter blocks light vibrating in all polarizations except one. The resulting polarized light is great for glare reduction and can be used in other applications such as stress-analysis tests in industry applications.

road, or snow. Such light is polarized. Navigating your way through some situations could be difficult without polarized sunglasses.

For example, think about light reflecting off the surface of a lake. The light polarized parallel to the water's surface is reflected with a greater intensity than the light polarized perpendicular to the surface. Sunglasses that pass vertically polarized light will reduce the glare from the water.

How can you check to see if a pair of glasses is polarized?

Polarized lenses are transparent only to light that has its electric field oriented in one direction. Therefore, if a pair of sunglasses is polarized, two pairs of the same sunglasses, when aligned perpendicular to each other, should not allow any light to pass through the lenses.

Light from the sky is also partially polarized due to the scattering of light off gas molecules in the atmosphere, so if you put on a pair of polarized sunglasses and tilt your head so that your ear is near your shoulder, you should see a change in the intensity of the sky on a clear, sunny day. If you see no such change, then the sunglasses are not polarized.

How does an LCD use polarized light?

When light enters the back of the LCD, it becomes polarized. As it passes through the crystal, its polarization direction is rotated through 90 degrees so that it passes through the second polarizer. Thus, light passes through the display, and it appears bright.

Also, each glass sheet is coated with a thin, electrically conductive layer. If a voltage is placed across the sheets, the molecules align themselves with the electric field.

Why does my smartphone screen have polarizing filters?

Most smartphone screens, computer monitors, and flat-screen televisions are examples of LCD devices. LCD stands for liquid crystal display. A liquid crystal is composed of long, thin molecules that are free to move like a liquid but organize themselves in a regular array like a crystal, so under the right conditions, it can act like a polarizer.

In an LCD display, the liquid crystal material is in a thin layer between two glass sheets. The bottom sheet is rubbed in one direction so that the molecules in the liquid crystal touching the surface align themselves with the rubbing. The top sheet is rubbed in the perpendicular direction, aligning the molecules touching it in that direction. As a result, over the thickness of the liquid crystal material, the direction of the long axis of the molecules can rotate from 0 degrees through 90 degrees. Polarizing filters are placed on the outsides of the glass sheets in the same direction as the rubbing.

The molecules no longer rotate in the direction of polarization of the light, so no light passes through the display. It appears dark. By varying the voltage, different degrees of darkness can be achieved. Different colors can further be created using different filters.

The entire LCD screen is composed of tiny pixels, each connected to a source of the control voltage. Thus, each pixel can be switched between light and dark. Color filters are placed over each pixel to produce a full-color display. There can be millions of pixels on a screen, and each pixel's brightness and color can be changed many times a second. That's how these screens display pictures and video.

OPACITY AND TRANSPARENCY

What is opacity?

Opacity is a measure of how much light can go through a material. A substance with low opacity, such as air or clear glass, is called transparent. Something with high opacity, which allows no light through it, is called opaque.

Concrete, brick, and metal are some examples of opaque materials. Some materials can be opaque to light but not to other types of electromagnetic waves. For example, wood does not allow visible light to pass through it, but it will allow microwaves and radio waves to pass through. The physical characteristics of the material determine what type of electromagnetic waves will and will not pass through it.

What's the difference between translucent and transparent?

Transparent media, such as air and water, allow light to pass through. Rays of light are not reflected, and closely spaced light rays are only bent together. Translucent materials, on the other hand, allow light to pass through but bend closely spaced light rays into different directions. For example, frosted glass and thin paper are translucent because they let light through, but they are not transparent because you cannot see clearly through them.

How does the opacity of Earth's atmosphere affect global temperatures and climate?

Carbon dioxide, methane, and water vapor are transparent to visible light and short-wavelength infrared radiation (IR) but are opaque to the long-wavelength IR emitted by warm objects. They are called greenhouse gases. These gases let through the IR that warms Earth, but they absorb the IR rays emitted by the warm Earth and keep them near to the ground. Thus, they act as an insulating blanket for Earth.

Over the past 200 or so years, the amount of carbon dioxide in the atmosphere has increased dramatically, with almost all of the increase due to human activity. As a result of this increased opacity to long-wavelength infrared light, the additional thermal en-

Is Earth's atmosphere transparent to invisible light?

Oxygen and nitrogen gas, the two principal components of Earth's atmosphere, are transparent to visible light and to most infrared (IR) and ultraviolet (UV) wavelengths of light. Infrared radiation from the sun helps warm Earth's surface, and long-wavelength ultraviolet radiation is necessary for our bodies to produce vitamin D. Too much UV, however, harms the skin and may damage our DNA.

A high-altitude layer of ozone in our atmosphere protects Earth from all but a small fraction of the UV radiated by the sun. Ozone molecules have three oxygen atoms as opposed to the two in oxygen gas. Late in the twentieth century, large and growing holes in this protective layer developed due to the emission of chlorofluorocarbon (CFC) molecules from things like escaped refrigerant gases and aerosol propellants. These gases are no longer being manufactured and their use has been curtailed worldwide, so the ozone layer's depletion has been greatly reduced. The U.S. Environmental Protection Agency has estimated that the ozone layer will be fully restored within a few decades if the world's governments continue their policies limiting the use of ozone-depleting gases.

ergy that is now in Earth's atmosphere compared to the end of the 1800s is equivalent to more than 100 million atomic bombs! This increase has been leading to a warmer Earth, which in turn is changing climate and weather patterns, disrupting food production, causing shifts in the locations of forests and animals, and increasing sea levels by melting ice in Earth's polar regions. The degree of global warming and its impact on Earth and humans is under intensive scientific study.

SHADOWS AND ECLIPSES

What are shadows?

Shadows are areas of darkness created by an opaque object blocking light. Whether they are observed when someone puts their hand in the light from a movie projector, stands outside in the sunshine, or sees the moon move between Earth and the sun during an eclipse, shadows have always intrigued us.

How is an eclipse a shadow?

An eclipse is created just as any other shadow: by the presence of an object in the path of incoming light. In a lunar eclipse, Earth blocks the light of the sun that illuminates the moon. In a solar eclipse, the moon keeps the sun's light from reaching Earth. Thus, the shadow of the moon on Earth is what we call a solar eclipse.

What is a lunar eclipse?

A lunar eclipse occurs when Earth is directly between the sun and the moon. Earth blocks the sun's light rays from hitting the moon, leaving it in complete darkness. For an observer on Earth, a shadow appears on the moon, causing it to become dark. As Earth moves out from between the sun and the moon, the moon is gradually illuminated until the entire moon is seen again.

What is a solar eclipse?

An eclipse of the sun occurs when the moon casts its shadow on Earth. As the moon moves between Earth and the sun, darkness falls upon the small part of Earth where the moon blocks the light. Earth is much larger than the moon, so the shadow cast during a solar eclipse is narrow and moves across Earth's surface at about one thousand miles per hour!

What are the umbra and penumbra of a shadow?

A shadow produced by a large light source has two distinct regions. The umbra is the area of the shadow where all the light from the source has been blocked, preventing any

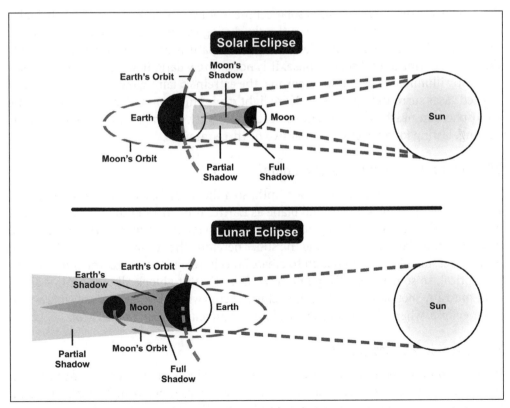

A solar eclipse occurs when the Moon casts a shadow on Earth; a lunar eclipse is when Earth casts a shadow on the Moon.

light from falling on the surface. The penumbra, or partial shadow, is a section where light from only part of the source is blocked, resulting in an area where the light is dimmed but not totally absent.

During a solar eclipse, the region of total darkness is the umbra. No direct sunlight can reach this area, resulting in a total eclipse. As Earth rotates, the moon's shadow races across Earth, producing the path of totality. Regions on either side of the path of totality experience the penumbra shadow and see the sun only partially covered by the moon.

How do the sizes of the umbra and penumbra vary?

An umbra and penumbra exist wherever there are shadows produced by a large light source. When an opaque object casts a shadow on a surface close to it, the shadow will be clear and distinct because it has a large umbra section and small penumbra. If, on the other hand, the object casting the shadow is closer to the light source, there will be regions on the surface where light from some of the source is not blocked by the object, producing a smaller umbra and a larger penumbra.

How dark does it get during a total solar eclipse?

The level of darkness during a total solar eclipse is at least as dark as typical late twilight. Depending on what objects are up in the sky during the eclipse, observers usually can see planets like Mercury, Venus, Mars, Jupiter, and Saturn, as well as some bright nighttime stars, during a total solar eclipse. It is not entirely dark, however, because the sky is slightly illuminated by the sun's corona, the extremely faint glow that is produced by ionized gases surrounding the sun. More light may come from reflection of light by the atmosphere in nearby areas not in the path of totality. Overall, a total solar eclipse is fleeting, lasting for an average of just two or three minutes.

Why don't eclipses happen every month?

The moon orbits Earth about once a month, so if the moon had a circular orbit around Earth that were in the same orbital plane as Earth's orbit around the sun, there would be a lunar eclipse and a solar eclipse every month. However, the moon's orbit is tilted with respect to Earth's orbit around the sun. Therefore, the sun-moon-Earth alignment is only occasionally perfect enough to create an eclipse. Furthermore, the moon's orbit is elliptical, so sometimes, it's too far away from Earth to block all of the sun's direct light onto any spot on Earth. If a solar eclipse occurs when it's that far away, it is an annular eclipse, where the moon covers all but a thin, circular ring of the sun.

How often do eclipses occur?

If you're counting partial eclipses as well as total eclipses, then over the entire Earth, there can be up to five solar eclipses or five lunar eclipses per year. Typically, though, there are at most one or two total solar or lunar eclipses in any given year. Total solar eclipses are especially rarely seen; they can only be observed in small areas and narrow paths on Earth's surface, while lunar eclipses can be seen over much larger areas. If a total solar eclipse passes over a given spot on Earth's surface, decades or even centuries can pass before another total solar eclipse happens there again.

When and where will total lunar eclipses occur in the next few years?

The following table shows the date and location of some future total lunar eclipses.

Date of lunar eclipse	Where eclipse will be visible
May 26, 2021	Asia, Australia, Pacific, Americas
May 16, 2022	Americas, Europe, East Asia, Australia, Pacific
November 8, 2022	Americas, Europe, Africa
March 14, 2025	Americas, Europe, Africa
September 7, 2025	Pacific, Americas, West Europe, West Africa
March 3, 2026	Europe, Africa, Asia, Australia
December 31, 2028	Europe, Africa, Asia, Australia
June 26, 2029	Europe, Africa, Asia, Australia, Pacific
December 20, 2029	Americas, Europe, Africa, Middle East

When and where will there be total solar eclipses for the next few years?

The following table shows the date and location of some future total solar eclipses.

Date of total solar eclipse	Where eclipse will be visible
December 14, 2020	Pacific Ocean, South America, Atlantic Ocean
December 4, 2021	Antarctica, Atlantic Ocean
April 20, 2023	Indian Ocean, Australia, Indonesia, Pacific Ocean
April 8, 2024	Pacific Ocean, Mexico, United States, Canada, Atlantic Ocean
August 12, 2026	Western Europe, Atlantic Ocean, Arctic Ocean
August 2, 2027	Atlantic Ocean, Spain, Northern Africa, Middle East, Indian Ocean
July 22, 2028	Indian Ocean, Australia, New Zealand, Pacific Ocean

When was the most recent total solar eclipse in the United States?

The most recent total solar eclipse in the United States occurred on August 21, 2017. The path of totality traveled across the North American continent from Oregon through Idaho, Wyoming, Nebraska, Kansas, Missouri, Illinois, Kentucky, Tennessee, Georgia, and the Carolinas. Many millions of people traveled from all over the country and the world to view totality, and tens of millions more people observed and experienced the partial solar eclipse that was seen across much of North America.

When will be the next total solar eclipse in the United States?

The next total solar eclipse in the United States will occur on April 8, 2024. The path of totality will travel roughly from south to north, going from Mexico into Texas all the way to New York State and into Canada. Mark your calendars; after that eclipse, there will not be a total solar eclipse over the continental United States for another 20 years.

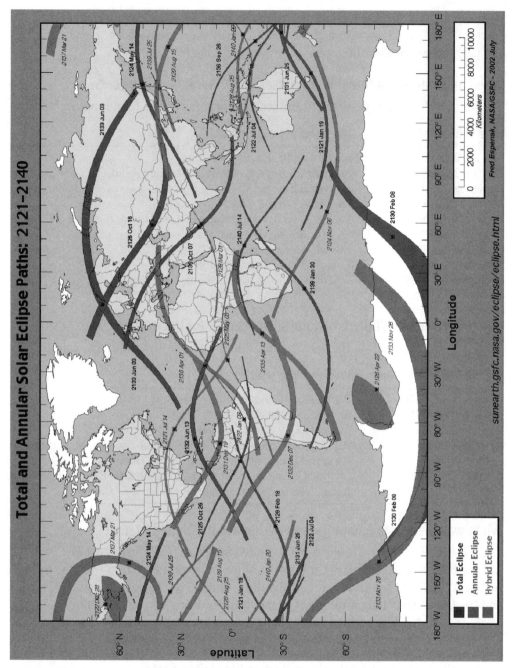

A map by NASA showing upcoming solar eclipses through the year 2040.

Is it safe to look at the sun during a total solar eclipse?

It is safe during totality—that is, when the moon is completely blocking the sun—to look at the eclipsed sun without eye protection. That, however, is the *only* time during an eclipse when it is safe to do so. Looking directly at even part of the sun's photosphere (the bright disk of the sun itself), even for just a few seconds, can cause permanent damage to the retina of the eye because of the intense visible and ultraviolet light that the photosphere emits. This damage can result in permanent impairment of vision up to and including blindness. The retina has no sensitivity to pain, and the effects of retinal damage may not appear for hours, so there is no warning that injury is occurring.

Normally, the sun is so bright that it is difficult to stare at it directly, so that is a sure sign that it would be dangerous to do so even during a partial eclipse. Viewing the sun's disk through regular sunglasses or any kind of optical aid (binoculars, a telescope, or even an optical camera viewfinder) is also extremely hazardous. Be sure to use certified eclipse glasses, solar filters, or welder's glass to view the sun, or use a pinhole camera to project the sun's image through a telescope and just view the projection.

REFLECTION

What is the difference between geometric optics and physical optics?

Geometric optics deals specifically with the path that light takes when it undergoes reflection and refraction—that is, when it encounters mirrors and lenses. Geometrical optics uses the ray model of light, in which an arrow represents the direction light trav-

---

How do solar eclipses aid in the study of world cultures and history?

Eclipses were often recorded or suggested in ancient documents and myths. Solar eclipses can thus be used to find dates and places of events recorded in those documents. For example, in the sixth century B.C.E., the Medes and Lydian armies were in a war when a solar eclipse occurred. The eclipse halted the battle and helped bring about a peace between the two armies because they thought the sun's disappearance was an omen. One possible date and place for that battle is May 28, 585 B.C.E., near the Halys River in modern Turkey, where a solar eclipse is calculated to have occurred.

In a story from an Indian epic, Arjun vowed to kill Jayadrath to avenge Jayadrath's killing of Abhimanyu. During a solar eclipse, Jayadrath came out of hiding to celebrate his survival, but when the sun reappeared, he was revealed, and Arjun killed him. Eclipses occurred in this region of the world in 3129 B.C.E. and 2559 B.C.E., providing two possible dates for this event—if it ever happened, that is.

els. The ray model does not consider the wave nature of light. Ray diagrams trace the path that light takes when it reflects and refracts in different media.

Physical optics is the division of optics that depends on the wave nature of light. It involves polarization, diffraction, interference, and the spectral analysis of light waves.

What is reflection?

When light strikes an object, it can either be absorbed, transmitted, or reflected. Opaque objects absorb or reflect light. Transparent and translucent materials transmit light but can also reflect it. The energy carried by the light must be conserved. The sum of the energy reflected, transmitted, and absorbed must equal the energy that strikes the material. Light energy that is absorbed increases the thermal energy of the material, so the material becomes warmer. Reflection occurs when light "bounces" off of a surface such as a mirror or a sheet of paper.

What is the law of reflection?

A smooth surface, like that of a mirror, reflects light according to the law of reflection. This is a rule that the angle of incidence equals the angle of reflection. Light that strikes the surface at, for example, 30 degrees above the surface is reflected at 30 degrees above the surface. You can test this for yourself with a small flashlight, a mirror, a sheet of paper, and a helper. Hold the piece of paper. Have the helper hold the flashlight, aim it at the mirror, and move it around until the reflected light ray hits your piece of paper. Change the angle with which the light from the flashlight strikes the mirror and see that you need to move the paper closer to the mirror to have the light hit it.

What is diffuse reflection?

What happens when light hits a sheet of paper? Try this experiment. Darken the room. Aim a flashlight at a piece of white paper and use a second sheet to catch the reflection. You'll see that the second sheet is illuminated in many different locations. The light is reflected from the paper but into many directions. This kind of reflection is called diffuse reflection.

Polished or smooth surfaces that do not absorb light are the best reflectors. Examples of reflective materials are shiny metals, whereas nonreflective materials may be dull metals, wood, and stone.

MIRRORS AND IMAGES

How were the first mirrors made?

People have viewed their own reflections in water for millennia, but some of the earliest signs of human-made brass and bronze mirrors have been mentioned in ancient Egyptian, Middle Eastern, Greek, and Roman literature. The earliest glass mirrors, backed with shiny metal, appeared in Italy during the fourteenth century. The original process for creating a glass mirror was to coat one side of glass with mercury and polished tinfoil.

What is the process of silvering a mirror?

A piece of glass is "silvered" when a layer of metal is deposited on one side, making the glass reflective enough to be a good mirror. In 1835 the German chemist Justus von Liebig (1803–1873) developed a method for silvering a mirror. His process consisted of pouring a compound containing ammonia and silver onto the back of the glass. Formaldehyde removed the ammonia, leaving a shiny, metallic-silver surface that reflected the light.

Today, most common mirrors are made by depositing a thin layer of metallic aluminum directly onto pieces of glass. The large mirrors used in modern telescopes, some of them more than 30 feet across, are coated in even larger chambers in layers that are sometimes only a few atoms thick; to increase reflectivity, they are often coated with silver or gold.

Why can't you always see yourself in a mirror?

The answer is a matter of angles. According to the law of reflection, the angle of incident light on a mirror (that is, the angle between the direction the incoming light is traveling and the mirror surface) must be equal to the angle of the reflected light. If you're standing directly in front of a mirror, the angle of incidence is close to 90 degrees, so the reflection comes out at about 90 degrees as well, or directly back to your eyes. If, however, you stand off to one side of the mirror, then the angle of reflectance is much smaller, so the incident light won't come from your face but from another part of the room.

How does a mirror reverse my image?

If you look in a mirror, you seem to see yourself reversed left to right. Your left eye appears to be your right eye in the mirror. Yet, the vertical direction is not reversed; your chin is still at the bottom of your face. What happens if you look at a mirror while lying down? Now left and right are not reversed—your chin is on the same side of both you and your image. But if your right eye was higher than your left, then in the image, the left eye will be higher.

So, what, indeed, is reversed in a mirror image? Think about what makes a glove right-handed or left-handed. If you

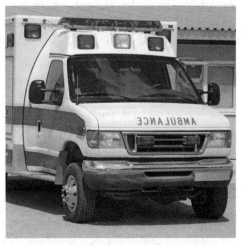

The word "AMBULANCE" is printed backwards on these emergency vehicles so that drivers will see the word correctly in their rearview mirrors.

207

examine a vinyl or latex glove, you'll notice that it can fit on either hand. How does such a glove differ from most other gloves? The thumb is in line with the fingers. In other words, there is no front or back! A mirror reverses not left and right or up and down but front and back. When you look at yourself, you are facing the opposite direction from your image. It's just like facing another person—his or her left and right sides are also reversed in comparison to yours.

Why is the word "ambulance" printed backward on the front of some ambulances?

The word "ambulance" is printed backward (left and right reversed) so that when viewed in a mirror—specifically, the rearview mirror of a car—it will appear correctly (that is, forward), ensuring that the driver can read it and respond as quickly as possible.

How do day/night rearview mirrors work in a vehicle?

When drivers traveling at night encounter a bright light from the vehicle behind them shining in their eyes, many will flip a tab on the underside of the rearview mirror to deflect the light up toward the ceiling of the car. The silvered surface of the mirror reflects approximately 85–90 percent of the incident light on the mirror, which is now directed toward the ceiling. The remaining 10–15 percent of the light is reflected by the front of the glass on the mirror. The mirror is wedge-shaped, thicker at the top than at the bottom. Thus, the front surface is angled downward to allow the much smaller amount of light to be reflected into the driver's eyes.

How do one-way mirrors work?

A one-way mirror seems to be a mirror when seen from one side but a window when seen from the opposite side. Thus, the window is disguised as a mirror to allow secret surveillance. Physically, though, there is no such thing as a one-way mirror. That is, the amount of light reflected from one side is the same as that reflected from the other. The light transmitted in one direction is equal to that transmitted in the opposite direction.

How, then, does a one-way mirror work? First, the mirror isn't totally reflecting. It transmits half the light and reflects the other half. The second requirement has to do with lighting. It is imperative that the observation room remain dark because if a lamp were turned on in it, some of that light would pass through the window into the interrogation room as well, and anybody in that room would be able to see through the glass into the observation room.

What is an image?

Suppose light falls on your face. Light is reflected diffusely from this surface; that is, it is reflected into a wide variety of directions. Consider a very tiny object like the end of an eyelash. Light from that point will leave in many directions. That point is called an object. An optical device can cause the light rays from the object to converge again to a single point. That point is called the image.

What is a real image?

A concave mirror or a convex lens can redirect light from an object so that it converges to a single point. That point is a real image. It is in front of a concave mirror and on the other side of a convex lens from the object. If you place a screen, piece of paper, or wall at the location of the image, you'll see it on that surface.

What is a concave mirror, and how does it work?

A concave mirror is curved inward so that the incident rays of light are reflected and can be brought together. If the rays striking the mirror were parallel to each other, as they would be from a distant source, then the reflected rays converge at the focal point. Concave mirrors are typically used to focus waves, whether it is a microwave signal to a receiver or visible light to an observer. They can produce real images.

If the object is closer to the mirror than the focal point, as is often the case with a concave bathroom mirror, the image is virtual and upright. If you look at the reflection of a more distant object, like the bathroom wall, you'll notice that it is upside down, or inverted. The wall is farther away from the mirror than the focal point, and its image is real.

What is a convex mirror and how does it work?

A convex mirror is the opposite of a concave mirror in that it is curved outward. The reflected light spreads out rather than converging at a point. Therefore, convex mirrors form virtual images. Convex mirrors are used for security purposes in stores because they broaden the reflected field of vision, allowing clerks to see a large section of the store. The images are smaller than the objects, but the mirrors help to see a wide area.

Why is it printed on sideview mirrors: "Objects viewed in mirror may be closer than they appear"?

This statement is actually an important safety message that warns the driver that the mirror may be deceptive because it is convex. Why would an automobile manufacturer put a deceiving mirror on a car? A flat-plane mirror would only show the driver a small,

What is a virtual image?

The light reflected from an object by a flat, or plane, mirror is not redirected so it will converge again. What you see when you look in a mirror is a virtual image. It is located behind the mirror. Some of the rays from the object are reflected by the mirror and enter your eye. Your eye believes that all these rays came from a single point, and that point is behind the mirror. The image isn't real; it is virtual. Virtual images can also be formed by lenses, but they cannot be focused onto a screen.

narrow section of the road behind the car. If a convex mirror is used, the driver can see not just behind the car but also to the side, reducing any "blind spots" that would not show a car in the reflection. Convex mirrors, however, make objects appear smaller and therefore farther away, so the message is there to serve as a reminder that the image is not exactly as it might appear.

REFRACTION

What is refraction?

Refraction is the bending of light as it goes from one medium to another. The most common use of refraction is in lenses. Eyeglass lenses refract light so that the wearer's eyes can focus the light properly. A magnifying glass is used to enlarge images. Lenses in cameras produce an image on film or on an image sensor, called a CCD sensor. Refraction also occurs when sunlight strikes Earth's atmosphere and when it goes through water.

How do we measure the refraction of light?

The extent to which a beam of light bends when it hits a different medium depends on the indices of refraction of the medium, the medium from which it came, and the angle at which the light strikes the boundary between the two media.

All materials have an index of refraction that depends on the speed of light in the material. The index of refraction is the speed of light in a vacuum divided by the speed of light in the material. A vacuum has a refractive index of 1, water is 1.33, and most types of glass are around 1.5. The higher the index of refraction a medium has, the more slowly light travels through it.

What is the index of refraction of some common media?

The following are some sample indices of refraction. The index of refraction, represented by n, is the ratio of the speed of light in a vacuum divided by the speed of light in the material. The larger the index of refraction, the greater the bending that takes place.

Medium	Index of Refraction (n)
Vacuum	1
Air	1.003
Water	1.33
Ethyl alcohol	1.36
Crown glass	1.52
Quartz	1.54
Flint glass	1.67
Diamond	2.42

What is Snell's Law of Refraction?

Snell's Law of Refraction, named after Dutch physicist Willebrord Snellius (1580–1626), tells us how light behaves at a boundary between two different media. Consider the interface between two media where the refractive index of the top medium is lower than that of the bottom medium. According to Snell's Law, when light hits the boundary between two materials, it is bent from its original path to a smaller angle with respect to the line perpendicular to the surface of the interface. As the incoming, or incident, angle increases, so does the refracted angle. When

Magnifying glasses make images larger by bending light through a glass medium.

light goes from a medium with a higher refractive index into one with a lower index, then it is bent away from the line perpendicular to the surface.

How does atmospheric refraction affect how we see celestial objects?

The index of refraction is close to, but slightly larger than, that of the vacuum of space. Thus, light from stars, planets, and other celestial objects refracts slightly as it enters Earth's atmosphere. Refraction is largest nearest the horizon. As a result, the true position of the stars is a bit off from where we observe them. Astronomers pay careful attention to this effect when making detailed measurements.

How does atmospheric refraction affect our view of the moon and sun?

Earth's atmosphere refracts a very small amount of reddish-colored sunlight onto the moon's surface. This red light is drowned out by ordinary sunlight reflected from the moon onto Earth, but it becomes visible during a total lunar eclipse when the moon is in Earth's umbral shadow. That's why a total lunar eclipse is sometimes called a "blood moon."

Refraction of light from the sun results in sunrise being slightly earlier than it would be without the atmospheric refraction and sunset being slightly later. Refraction also distorts our view of the sun and moon when they are very close to the horizon.

What is a mirage?

Mirages typically occur on hot, summer days when surfaces such as sand, concrete, or asphalt are warm. Mirages look like pools of water on the ground, along with an upside-down image of a building, vehicle, or tree in the distance. As you approach a mirage, the puddle of water and the reflection seem to disappear.

A mirage occurs because of a temperature difference between the air directly above the surface, which is hot and thus less dense, and the cooler, denser air a few meters above the surface. The denser air has a higher refractive index, and that causes the light from an object to bend up toward the observer. As a result, the object is right-side up while the refracted image is inverted and underneath the original object. The illusion of water is also a refracted image—the image of the sky. Mirages can only occur on hot surfaces and objects that are at relatively small angles in relation to the observer. Therefore, a person cannot see a mirage from an object that is just a few meters away. A mirage is not a hallucination but a true and well-documented optical phenomenon.

LENSES

When were lenses first made?

The word "lens" comes from the name of the lentil bean because of the similarity in shape of the bean and a converging lens. Lenses have been used for over three thousand years. It's possible that ancient Assyrians used them as "burning glasses" to start fires. A "burning glass" is mentioned in a play written in 424 B.C.E. by ancient Greek playwright Aristophanes.

Roman emperors used corrective lenses and knew that glass globes filled with water were able to produce magnified images. Ibn Al-Haitham (Alhazen; c. 965–1038 C.E.) wrote the first major textbook on optics, which was translated into Latin in the twelfth century and influenced European scientists. Shortly thereafter, in the 1280s, eyeglasses were used in Italy. The use of diverging lenses to correct nearsightedness (myopia) was documented in 1451.

What is a converging lens?

A converging lens has at least one convex side. Its shape causes the entering light rays to converge or come closer together. A converging lens can create a real, inverted image that may be projected on a screen. When used as a magnifying glass, it creates a virtual, upright image.

What is a diverging lens?

A diverging lens has at least one concave side. The shape of the lens causes the entering light rays to spread apart when they leave the lens. A diverging lens is often used in combination with converging lenses. Eyeglasses to correct nearsightedness use diverging lenses.

What is the focal length of a lens?

The focal length is a measure of the strength of a lens. Consider a converging lens. Rays from a very distant object come together at the focal point. The focal length is the distance from the lens to the focal point. Lenses with very convex surfaces have shorter focal lengths, while flatter lenses have longer focal lengths. For diverging lenses, the focal point is virtual. That is, it is the point from which the diverging rays leaving the lens would have come if a point object were placed there.

FIBER OPTICS

What is total internal reflection?

Total internal reflection occurs when a ray of light in a medium with a higher index of refraction strikes the interface between that medium and one with a lower index of refraction. When the incident rays are at a small angle with respect to the perpendicular to the interface, the rays pass through the interface being refracted to a larger angle. As the angle of incidence increases, so does the refracted angle. At the "critical" angle, the refracted angle is 90 degrees; that is, the ray's direction is along the interface. If the angle of incidence is increased any more, there is only a reflected ray; the refracted ray no longer exists. Because all the light is reflected, it is called total internal reflection.

If you open your eyes underwater, can you see out of the water?

One common example of total internal reflection can be observed when you are underwater. If you look straight up out of the water, you will see the sky and any other visible surroundings directly above the water. If, however, you look out of the water at an angle of 48 degrees or more from the vertical, you will not see out of the water but instead will see a reflection from the bottom of the water. The next time you are in a pool or a lake, try to look up out of the water, and you will see a

Fiber optics are glass fibers that transmit laser light that can contain digital information more quickly and efficiently than metal cables.

213

point on the surface where you no longer can see out of the pool, for the light has reached its critical angle.

Why are cut diamonds so sparkly?

Diamond has a very large index of refraction and thus has a small critical angle of only 25 degrees. If a diamond is expertly cut, light striking the top of the diamond will be internally reflected and will emerge again from the top rather than the sides of the diamond, creating a striking, sparkling shine.

How do optical fibers use total internal reflection?

Strands of carefully manufactured glass fiber, commonly known as optical fibers, use the principle of total internal reflection to transmit information near the speed of light. The fiber has an inner core of glass with a high refractive index surrounded by a cladding of glass with a lower index. A laser sends light into the end of a strand of fiber. When the light strikes the interface between the core and the cladding, the light is reflected back into the cable, continuing to move down the length of the fiber.

What are some advantages of optical fibers for transmitting information?

Visible light and near-infrared radiation can travel kilometers through fibers without significant energy loss. One reason is the total internal reflection. The second reason is that they are made from materials designed to absorb as little as possible of the radiation. Another advantage of optical fibers is that information sent through the fibers is more secure because, if it doesn't escape the fiber, it is not accessible to those trying to intercept the information.

How did fiber optics originate?

The idea that light could travel through glass strands originated as far back as the 1840s, when physicists Daniel Colladon (1802–1893) and Jacques Babinet (1794–1872) demonstrated that light could travel through bending water jets in fountains.

The first person to display an image through a bundle of optical fibers was a medical student in Germany named Heinrich Lamm (1908–1974), who in 1930 used a bundle of fibers to project the image of a light bulb. In his research, Lamm ultimately used optical fibers to observe and probe areas of the human body without making large incisions. From that point on, serious research in optical fibers ensued, expanding later with the development of the laser.

How are fiber optics used for imaging today?

The transmission of information through optical fibers has had a huge impact on the world. Modern medicine, for example, has benefited greatly from the use of flexible fiber-optic bundles that enable the viewing of areas of the body that might otherwise be inaccessible. It is possible to use fiber optics to look down into your digestive tract, in blood vessels, and even into the skull without surgery.

In astronomy, fiber optics are used to make telescopes much more powerful. In the past, only a few spectroscopic measurements could be made at one time with any single telescope; today, fiber-optic systems can create spectroscopic masks and plates that allow astronomers to measure the spectra of more than 1,000 objects at a time. This hundredfold increase in data collection ability has helped revolutionize our understanding of the universe.

How are fiber optics used for communication today?

Communications has been transformed tremendously thanks to fiber optics. Long-distance internet, telephone, and television signals use fiber-optic cables. Some systems use fiber optics to transmit information directly to the home or business. Fiber optics can transmit large amounts of data at high speeds at low cost, permitting hundreds of television broadcasts, very high-speed internet, text messages, streaming connections, and telephone conversations to be sent and received at the same time.

DIFFRACTION

What is diffraction?

Diffraction describes how waves move, spread, and interact in ways different from rays. In geometric optics, light is treated as a series of rays that can be reflected and refracted. But, for example, when light goes through a very small opening, the ray model is no longer sufficient, and the wave properties of light become important. Suppose you pass light through a round aperture whose diameter you can change. Let the light fall on a screen. As you first begin reducing the size of the hole, you will see a bright spot that is shrinking in size. But when the hole becomes very small, a strange thing happens: the spot no longer shrinks, but its outer edge becomes fuzzy. Light begins to bend around the edge of the hole. This is an example of diffraction. Diffraction also occurs when light is sent through narrow slits or if there are small objects that cast shadows in a broader beam of light.

Who conducted early research on diffraction?

In 1665, a publication by Francesco Grimaldi (1618–1663) described how light could be diffracted when passing through thin holes or slits or around boundaries. In 1803, Thomas Young's (1773–1829) experiments with sending light through narrow slits demonstrated the diffraction and interference of light.

How does the interference of light work?

Waves, as we discussed in an earlier chapter, can pass through and around each other, creating interference. The key property of the interference of light is that the two or more interfering waves must have the same wavelength and phase.

Have you ever heard a sound from around the corner of a building, even though there was no wall to reflect the sound toward you? That is an example of diffraction as well. Diffraction occurs with all types of waves. You can often see it in water waves, and it is one reason that sound waves spread when they go through a door or window.

While interference can be seen with ordinary light, it can be seen more easily using laser light. If light from a laser passes through two slits that are a small distance apart, the diffraction patterns from the two slits will overlap. When they do, a pattern of light and dark bands will be seen. The bright bands are where waves from the two slits are constructively interfering. The distance the light has traveled from the two slits will either be equal or will differ by an integer number of wavelengths.

The dark bands occur when the waves destructively interfere. In this case, the difference in distance that light from the two slits will have traveled will be an odd number of half-wavelengths—that is, one-half, one-and-a-half, two-and-a-half, three-and-a-half, and so on.

Why can soap bubbles sometimes be so colorful?

The bright spread of colors that are produced when light hits a thin film like a soap bubble is called iridescence. It is caused by the interference of light waves resulting from reflections of light off the top and bottom surfaces of the thin film. A soap bubble displays an iridescent pattern because light reflects off the front and back (inside and outside) surfaces of the soap bubbles. As the thickness of the layer varies, the interference between the two reflections will vary, causing the colors to vary as well.

Why is there sometimes iridescence on roads and parking lots?

Sometimes on roads or parking lots, especially when the roads are wet, a thin film of gasoline or oil can be seen iridescing on the road surface. These films are easier to see on wet surfaces because of the iridescent patterns that result from light reflecting off the top of the gasoline and off the boundary between the gasoline and water. The resulting pattern appears as the colors of the

Soap bubbles have an iridescent shine when light hits them because light waves reflect off the two surfaces of their thin film.

visible light spectrum in the thin film of gasoline—something that is perhaps surprisingly beautiful for having been caused by pollution.

What is a diffraction grating?

If you take a surface of shiny, reflective material and cut tiny grooves close together across its surface, you can create a diffraction grating. There must be hundreds or even thousands of grooves per inch; but if that happens, light that strikes the grating gets reflected back, split into the colors of the rainbow. This is because the grooves cause interference and diffraction at the grating's surface, which serves to break up a single beam with many wavelengths of light into a spread-out series of beams at individual wavelengths. Diffraction gratings are very important tools that scientists use to study the spectrum of light sources.

COLOR

What is white light?

White light is the name given to light that is a combination of all the colors in the visible light spectrum. When separated from each other, the different wavelengths have different colors. The longest-wavelength light is the color red, and decreasing wavelengths result in orange, yellow, green, blue, indigo, and finally, the shortest-wavelength visible color, violet.

What is indigo?

Indigo is the color between blue and violet in the spectrum, but almost no one can distinguish that color. Why, then, are there seven colors rather than six? Historians of science think it may be because Isaac Newton, the scientist who wrote the first book about optics, was drawing an analogy between color and musical tones. There are seven notes in the familiar European scale—A, B, C, D, E, F, and G. Therefore, Newton decided there should be seven colors in the spectrum, and he identified indigo as the seventh. So, don't be too disappointed if you can't tell indigo apart from blue and violet.

How do humans see objects?

For us to see an object, light from the object must enter our eyes. We can see stars, lightning, and light bulbs because they are emitting or giving off light. We depend on the light emitted from these sources to see objects that don't emit light: we see those objects because they reflect light into our eyes. The paper on which this book is printed, for example, does not emit light. We see it because the paper reflects light into our eyes.

How do we see specific colors?

When we "see" colors, we are seeing only part of the spectrum of colors that make up white light. The rest of the visible light spectrum is missing.

Selecting the color seen can be done in three ways. (1) You can view the object through a transparent material that transmits only part of the spectrum while absorbing the rest. (2) The object may itself have color—that is, it reflects part of the spectrum while absorbing the rest. (3) Finally, the spectrum can be physically separated, as it is by interference or by refraction by a prism or rainbow. For example, theatrical "gels" on spotlights produce colored effects for a stage show. Snow reflects the entire visible light spectrum, so it appears white, but green paper reflects only the green part of the spectrum. A black cloth absorbs all visible light, so it appears black.

Who discovered the basic properties of light and the color spectrum?

By the seventeenth century, glassmakers worldwide had learned to make gem-shaped pieces that were used in chandeliers. Candlelight was refracted in these pieces. When viewing the candlelight, different colors were seen, depending on the angles made by the light, the glass, and the viewer's eye. The British scientist Isaac Newton (1642–1727), who was intrigued by the colors that were produced by those chandeliers, decided to examine how a piece of glass shaped as a rectangular prism could create a spectrum of colors.

In Newton's own words, as he wrote in his 1704 book *Opticks*: "In a darkened Room make a hole in the [windowshade] of a window, whose diameter may conveniently be about [one-third] of an inch, to admit a convenient quantity of the sun's light: And there place a clear and colourless Prisme, to refract the entering light toward the further part of the Room, which, as I said, will thereby be diffused into an oblong [spectrum]."

How did Isaac Newton show that many colors combine to create white light?

To prove that the colors did not come from the prism itself, Newton expanded his prism experiment by reversing the procedure and forming white light from the spectrum of colors. He accomplished this by placing a lens in the middle of the spectrum to converge the colors on a second prism in the path of the colors. Sure enough, a beam of white light emerged out of the second prism.

How does a prism separate light into a spectrum of colors?

If all wavelengths of light were refracted by the same amount when entering and leaving a prism, then there would be no separation of colors. The refractive index of all materials depends on the wavelength of light. In diamond, the refractive index for blue is 1.594, for red 1.571. In flint glass, the index varies from 1.528 to 1.514. In crown glass, the variation is from 1.528 to 1.514, while in water, it is only from

Isaac Newton used a prism to separate white light into a spectrum, leading to the publication of his findings in his *Theory of Colors* (1672).

1.340 to 1.331. In all cases, the refractive index of blue is larger than that of red, so the blue light is refracted through a larger angle than the red. The large difference in diamonds accounts for their "flash."

If white is all colors together, what is black?

Black, the exact opposite of white light, is the absence of light or the absorption of all light. It may seem obvious to us, but Newton was the first to recognize this fact. A black piece of paper appears black because all the light is being absorbed in the paper—none is reflected back out to our eyes.

What are the primary colors from a light source?

When mixing light (or "additive color mixing"), the three primary colors are blue, green, and red. Computer monitors and both cathode ray tube (CRT) and flat-panel television sets use these colors. The combination of these primary colors results in other colors, and when all three colors are combined with equal intensity, white is formed.

What are the secondary colors for additive color mixing?

When any two of the three primary colors are mixed, secondary colors are formed; they are called secondary because they are by-products of the primary red, green, and blue colors. Red light mixed with green creates yellow light. Red and blue produce magenta. Finally, cyan is formed when blue light and green light are added together.

What are complementary colors?

Complementary colors are pairs of one secondary and one primary color that, when mixed, form what is close to white light. For example, yellow and blue light are complementary because, when combined, they form white light, as will magenta and green and cyan and red.

What is subtractive color mixing?

As opposed to the mixing of light ("additive color mixing"), subtractive color mixing only occurs when combining dyes, pigments, or other objects that absorb and reflect light. For example, you could shine white light through two colored filters or gels. Or, you can reflect light from colored surfaces. If you shine white light on a blue wall, the wall absorbs red and green but not blue. So, if we know what colors are reflected, we know the color of the object.

What are the primary pigments?

The primary pigments or dyes are magenta, which reflects blue and red light but absorbs green; cyan, which reflects blue and green but absorbs red; and yellow, which reflects red and green but absorbs blue. These are the same colors as the secondary colors obtained when mixing light. Note that if we combine magenta, cyan, and yellow, the red, green, and blue are all removed, leaving nothing—that is, black.

What are the wavelength ranges of some colors?

Blue is light with wavelengths in the range of about 400 to 500 nanometers. Green is roughly 500 to 560 nanometers. Yellow is 560 to 590 nanometers, orange 590 to 620 nanometers, and red beyond 620 nanometers. It may seem surprising that a yellow filter transmits green, yellow, and red. But the eye is not very sensitive to red in comparison to green and yellow. That's the same reason that the transmission of the blue and indigo filters beyond 660 nanometers does not impact what the eye sees. Finally, in subtractive color mixing, purple is a mixture of blue and red; purple (or violet) light, however, has wavelengths that are actually a little shorter than 400 nanometers.

What are the secondary colors for subtractive color mixing?

The secondary colors for dyes and pigments are the same as the primary colors in additive color mixing. Red, green, and blue dyes or pigments reflect their own color while absorbing the other two colors. For example, red would reflect red but absorb the green and blue light.

Why do most color printers use four colors to print?

It would seem that a color inkjet printer mixing the three primary pigments of yellow, magenta, and cyan should be able to produce all the other colors, including black. When all three primary colors are combined, however, the mix looks more like a muddy brown color than black. Although these are the primary colors of which other shades can be created, they do not represent all the colors of the spectrum needed to form black. That is, there are gaps in the wavelengths that these pigments absorb. Therefore, most color inkjet printers have a cartridge with yellow, cyan, and magenta ink and another separate ink cartridge of just black.

What is colorimetry?

Because the perception of color is mostly a neurophysiological function between the eyes and the brain, it can vary slightly from person to person. Also, the subtractive color seen depends on the light source. If you plan to paint a room, for example, you should examine the color when it is illuminated by several different light sources—for example, incandescent lamps, fluorescent lights, LEDs, and sunlight. The colors of the room may look very different in each case. Scientists, artists, advertisers, and printers need an objective method of specifying color as it relates to the frequencies of light.

Color inkjet printers use black, along with yellow, cyan, and magenta, because the colored pigments do not cover the full spectrum. Therefore, they cannot produce a dark black but only a muddy dark brown.

This technique for measuring the intensity of particular wavelengths of light is known as colorimetry.

What are hue and saturation?

Hue is related to the wavelength of a color. Saturation is the extent to which other wavelengths of light are present in a particular color. For example, for the hue red, deep red is saturated, while pink is a mixture of red and white.

On humid summer days, why does the sky look white or grayish?

When high amounts of humidity are in the air, water molecules are more prevalent than on a cool, dry day. Water molecules, which have two hydrogen atoms and one oxygen atom, are larger than oxygen and nitrogen found in the air, and the size of a molecule plays a significant role in what frequencies of light are scattered. When white light encounters a larger molecule or dust particle, larger-wavelength light will be scattered, whereas if a smaller molecule is struck by white light, smaller-wavelength light will be scattered. Snow, beaten egg whites, and beer foam all look white for the same reason.

If the water or smoke is dense enough, all visible light waves are scattered many times, and the cloud looks gray.

Why are sunrises and sunsets often orange or red?

During the evening and early morning, when the sun is lower in the horizon, the light that the sun emits has to travel through more of the atmosphere to reach us than it does during midday, when the path of sunlight through the atmosphere is shorter. Since the distance through the atmosphere is much larger for sunlight in the morning and evening than during midday, more of the shorter wavelengths of light are scattered out of the direct light from the sun. Thus, the sun's color goes from yellow to orange and finally to red. Dust and water vapor in the atmosphere enhance this effect, making sunsets even redder.

Why is the ocean blue?

There are two major reasons why the ocean and most bodies of water appear blue. The first can be observed by looking at the water on a cloudy day and then on a sunny day. There is a rather large difference in how blue the water looks on the two different days. The water acts as a mirror for the sky, so on a sunny day with a blue sky, the water will have a richer blue color than on a cloudy day.

The second, and usually more significant, reason why bodies of water have a blue appearance is that water scatters shorter-wavelength (bluer) light more than longer-wavelength (redder) light. In fact, water absorbs some orange, red, and very-long-wavelength infrared. As a result, it absorbs more energy in the sunlight, increasing its temperature. The much larger amounts of reflected, short-wavelength light results in a crisp-blue-colored body of water.

Why do bodies of water sometimes not look blue?

Some bodies of water may take on a more greenish or at times a brownish or black color. Usually, this is due to other elements in the water such as algae, silt, and sand. Runoff water from glaciers is very white due to the tiny grains of silt in the water. Still, in the majority of cases, water looks blue.

Why do we see blue sky and not a blue-indigo-violet sky?

The British scientist John Strutt, Lord Rayleigh (1842–1919) determined that the nitrogen and oxygen molecules in the atmosphere scattered sunlight, allowing us to see the sky. The scattering is strongest at the lowest wavelengths. So, why don't we see a violet sky? It's because of our eyes. Human eyes are most sensitive to color in the middle of the visible light spectrum—

John William Strutt, 3rd Baron Rayleigh, was a British physicist who figured out that the sky looks blue to us because of how nitrogen and oxygen molecules scatter sunlight.

that is, around a wavelength of 550 nanometers. Blue is closer to this wavelength than indigo or violet, so we are more sensitive to the blue color. So, even though all three colors are scattered by the molecules in the air, humans see a predominantly blue sky.

Why does the sun itself look yellow?

In the visible light spectrum, the sun emits white light. Because of Rayleigh scattering, the violet, indigo, and blue wavelengths of light from the sun are scattered throughout the sky. The remaining light that gets through is thus primarily the longer wavelengths of yellow, orange, and red light. The closer to the horizon the sun gets, the more short-wavelength light is scattered; that's why, as sunset approaches, the sun looks more yellow, then orange, then red.

RAINBOWS

How are rainbows produced?

A rainbow is a spectrum of visible light formed when sunlight interacts with droplets. Upon entering a water droplet, the white light is refracted and dispersed—that is, spread apart into its individual wavelengths—just as in a prism. The light inside the droplet then reflects against the back of the water droplet before it refracts and disperses as it exits the droplet. The angle between entering and leaving is 40 degrees for blue light, 42 degrees for red.

What conditions must be met to see a rainbow?

There are two main conditions for witnessing a rainbow. The first is that the observer must be between the sun and the water droplets. The water droplets can either be rain, mist from a waterfall, or the spray of a garden hose. The second condition is that the angle between the sun, the water droplets, and the observer's eyes must be between 40 degrees and 42 degrees. Therefore, rainbows are most easily seen when the sun is close to the horizon, so the rays striking the droplets are close to horizontal.

Is there such a thing as a fully circular rainbow?

All rainbows would be completely round except that the ground gets in the way of completing the circle. However, if viewed from a high altitude, such as an airplane, circular rainbows can be seen when the angle between the sun, the water droplets, and the plane is between 40 degrees and 42 degrees. In this case, the rainbow is horizontal, meaning that it is parallel to the ground and therefore not blocked by the ground. It's quite a sight.

What is the order of colors in a rainbow?

The order of colors in a rainbow goes from longest-wavelength red on the outer arc to shortest-wavelength blue on the inside of the arc. The full order from outer to inner is: red, orange, yellow, green, blue, indigo, and violet.

Who first explained how rainbows are formed?

The British physicist Sir Isaac Newton (1642–1727), a great pioneer of optics, was *not* the first person to understand the optical characteristics of a rainbow. In fact, it was a German monk, Theodoric of Freiberg (c. 1245–c. 1311), who in 1304 discovered that light refracted and reflected inside a water droplet. To demonstrate his hypothesis, the monk filled a sphere with water, sent a ray of sunlight through the sphere, and observed the separation of the white light into colors along with the reflection on the back of the water droplet. It wasn't until centuries later that the French mathematician René Descartes (1596–1650) rediscovered and reframed some of Theodoric's results.

Who is Roy G. Biv?

The fictitious name "Roy G. Biv," which is an acronym for "Red, Orange, Yellow, Green, Blue, Indigo, Violet," is a handy mnemonic device for remembering the order of colors in a rainbow from longest to shortest wavelength.

Does everyone see the same rainbow?

A rainbow is dependent on the position of the sun, the water droplets in the sky, and the location of the observer. So, everyone watching a rainbow is watching their own personal rainbow.

What is a double rainbow?

Sometimes, a secondary rainbow shows up, creating a "double rainbow." The secondary rainbow has its color spectrum reversed in order, is outside of the original rainbow, and is significantly dimmer than the primary rainbow. A secondary rainbow occurs because an additional reflection of the light takes place inside the water droplets. Instead of reflecting once in the water droplet, the light reflects twice inside the water, reversing the order of the colors. The secondary rainbow appears between the angles of 50 degrees and 54 degrees.

EYESIGHT

How does the human eye work?

The eye is essentially an optical extension of the brain. It consists of a lens to focus the image, an iris to regulate the amount of light entering the eye, and a screen called the retina. The cells of the retina do some preliminary processing of the information they receive, then send signals along the optic nerve to the brain.

What is the outer part of the human eye like?

The cornea is a transparent membrane on the outer surface of the eye. Between the cornea and lens is a fluid. Light refracts when going through the convex surface of the cornea into the fluid. In fact, most of the focusing of light in the eye occurs at the cornea. Light passes through the iris, which opens and closes in response to the amount of light entering the eye. The iris can only change the amount of light going through it by a factor of 20, while our eye can respond to differences in light level of 10 trillion! The major task of the iris, then, can't be to control light intensity. In addition, when the opening in the iris shrinks, the eye can keep objects in focus from a wider range of distances. After passing through the iris, the light goes through the lens.

What is the middle of the human eye like?

The lens consists of layers of transparent fibers covered by a clear membrane. To focus in on objects that we want to see, our eye changes the shape, and thus the focal length, of the lens and cornea by contracting or relaxing the ciliary muscle around the eye. Light then passes through a liquid called the vitreous humor, which fills the major volume of the eye, and falls on the retina. The cornea and lens have created an inverted image on the surface of the retina.

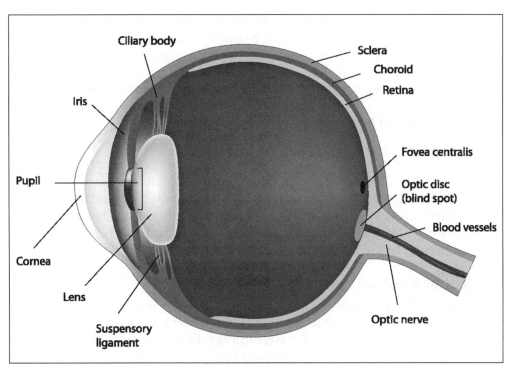

The human eye is quite a remarkable organ. It manages to allow light and images in, capture them on the retina, and send the information through the optic nerve to the brain for processing.

How does the back of the human eye work?

The retina is composed of a layer of light-sensitive cells, a matrix of nerve cells, and a dark backing. There are two kinds of light-sensitive cells: cones and rods. The seven million cones are sensitive to high light levels and are concentrated around the fovea, the part of the retina directly behind the lens. Surrounding it are some 120 million rods that are sensitive to low light levels. The entire retina covers about five square centimeters. A system of nerves in the retina do some preliminary processing of the electrical signals produced by the rods and cones before sending the results through the optic nerve to the brain.

What is the difference between rod cells and cone cells?

Cone cells, as their name suggests, are cone-shaped nerve cells on the retina that can distinguish fine details in images. They are located predominantly around the center of the retina, called the fovea. The cones are also responsible for color vision. Some cones respond to blue light, being most sensitive to 440-nanometer wavelengths. A second kind has peak sensitivity in the green, 530-nanometer light. The third is sensitive to a wide band of wavelengths from cyan through red. Its sensitivity peaks in the yellow, 560-nanometer light.

As the distance grows from the fovea, rod-shaped nerve cells replace the cones. The rods are responsible for a general image over a large area but not fine details. This explains why we look at objects straight on when examining something carefully. The image will be focused around the fovea, where the majority of cones pick up the fine details of the image. The rods, being much more sensitive in low light levels, are used a lot for night vision.

What wavelengths of light are our eyes most sensitive to?

Human eyes are most sensitive to the wavelengths corresponding to the yellow and green colors of the spectrum. Flashy signs and some fire engines are painted in a yellowish-green color to attract our attention. Even simple objects such as highlighter markers, used to emphasize words or phrases while taking notes, are typically bright yellow and green. When we glance over something or see an object out of the corner of our eyes, we are more likely to notice bright yellowish-green objects than red or blue objects because the eye is less sensitive to these wavelengths.

How does the lens in our eye change shape when focusing?

The ciliary muscle, responsible for changing the shape of the lens, adjusts its tension to focus on different distances. When focusing on objects far away, the lens needs a large focal length, so the muscle is relaxed in order to make the lens relatively flat. When an object is closer to the eye, however, a shorter focal length is needed. The ciliary muscle then contracts, reducing the focal length of the lens by making it more spherical. The process of adjusting the shape of the lens to focus in on objects is called "accommodation."

When swimming underwater, why is my vision blurred unless I'm wearing goggles?

Although your eye's lens changes shape to focus images on the retina, most of the refraction of light takes place during light's transition from air to the cornea. When water is substituted for air, the angles through which light is refracted are reduced, producing a blurred image on the retina. Swim goggles preserve a layer of air that stays in contact with the cornea, so the refraction is more normal.

What is color blindness?

Some people are born with an inability to see some colors due to an inherited condition called color blindness. The British physicist and chemist John Dalton (1766–1844) described color blindness in 1794. He was color-blind himself and could not distinguish red from green. Many color-blind people do not realize that they cannot distinguish colors. This is potentially dangerous, particularly if they cannot tell apart the colors of traffic lights or other safety signals. Some color-blind people are only able to see black, gray, and white. It is estimated that 7 percent of males and 1 percent of females are born with some level of color blindness.

How close can an object be viewed before it appears blurry?

There is a limit as to how close an object can be to the eye before the lens can no longer adjust its focus. Up to about 30 years of age, the closest an object can be focused is approximately 10 to 20 centimeters (4 to 8 inches). As one grows older, the lens tends to stiffen, and it becomes more difficult for the person to focus on close objects. In fact, by the time a person reaches the age of 70, their eyes may not be able to focus on objects within almost one meter of their eyes. As a result, most aging adults need some kind of vision correction to focus on close objects.

What is farsightedness, and how is it corrected?

Farsightedness, or hyperopia, occurs when the lens of the eye can see objects far away but cannot focus in on objects at closer range. The cornea and lens of the person with farsightedness cause the image to focus behind the retina, resulting in the images from objects close to the eye to be blurred. To correct for farsightedness, a convex lens is used to converge the light rays closer together, permitting the image to fall on the retina. The rigidity in the eye's lens that affects older people, making them unable to focus on close objects, is called presbyopia.

What is nearsightedness, and how is it corrected?

Nearsighted vision means that a person can only clearly see objects that are relatively near the eye. Images from distant objects are focused in front of the retina. Nearsight-

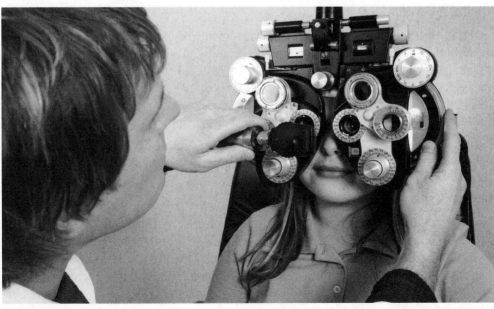

Farsightedness and nearsightedness are common eyesight problems. The former occurs when images entering the eye are focused behind the retina, and in the latter, images focus in front of the retina.

What allows some animals to see in the dark better than humans?

There are three main reasons why some animals can see better than humans can at night. The first reason is that their eyes, relative to body size, are larger and can gather more light than human eyes can. More light results in a brighter image.

The next reason has to do with the rods and cones in the nocturnal animal's eyes. Cones are used for detail and work best in bright light. A nocturnal animal has little need for the color vision provided by the cones and therefore has more room for the rods that detect general information such as motion and shapes.

The third reason why nocturnal eyes excel in the absence of light is due to the *tapetum lucidum,* a membrane on the back of the retina that reflects light back to the retina, effectively doubling the retina's exposure to light. The reflective tapetum can be seen in the light reflected back out of animals' eyes at night when you shine a flashlight on them.

edness, or myopia, is most often caused by a cornea that bulges outward too much. The lens cannot be flattened enough to compensate, and so distant objects appear fuzzy. To correct for the short focal length of the lens, a concave lens is used to make the light rays diverge just enough so that the image will fall on the retina. So, contact lenses to correct for myopia are thicker at the edges than at the center.

What is 3-D vision?

Seeing in three dimensions, which is how a person with normal eyes sees, means that in addition to perceiving the dimensions of height and width (such as seen on a piece of paper, a poster, or a TV or movie screen), one can see the third dimension of depth. We see real objects in 3-D because we have two eyes that see slightly different perspectives of the same view. The combination of these views, when interpreted by our brain, gives us the ability to perceive depth, the third dimension.

If you close one eye, your ability to perceive depth is eliminated. With only one eye, the world won't look very different to you, but you'll experience difficulty in judging distances.

How do 3-D movies and television work?

The original way to create a 3-D movie was through creative filming. When a 3-D movie is filmed, two cameras film the movie from slightly different positions. When the film is projected onto the screen, each projector uses a separate polarizing filter. The left projector might use a horizontally polarized filter, while the right projector might use a vertically polarized filter. The viewer then wears so-called 3-D glasses, which are also polarized; the left eye might let through only the horizontally polarized image produced by the left projector, while the right eye might let through only the vertically polarized images produced by the right projector. This arrangement simulates the different perspectives that each eye sees when looking at a real-life 3-D scene, allowing the brain to interpret the difference as the third dimension, depth.

How can 3-D movies be digitally produced?

Newer methods of producing three-dimensional views use digital methods rather than film. In the method best suited to 3-D television, the images from the left and right camera lenses alternate at a rapid rate. The viewer wears glasses in which each lens can be switched from transparent to opaque on command from an infrared signal sent from the television set. Thus, each eye sees only the frames captured by the appropriate camera lens. Another method, which is more suitable to movies shown in theaters, uses digital images that alternate between those captured by the right and left camera lenses. A device placed in front of the projector lens switches the polarization of the light coming from the projector so that the left-side images are polarized one way and the right-side images are polarized the other way. The movie viewer wears polarized glasses so that each eye sees only the appropriate frames.

CAMERAS

How does a camera work?

A camera consists of a series of lenses and apertures (holes that light can travel through) and other optical elements to create a visual reproduction of an object. One of the simplest cameras is a pinhole camera, and it can be used to illustrate how cameras work. Figure 21 shows how pinhole cameras work.

A pinhole camera is typically made from a box with a small "pinhole" in one side of the box and a screen on the other side. The pinhole is so tiny that only a very small number of light rays can go through it. The diagram shows how a pinhole creates a reproduction of the object on the screen. Note that it is not a real image because the light rays do not converge on the screen.

Pinhole cameras are easy to make and are often used during solar eclipses because it is very dangerous to look directly at the sun except during totality. With the sun at your back, point the hole up toward the sun and view the image on the screen of the moon passing in front of the sun.

How does a camera create an image?

Figure 21 shows what would happen if there were three pinholes, each creating an inverted reproduction of the object. Now if a converging lens is placed just behind the pinholes, it will bend the rays going through it. If the focal length of the lens and the distance between the lens and screen are chosen correctly, then the three reproductions from the pinholes will all be at the same location. Light rays from the top of the object

229

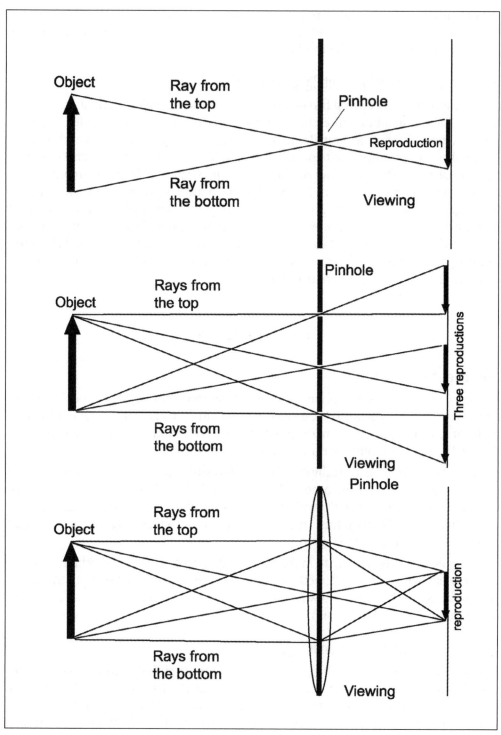

Figure 21.

will converge on the appropriate point on the image. Note that the image is inverted and the same size as the reproductions.

What would happen if you had a multitude of pinholes at the location of the lens? The reproductions from all the pinholes would be at the same location, and many more rays from the object would end up at the same place on the image. The image would be much brighter. So, you can model the formation of an image by a lens as a collection of reproductions of pinholes. This is shown in the bottom panel of the diagram. The larger the lens, the brighter the image.

How is a camera similar to the human eye?

A camera performs many of the functions of the eye. It has a lens to form an image on a photosensitive surface. The lens must be able to form sharp images of objects both close and far away. The amount of light reaching the photo detector must be controlled to make the exposure correct. In older cameras, the photosensitive surface was film. Today, cameras use a digital sensor. These sensors are small—most are between 3/8 of an inch and 1 inch in size—but contain as many as 10 million or more separate light detectors called pixels. Each pixel is covered by a red, green, or blue color filter, so the camera can produce full-color images.

How do you control focus and light for a camera?

A camera's lens isn't flexible like the one in the eye, but the distance between the lens and the sensor can be varied. Bringing a distant object into focus requires that the lens be closer to the sensor. A close object requires that the lens be moved further away. The amount of light is controlled two ways. One is to have an aperture that can be opened to admit more light or closed down to reduce the amount of light. The second is a shutter that controls the amount of time light is allowed to reach the sensor. While leaving the sensor exposed for a longer amount of time is needed when the light is dim, it also will cause a blurred image if the object is moving. Thus, it is important to select the correct combination of aperture and shutter speed to take good pictures.

What causes red eye in photographs?

Red eye occurs when a flash is used because there is not enough natural light for a good exposure. Under normal conditions, to allow enough light to enter the eye, the pupil dilates, and to prevent too much light from entering the eye, the pupil constricts. But when a camera flash

Camera lenses are similar to human eyes in that both have lenses that focus images onto a surface, but camera lenses are not flexible like an eye lens.

231

is fired, the pupils are not expecting the bright light and do not have a chance to constrict. As a result, a large amount of light enters the eye and reflects off the blood vessels that supply the retina in the back of the eye. The redness on the pupil is actually the reflection off these blood vessels that has been recorded by the camera.

How can red eye be reduced?

Red-eye reduction is a feature found on many modern cameras. It simply tries to have the pupils of the photographic subject's eyes contract so that not as much light will be reflected back from the retina. There are several ways to achieve this. One method is to have a smaller light illuminate before the real flash; another method is to have a quick burst of five or six miniflashes that cause the pupil to contract before the picture is taken.

TELESCOPES

Who invented the telescope?

There are several conflicting claims for the first person to combine two lenses to "see things far away as if they were nearby." The Dutch eyeglass makers Hans Lippershey (c. 1570–1619), Zacharias Janssen (c. 1585–1638), and Jacob Metius (c. 1571–c. 1628) were among the first. Lippershey described the design and applied for a patent on October 8, 1608, but he was turned down. Copies of Lippershey's device, which was constructed from a convex and a concave lens and had a magnifying power of 3, were common in the Netherlands that year.

How did Galileo Galilei use a telescope to study the universe?

The Italian astronomer and physicist Galileo Galilei (1564–1642) heard about the invention of the telescope in June 1609, while he was in Venice. He quickly figured out how it worked, and as soon as he returned home to Padua, he constructed one. A few days later, he demonstrated it to the leaders in Venice, who agreed to support his construction of more telescopes for practical and research purposes. Over the next year, Galileo improved his instruments and began studying the sky with them, and in March 1610 he published his first revolutionary discoveries in his book *Siderius Nuncius*. He outlined his discoveries of the moons of Jupiter, the rotation of the sun, the phases of Venus, sunspots, individual stars in the Milky Way, and mountains on the moon. Galileo's telescopes, with convex and concave lenses, produced an upright image. A few of his original telescopes still survive today.

What is a refracting telescope?

The refracting telescope was the first kind of telescope ever invented, and it is the kind Galileo used. It employs one lens to gather, refract, and focus light toward an eyepiece. The eyepiece contains one or more lenses that create an image that the eye can see. The

How were refracting telescopes improved after Galileo?

Based on the ideas of astronomers Johannes Kepler (1571–1630) and Christoph Scheiner (1573–1650), the telescope was improved by using two convex lenses separated by a distance equal to the sum of their focal lengths. Such a telescope inverts the image. To achieve high magnifications, one of the focal lengths had to be very large.

Refracting telescopes eventually proved to be cumbersome and difficult to use. The prism-like shape of the lenses introduced colors into the images. This defect, called chromatic aberration, was eliminated 120 years later using a lens made of a combination of two glasses. But this invention did not stop the weight of the large lens from causing it to sag, creating distorted images.

larger the diameter of the lens, the more light the telescope can gather. The weight of the lenses then limits the practical size of refractor telescopes.

What is a reflecting telescope?

Telescopes that use mirrors to focus light were first invented by Isaac Newton (1642–1727) in 1668. A reflecting telescope uses mirrors rather than lenses to gather light and focus the light to make images. It usually consists of at least two mirrors: one large, curved mirror at the end of the telescope to gather light and a smaller mirror used to direct the light to the eyepiece. Mirrors are much easier to manufacture in large sizes than lenses, and they weigh much less; however, part of the light coming into the telescope is usually blocked by one or more of the smaller mirrors in the telescope's optical path, so such a telescope must be very carefully constructed and aligned.

What is a segmented-mirror telescope?

When mirrors exceed several meters in diameter, they are typically no longer rigid enough to maintain their shape perfectly when they are tilted. One way to solve this problem is to use smaller mirrors and link them together with computers and electronics so that they work together as if they comprised one single, large mirror. This method results in a multiple-mirror or segmented-mirror telescope. The first large, multiple-mirror telescope was the MMT on Mount Hopkins in southern Arizona. The first large, segmented-mirror telescopes were the Keck I and Keck II telescopes on Mauna Kea on the big island of Hawai'i.

What are active optics and adaptive optics?

Another way to improve the image quality of a large telescope mirror that may deform when tilted or exposed to temperature variations is to use computer-controlled systems to adjust the shape of a telescope mirror's surface ever so slightly in real time so that it can maintain maximum image quality and focus. If it is used on a large primary mirror of a telescope, and the shape is adjusted a few times per second, it is called an active optics system.

If computer control systems are used to change the shape of a mirror dozens, hundreds, or thousands of times per second, it is called an adaptive optics system. These kinds of systems can be used to remove the blurring effects of Earth's atmosphere and allow the telescopes to take pictures that are almost as perfect as if they were taken with a space telescope. Adaptive optics systems are very challenging to build and operate, and they sometimes use lasers fired into the air next to the telescope to guide the corrective actions. When they succeed, the resulting images are spectacular, and they can be used to detect very difficult astronomical targets, including planets far beyond our own solar system.

What is the most famous space telescope?

The Hubble Space Telescope (HST), launched in April 1990, has proven to be the most influential scientific instrument of its time. The HST's primary mirror is about 2.4 meters (94 inches) across, much smaller than modern, large, ground-based telescopes like Keck I and Keck II, but being in space, it is not limited by the image distortions caused by variations in the refractive index of air above the telescope. In addition, a space telescope can detect the infrared and ultraviolet rays blocked by Earth's atmosphere. Many other highly successful space-based telescopes have been launched over the decades, but none of them have yet surpassed the HST in the number and importance of cosmic discoveries that have been made with them.

What was wrong with the Hubble Space Telescope when it was first launched?

An error in the shape of the Hubble Space Telescope's primary mirror seriously threatened the success of its entire mission. The error was only 1/50th the width of a human hair, but it created a spherical aberration in the optics of the HST. As a result, it could only focus its images about as well as a typical ground-based telescope. Furthermore, electronic noise in the HST's control systems and mechanical jitter in its solar panels caused the telescope to operate far below optimal expectations.

What was done to correct the Hubble Space Telescope's optical problems?

Luckily, the Hubble Space Telescope was placed into an orbit around Earth that al-

The Hubble Space Telescope can see deeper into space than earthbound telescopes because there is no atmospheric distortion to interfere with how it sees images.

lowed astronauts to service and repair it after its launch and deployment. After physicists and engineers worked for three years to examine the HST's problems and find solutions to them, a team of astronauts rode the space shuttle *Endeavor* to the space telescope in December 1993. Among other repairs and upgrades, they installed a new camera that could compensate for the primary mirror's spherical aberration, and they also installed a corrective optics system that acted somewhat like a pair of glasses to correct nearsightedness or farsightedness. The servicing mission went perfectly, and the HST began operating even better than the way it was originally designed.

What has happened with the Hubble Space Telescope since its launch?

Since it was launched in 1990, the Hubble Space Telescope has been serviced five times in all by NASA astronauts that reached the telescope aboard the space shuttle. The last servicing mission occurred in 2009. With these servicing missions, the HST has been upgraded to be hundreds of times more powerful than it was initially designed to be, and it remains fully operational after three decades in space. NASA has plans to continue operating the HST until at least 2021.

Astronomers worldwide have used the HST to conduct research in just about every area of cosmic discovery. More than ten thousand scientific papers have been published using data from the HST. Equally important, the beautiful photographs and spectra taken with the Hubble Space Telescope have fascinated the public and greatly broadened all of humanity's understanding of our wondrous and amazing universe.

What is the planned successor to the Hubble Space Telescope?

The James Webb Space Telescope (JWST), scheduled for launch in 2021, is a next-generation space telescope that is planned to succeed the Hubble Space Telescope. The JWST will have about 10 times the light-gathering power of the HST, and it will be sensitive to infrared as well as visible light. Instead of low Earth orbit, the JWST will orbit at the sun-Earth L2 Lagrange point (a point of gravitational equilibrium between the two bodies), about 1.6 million kilometers (1.0 million miles) from Earth. From there, it will be able to make observations 24 hours a day, seven days a week for the duration of its mission.

ELECTRICITY AND MAGNETISM

Why is electricity so important?

Electricity is an extremely important tool to generate a large amount of useful power. Before the use of electricity, humans relied on resources such as water wheels, windmills, and wood stoves to generate power and heat. Electric power, compared to those early methods, can be far greater in quantity and far more easily distributed to cities, homes, businesses, and even small, portable machines from medical devices like pacemakers to the phone in your pocket.

How did scientists first study electricity?

Prehistoric people valued and traded amber, a gemlike material that is petrified tree sap. Surely more than once, a person would have rubbed amber on his or her fur clothing and noticed that fur was attracted to the stone. Perhaps she rubbed it hard enough to produce sparks. The Greek philosopher Thales of Miletus wrote about these effects around 600 B.C.E.

But it wasn't until 1600 C.E., some 2,200 years later, that William Gilbert (1544–1603), an English physician, named this effect "electricity" after the Greek name for amber: "elektron." Gilbert showed that sulfur, wax, glass, and other materials behaved the same way as amber. He invented the first instrument to detect what we now call the electrical charge on objects. He called it a versorium, and it was a pointer that was attracted to charged objects. Gilbert also discovered that a heated body lost its charge and that moisture prevented the charging of all bodies.

In 1729, the English scientist Stephen Gray (1666–1736) determined that charge, or what he called the "electric virtue," could be transmitted over long distances by metals, objects that couldn't be charged.

How was electricity used as entertainment in the 1700s?

In the mid-1700s, demonstrations of electrostatics were extremely popular, especially in Parisian salons, where wealthy men and women gathered to discuss events of the day. Benjamin Franklin (1706–1790) was a popular guest. In Stephen Gray's most famous demonstration, called the Flying Boy experiment, a boy was suspended horizontally using two silk threads hung from hooks placed on the ceiling. When a charged tube was held near his foot, pieces of metal foil were attracted to his face and to his outstretched hands.

How did early scientists explain the difference between moving and static electric charges?

The French physicist Jean-Antoine Nollet (1700–1770) attempted to measure the speed of electricity by having a circle of 200 monks hold hands. The monks at the ends of the line touched a machine that produced charge. They all jumped simultaneously when they felt the painful shock, so he concluded that electricity moved instantaneously. In a similar demonstration, Louis-Guillaume le Monnier (1717–1799) discharged a Leyden jar through a chain of 140 courtiers in the presence of the king of France.

How could these results be explained? Charles-François du Fay (1698–1739) concluded that there were two types of electricity. He said that "vitreous" electricity is produced by friction from glass or precious stones and that "resinous" electricity is produced by rubbing amber, sealing wax, or silk; when the two types are combined, they neutralize each other. Nollet modeled these types as two fluids, each composed of particles that repelled each other. Charging amber gave it an excess of resinous fluid, he said, and charging glass with silk gave it an excess of vitreous fluid. When the two were touched together, the fluids combined with each other, leaving the objects uncharged.

STATIC ELECTRICITY

What is electrostatics?

Electrostatics is the study of the causes of the attractive and repulsive forces that result when objects have an excess or a deficit of electrons, causing them to be electrically charged. Electrostatics is the study of what is often called static electricity.

What can you discover about static electricity with a roll of tape?

You can learn something about attractive and repulsive charges with a roll of cellophane tape. Any brand will do—the cheaper the better. Pull off a strip about 5 inches long, then fold over about ¼ inch at one end to serve as a handle. Press the tape on your desk or a table. Mark the strip with the letter "B." Make a second identical tape and press it down next to the first.

Holding the two tapes by their handles, quickly pull them off your desk. They'll probably be attracted to your hands, so shake them until they hang free. Then, bring

What did Benjamin Franklin think about the fluid nature of electricity?

Benjamin Franklin (1706–1790) believed there was only one kind of electric "fluid." When glass was rubbed, the fluid filled the glass. When amber was rubbed, the fluid left the amber. He called an object with an excess of fluid "positive" and one with too little fluid "negative." When they were touched, the fluid flowed from the glass to the amber, leaving each with the proper amount of fluid. The flow was likened to water in a river. The "electrical tension" (difference in potential) and "electrical current" were analogous to the difference in water levels between two points and to the amount of water transferred.

them closer together. What do you see? You should see them bending, evidence that there is a force between them. If they don't bend, stick them on the desk and pull them off again. Do they provide evidence that there is an attractive or repulsive force between them?

We'll say that pulling them off the table caused them to be "charged," although we have no evidence of what they are "charged" with. They were obviously charged in the same way, so we can conclude that objects with like charges repel each other. By the way, we'll work toward an explanation for why they're attracted to your hands.

Press the two strips back on your desk. Now make two more strips the same length and press them on top of the first two strips. Mark these strips "T" to identify them as the top tapes, as opposed to the "B" or bottom tapes.

Slowly pull the T + B pair of tapes off the desk together. If they are attracted to your hand, then use the other hand to gently pat both sides of them over their entire length. That should remove any residual charge from the pair of tapes. If not, pat them down again.

You have a pair of objects with no charge. Holding the two handles of the pair, rapidly pull them apart. Again, if they are attracted to your hands, shake them until they hang freely. Bring them closer together. Is there evidence of a force between them? Is it attractive or repulsive?

How did these pieces of tape get charged by static electricity?

You started with a pair of objects with no electric charge. Pulling them apart caused them to be charged, but not in the same

Electrostatics—or static electricity—involves the attractive and repulsive forces that result when objects made of two different materials are rubbed together.

239

way, because they didn't repel but attracted each other. Thus, you can conclude that they must be charged differently, and objects with different charges attract.

To keep your charged tapes, you can hang them from the edge of your desk or a desk light. Make a second T + B pair and see if the two T (top) tapes are charged alike or differently. If the tapes stop interacting, you can repeat the charging procedure as often as you like.

Hang a T and a B tape so you can bring objects near them to see if there are forces between them. Make a list of the objects you tried and whether they attracted or repelled the T tape and the B tape. Try your finger. Then try rubbing a plastic pen on a piece of wool. Try plastic rubbed by silk or polyester. Try glass and metal.

Do some objects attract both tapes? Repel both tapes? Attract one and repel the other? If they do the latter, you can characterize them as being charged like the T tape or like the B tape.

Clearly, some objects can attract both kinds of charge, but no objects can repel both.

What combinations of charges cause attraction and repulsion?

Unlike gravitational forces, which only attract masses to each other, electrostatic forces can either attract or repel charges. Like charges (positive–positive or negative–negative) repel each other. Unlike charges (positive–negative) attract each other. A common phrase describing many human social relationships, "opposites attract," holds true for electrostatic forces.

What does it mean to charge an object with static electricity by contact?

When a rubber rod is rubbed with fur, the fur transfers electrons to the rubber rod. The rod and fur, originally neutral, are now charged. If an object touches the rod, some of the excess electrons on the rod can move to the object, charging it. The rod, which is now negatively charged because it has excess electrons, can attract positive charges. This method is called charging by contact.

But, as you observed with the cellophane tapes, your hand and other neutral objects attract both positively and negatively charged objects. How does this happen? The rod attracts positive charges and repels negative charges. Neutral objects contain equal numbers of positive and negative charges. In a conductor, the charges are free to move, so the electrons can be pushed to the far end of the object, making it negatively charged and leaving

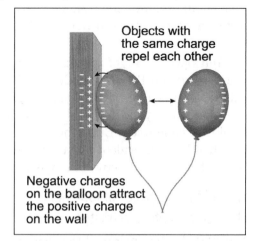

Kids often discover that when they rub a balloon on, say, a sweater, static electricity buildup causes the balloon to stick to stuff like walls or a nearby friend.

the close end positively charged. An object that is neutral but has separated charges is polarized. Is there a net force on a polarized object? And can it exert a net force on the charged object like the cellophane tape? Yes, because the electrostatic force is stronger at closer distances. Thus, the attractive force between the unlike charges is stronger than the repulsive force between the like charges, and there is a net attractive force.

In nonconducting materials, the charges cannot be widely separated, but they can move within the atoms or molecules. So, insulators, like pieces of paper, dust, or hair, can also be attracted, even though they are neutral.

What is charging by induction?

Did you ever see a piece of paper attracted to a charged rod, touch it, and then jump away? How would that happen? If it touched the rod, it became charged with a charge like that of the rod, so it would now be repelled. A conductor can also be charged after being polarized but without touching the charged object. If you bring a large, metal object, like a pie plate, near a charged rod, the positive charges will move to the far end of the plate. If you now touch this end briefly with your finger, the positive charges will be pushed even further away into your finger. When you remove your finger, the pie plate is negatively charged. This process is called charging by induction.

Rubbing a glass rod with silk will achieve the same effect. The glass rod is positively charged, while the silk receives the excess negative electrons. The glass rod can still pick up small objects, but it attracts the negative charges in those objects instead of the positive charges. When the pie plate is charged by induction, it will be positively charged.

Why does a charged balloon stick to a wall?

The attraction between a charged balloon and a wall is the result of electrostatic forces. When rubber is rubbed on human hair or a wool sweater, electrons transfer easily to the rubber balloon. The balloon is charged when it is rubbed. The hair or sweater fuzz may stand up as a result of the excess positive charges repelling each other. When the balloon is brought near the wall, it polarizes the wall, moving the positive sources toward it and repelling the negative charges away. The negatively charged balloon is attracted to the many positive charges in the wall. As long as the electrostatic force and frictional force between the balloon and the wall are stronger than the gravitational force pulling the balloon down, the balloon will remain on the wall.

Why do you sometimes get a shock when touching a doorknob?

This annoyance happens usually on dry days after walking on carpeted floors. The friction between the carpet and your shoes or socks causes charges to be moved between your body and the carpet. Usually, your body becomes negatively charged. When your hand approaches a doorknob, the negative charges in your hand are attracted to the positive charges in the doorknob (created by polarization), causing an electrical spark when the two charges meet.

What are some good conductors of electricity?

To be an effective conductor, a material must allow the electrons to move easily throughout it. The atoms in good conductors, such as most metals, have one or two electrons that can be freed easily from the nucleus to move through the material. Water is a fair electrical conductor, but when salt is added, it becomes a better one.

What are some good insulators of electricity?

In an insulator, the electrons are strongly bound to their nuclei and thus cannot move through the material. Many nonmetals, such as plastic, wood, stone, and glass, are good insulators. Your skin is a good insulator, unless it is wet.

How is the strength of an electrical force measured?

British philosopher, theologian, and scientist Joseph Priestley (1733–1804) suggested that the force caused by static electricity might depend on distance the same way gravity does. Using Priestley's idea, the French physicist Charles Coulomb (1736–1806) made quantitative measurements of the force of attraction and repulsion between charged objects using an apparatus shown in the accompanying illustration. He found that the force depended on the charge of the two objects and the distance between them. The relationship he found is called Coulomb's Law, and the unit of measurement of charge is the coulomb (C).

What is Coulomb's Law?

Coulomb's Law describes the strength of the electrical force between two charged objects. The formula is $F = k \, (Qq/r^2)$, where k is a constant equal to 9.0×10^9 Nm^2/C^2 (newton-meters squared per coulombs squared). The charges Q and q, measured in coulombs, represent the charges on the objects that cause the force F, measured in newtons. Finally, r is the distance between the centers of the two charged objects. A negative force is an attractive force, while a positive force is repulsive.

What is a coulomb of charge?

A coulomb of charge is equal to the charge of 6.24×10^{18} electrons (negative) or protons (positive). A coulomb is a very large charge. Objects that are charged by rubbing or induction have typically a microcoulomb (10^{-6} C) of charge.

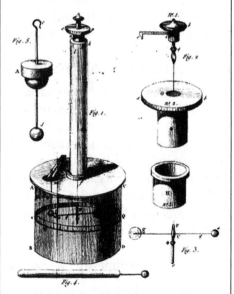

Coulomb used a torsion balance to calculate electrical charges: a rod hanging from a silk thread to a ball with a known electrical charge and a second ball with an unknown charge. The balls would twist the fiber at a certain angle, and the results would be measured to figure out the force between the balls.

What is an electroscope?

An electroscope is a device used to measure the charge on an object. It consists of two metal leaves (either thin aluminum foil or gold leaf) attached to a metal rod. If you touch a charged object to the metal rod, the two leaves will be charged with like charges, and so they will repel each other. The larger the charge, the greater the angle will be between the leaves.

What is an electric field?

As discussed before, a gravitational field surrounds Earth or any object with mass. Another object with mass that is placed in this field will experience a gravitational force on it. In the same manner, an electric field surrounds a charged object. Another charged object placed in that field would experience a force. If a positive force creates the field, then the force caused by the field on a negative force will be toward the source of the field. A positive charge will experience a force away from the source. The English physicist Michael Faraday (1791–1867) was the first to use the concept of a field to describe the electrostatic force.

CAPACITORS

What is a Leyden jar?

In November 1745, Ewald Jürgen von Kleist (1700–1748), dean of a cathedral in Pomerania, put a nail into a small medicine bottle and charged it with an electrical machine. When he touched the nail, he received a strong shock. In March 1746, Pieter van Musschenbroek (1692–1761), a professor at the University of Leyden in Holland, performed a similar experiment with the device, now called the Leyden jar. A Leyden jar is an early kind of electrical capacitor.

How does a Leyden jar work?

A Leyden jar is an insulating container with conductors on the inner and outer surfaces. When charging the Leyden jar, the source of charge is connected to a rod touching the inner conductor while the outer conductor is connected to ground. The inner and outer conductors become

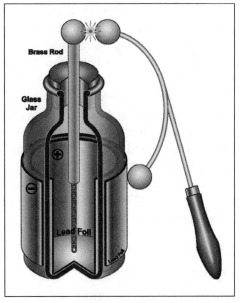

A Leyden jar is an early type of capacitor invented in the eighteenth century. It is simply a jar with a metal electrode that has been charged and maintains the charge until it touches another conductor.

243

oppositely charged. It takes energy to move additional charges to the jar as the charges overcome the repulsive forces of the charges already on the conductors. The jar stores this electrical energy. If the inner and outer conductors are connected by a wire, the charges flow and make the two conductors neutral again.

What were the uses of the Leyden jar?

In the late eighteenth and nineteenth centuries, people attempted to use the Leyden jar in a variety of ways. Some believed that it could cure medical ailments, and many doctors used the jar as primitive electroshock therapy. Others used it as a demonstration device and for entertainment purposes. Still more people felt that it could be used in cooking. (Try cooking a turkey with an electrical spark!)

How can I make a Leyden jar?

You can use either a glass or plastic container that has a tight-fitting cap. Use a small nail to make a hole in the center of the cap. Straighten a paper clip and push it through the hole. Make sure the end of the clip reaches the bottom of the jar. Cover the outside of the jar with aluminum foil and fill the jar about two-thirds full of water. Make sure that the jar cap is dry. Now rub a plastic pen with wool and touch the pen to the paper clip. Repeat the rubbing and touching several times. Then, touch the clip with your finger. You should feel a very tiny shock. The jar has stored the charge that you gave the pen when you rubbed it.

What is a capacitor?

The capacitor is the modern version of the Leyden jar. Like the jar, it consists of two conductors separated by an insulator. The insulators used can be air; a thin, plastic film; or a coating of oxide on the metallic surface.

One use of a capacitor is to store the energy needed to fire a flash lamp on a camera. A battery-powered circuit slowly charges the capacitor. When the flash lamp is triggered, the capacitor's energy is quickly transferred to the lamp, creating a brief, intense flash of light. Capacitors are also used in electronic devices from telephones to televisions to store energy and reduce changes in voltage.

What did Benjamin Franklin's kite experiment prove?

American statesman, writer, and inventor Benjamin Franklin (1706–1790) is probably most famous for flying kites in thunderstorms. In the mid-1700s, there were

Unlike what some people believe, Benjamin Franklin did not discover electricity. His kite experiment was designed simply to show that lightning consists of electricity.

three different phenomena that had similar effects. You could draw sparks with frictional or static electricity. Lightning appeared to be a giant spark, and electric eels could cause shocks like static electricity. But no one knew if these three had the same or different causes. Franklin touched a Leyden jar to a key tied to the string of his kite. When sparks jumped from the cloud to the kite, the charges went down the string and charged the Leyden jar. Thus, Franklin showed that lightning and frictional electricity were the same.

Did Benjamin Franklin's definition of positive and negative agree with what we use today?

Franklin decided that sparks given off by an object charged by a glass rod (vitreous electricity) looked more like fluid leaking out than did the sparks from an object charged by a rubber rod (resinous electricity). Thus he decided that glass had an excess of electrical fluid.

Today, we know that electric charge is carried mostly by electrons. Electrons are charged the same way that rubber or plastic is (negatively). Thus, we say that they have a negative charge. Because they are transferred much more easily than are the more massive, positively charged nuclei, when there is an excess of electrons, the object is negatively charged; when there is a lack of electrons, it is positively charged. So, even though Franklin made the wrong choice, we still follow his convention.

VAN DE GRAAFF GENERATORS

What is a Van de Graaff generator?

Named after its American creator, Robert Jemison Van de Graaff (1901–1967), the Van de Graaff generator has been the highlight of many electric demonstrations in both physics classrooms and museums around the world. The device, created in 1931, consists of a hollow, metal sphere that stands on an insulated, plastic tube. Inside the tube is a rubber belt that moves vertically from the base of the generator to the metal sphere. A metal comb attached to the base almost touches the belt. The rubber belt carries negative charges from the comb up the tube and into the metal sphere. There, a second metal comb captures the charges. They repel each other and spread over the exterior surface of the metal sphere. As more and more charge is carried upward, it takes more and more energy from the motor to move them up because of the repulsive force of the charges already there. The energy of the charges can reach up to a million joules per coulomb of charge—that is, up to one million volts.

What happens if you touch a Van de Graaff generator?

If you place your hands on the upper sphere while the generator is charging it, the electric charges accumulating on the sphere move onto your body as they are repelled by the other charges. When your body has enough charge, your hair may stand up on end because the electric charges on the hair repel each other. You won't be hurt because the current through your body is very small.

What happens if you get close to a charged Van de Graaff generator?

The sphere on the Van de Graaff is a conductor surrounded by an insulator (the air). While there are strong forces on the negative charges on the sphere, they're not strong enough to break down the insulating properties of the air. If, however, you bring another object with less negative charge close to the generator, the forces on the air molecules can become strong enough to rip them apart, separating their negative electrons from the positive nuclei. A spark will jump. If that object is your finger, you'll feel a shock when the charges carried through the spark move through your body. While the shock can be painful, one produced by the kind of Van de Graaff in most physics classrooms is not harmful.

How much charge is inside the sphere of a Van de Graaff generator?

None. When negative charges leave the rubber belt, they move immediately to the outer surface of the sphere. Negative charges like to be as far away from each other as possible, so they move to the outer surface of the Van de Graaff generator's sphere.

What is a Faraday cage?

A Faraday cage, named after British physicist Michael Faraday (1791–1867), is a cage, metal grating, or metallic box that can shield electrical charge. Charges gather on the outer shell of the cage because they are repelled by one another and can be further from each other if they are on the outside of the cage. This results in no charge within the Faraday cage. The metal sphere of a Van de Graaff generator is a Faraday cage. Cars and airplanes can be Faraday cages as well, and they may provide some protection from lightning during an electrical storm.

LIGHTNING

What is lightning, and how is it created?

Lightning is an electrical discharge in the atmosphere, like a giant spark. There is still debate about the cause of the separation of charges needed to create the discharge. Atmospheric scientists believe that

An old, damaged photograph of British physicist Michael Faraday, who was the first to describe an electric force in terms of a field. He also invented the Faraday cage, which permits electric charges on the outer shell but not within the cage itself.

strong updrafts in the clouds sweep droplets of water upward, cooling them far below the freezing point. When the droplets collide with ice crystals, the droplets become a soft mixture of water and ice. As a result of these collisions, the ice crystals become slightly positively charged and the water-ice mixture becomes negatively charged. The updrafts push the ice crystals up higher, creating a positively charged cloud top. The heavier water-ice mixture falls, making the lower part of the clouds negatively charged.

The ground under the cloud, meanwhile, is charged by induction. The buildup of negative charges on the underside of a thundercloud attracts the positive charges in the ground. The negative charges are repelled further into the ground, leaving a positively charged surface.

How do the clouds and ground act as a giant capacitor?

A capacitor consists of two conducting plates with opposite charge separated by an insulator. When a wire is connected between the two plates, a large electric current flows the charges rapidly from one plate to the other, neutralizing the capacitor.

The charged regions of the clouds act as conducting plates, while the air between them acts as the insulator. The same thing occurs between the lower section of the cloud and the ground. The air between these sections acts as the insulator, but when the forces exerted by the charges on the air molecules are large enough, they can rip the electrons from the molecules. The result is a positively charged molecule, called an ion, and a free electron. The air is changed from an insulator to a conductor. The mobile electrons gain more energy, creating more and more ions and additional free electrons. When the electrons and ions combine again, light is emitted. The tremendous amount of energy released rapidly heats the surrounding air, producing thunder.

Does lightning always strike the ground?

Although most people think of lightning when it goes between Earth and clouds, the most common type of lightning occurs inside and between thunderclouds. It is usually easier for lightning to jump between the clouds than it is for it to jump from the clouds to Earth. As a result, only one-quarter of all lightning strikes actually strike the ground.

How does air become a conductor?

When the charges have enough energy to begin to ionize the air, the free electrons will form a negatively charged "stepped leader" that will go from the cloud and make its zigzagged and often branched trip toward the ground. This process is slow, taking a few tenths of a second. The leaders are weak and usually invisible. The atoms in the air near the ground, feeling the attractive force from the electrons in the stepped leader, separate into ions and free electrons. The positively charged air ions from tall objects, such as trees, buildings, and towers, leave in streamers. When a stepped leader and streamer meet, a channel of ionized air is created, allowing large amounts of charge to move between the cloud and the ground. The return stroke of charge back to the cloud is the brightest part of the process.

247

Where in the world does lightning strike most frequently?

Satellite lightning detectors show that over the entire Earth, lightning strikes about 45 times each second, or 1.5 billion strikes each year. In the eastern region of the Democratic Republic of the Congo in Africa every year, each square mile, on average, has some 200 lightning strikes. A section of Florida known as "lightning alley" is a 60-mile-wide hotspot of lightning activity in the United States. On average there are 50 lightning strikes in each square mile per year.

What is a free electron?

In atoms and molecules, the negatively charged electrons are attracted to the positively charged nucleus. It takes a considerable amount of energy to remove an electron from an atom or molecule. The electric fields produced by thunderclouds have enough energy to do this, so they can pull an electron from an atom, creating a positively charged atom, or ion, and a free electron that moves without being attached.

LIGHTNING SAFETY

Is it true that lightning never strikes the same place twice?

This is absolutely false. The Empire State Building in New York City is just one example of where lightning has struck more than once. In some thunderstorms, the tower on the Empire State Building has been hit several dozen times.

How many people are hurt or killed by lightning each year?

Of the 40 million lightning strikes per year in the United States, about 400 of them hit people. About half of those people die as a result of the strike, while many of the others sustain serious injuries.

Why is a car often a good place to be when lightning strikes?

It is not because, as many people falsely believe, of the rubber tires. Many people think the rubber tires of a car provide insulation from the lightning striking the ground. If this were the case, wouldn't riding a bicycle do the same thing? The real reason why a car is

Studies in the 1980s showed that airplanes actually attract lightning. While lightning storms can still be danger-ous, passengers inside airplanes are safe from electric shock because they are actually inside a Faraday cage.

a safe place to be when struck by lightning is because most cars have metal bodies, and these act as Faraday cages, keeping all the electrical charge on the outside of the car. Since the charge is kept on the outside of the vehicle, the person sitting inside the car is kept perfectly neutral and safe. It is the shielding of the metal car body, and not the rubber tires, that protects people in automobiles.

What happens to an airplane when it is struck by lightning?

Airline pilots tend to avoid thunderstorms, but when a plane is struck by lightning, the passengers inside the plane are kept perfectly safe, for they are inside a Faraday cage, which shields them from the massive electrical charge. The lightning can, however, dis-turb and even destroy some of the sensitive electronics used to fly the plane.

Studies were performed by NASA in the 1980s in which they flew fighter planes into thunderstorms to see how the planes would react to lightning. The scientists quickly found that the planes actually encouraged lighting because the planes caused increases in the electric field of the cloud, which in turn caused the lightning to hit the plane's metal body.

What are some things you should do if caught in a lightning storm?

The safest place to be during an electrical storm is inside a building (where you should stay away from electrical appliances, such as the phone and television, as well as all plumbing and radiators) or car, but if you are unable to shield yourself in this way, the following precautions should be taken:

249

- Crouch down on the lowest section of the ground, but do not let your hands touch the ground. If lightning strikes the ground, the charges spread out sideways and can still reach you. If only your feet are on the ground (especially if you're wearing rubber-soled shoes), this might limit the amount of charge that passes through your body. If you must lie down because of an injury, try to roll up into a tight ball.

- Take off all metal items like jewelry or watches, and move away from all metal objects unless they act as Faraday cages (refer to the question about Faraday cages in the Van de Graaff Generators section).

- Move away from isolated and tall trees.

- Avoid the tops of hills or mountains and open areas such as water and fields.

If you are out on a lake or on the ocean, get back to shore as quickly as possible. If that is not practical, get down low in the boat and move away from any tall, metal structures like a mast or an antenna.

How do lightning rods work?

Lightning rods are pointed, metal rods that are installed above a tree or rooftop to protect the object. The rod, connected to the ground by a metal wire, both encourages and discourages a lightning strike. The rod discourages the lightning strike by "leaking" positive charges out of its pointed top to satisfy the need for positive charge in the clouds. If the rod cannot leak out enough charge to satisfy demand, the stepped leader from the cloud is instead attracted to the rod, and a flash of lightning occurs. Therefore, the rod attempts to discourage lightning, but if it cannot satisfy the negative charge, it attracts the lightning to the rod instead of to the tree or house.

If the lightning rod doesn't have a good connection to the ground through the wire, it can increase the danger to the building. Often, these heavy grounding wires come loose from the lightning rods, and if the rod is then hit by lightning, the charges will flow along the surface of the building to the ground and could cause a fire. The rods can become disconnected from lack of routine maintenance; it is wise to check these connections on a regular basis.

Who invented the lightning rod?

Although a Russian tower built in 1725 had what would now be called a lightning rod, credit is usually given to American inventor Benjamin Franklin (1706–1790). He invented the lightning rod in 1749 to protect houses and tall trees from being destroyed by lighting bolts.

ELECTRIC CURRENT

When did people think there were different kinds of electricity?

In the 1700s, Benjamin Franklin's kite experiment showed that lightning and static electricity were the same. Since ancient times, humans knew that certain fish, such as the electric eel, could shock a person. Was this "animal electricity" the same as static electricity? According to legend, the Italian physician Luigi Galvani (1737–1798) was making frog-leg soup for his sick wife. Whenever a nearby static electricity machine created a spark, the legs jerked. After completing several experiments, in a 1791 paper, Galvani reported that when one metal touched the muscle of a frog's leg while another metal touched the nerve, the muscle contracted. Thus, Galvani helped to show that there was a connection between static electricity and electric effects in animals.

How did Luigi Galvani's experiments help develop the idea of electric current?

Galvani believed that the flow of charge from the nerve into the muscle caused the contractions. His fellow scientist at the University of Bologna, Alessandro Volta (1745–1807), recognized that Galvani's frog leg was both a conductor and a detec-

You can use the acidic juice within a lemon—or a potato, if you're short on lemons—to create a simple battery. Insert two different metals (zinc and copper), and electrons will flow from the zinc to the copper to create a weak current. Connecting multiple lemons gives you a stronger current.

tor of electricity. In 1791 he replaced the leg with paper soaked in saltwater, a conductor, and used another means of detecting the electricity. He found that charges flowed only if the two metals touching the paper were different. The combination of two different metals separated by a conducting solution is called a galvanic cell after Galvani.

Volta went further. He found that the two metals that produced the greatest electrical effect were zinc and silver. In 1800 he stacked alternating disks of zinc and silver, separated by a card wet with saltwater. He found that this device, called the voltaic pile, was a continuous source of charge flow. English chemist Humphry Davy (1778–1829) showed that the charge flow was due to a chemical reaction between the metals and the conductive solution in the cards.

What causes the flow of charges in a voltaic pile?

Volta invented the term "electromotive force" (emf) to describe what causes the separation of charges. The more disks there were in the voltaic pile, the greater the emf. Unfortunately, the word "force" is an incorrect use of that term because there is no mechanical push, measured in newtons, on the charge. The correct term is potential difference or voltage, the energy change per unit charge separated. Both the term "voltage" and the unit in which it is measured, the volt (V), are named after Volta.

What is potential difference, or voltage?

In a voltaic pile, today more commonly called a battery, chemical energy is converted into increased energy of electric charges. Positive charges at the positive terminal of

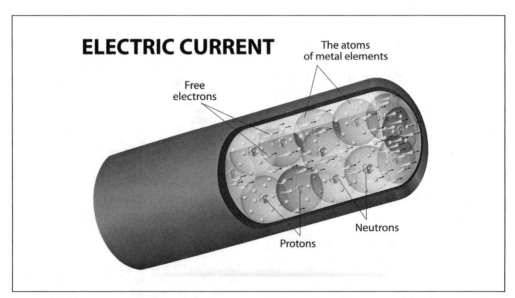

An electric current occurs when electrons in atoms break away from the nuclei and become free electrons; the flow of free electrons is the current. For example, copper atoms in electrical wire will provide one free electron per atom, making copper a useful metal for this purpose.

the battery have a greater energy than those at the negative terminal. The quantity that is important is not the total energy difference of charges at the two terminals of the battery but the energy difference divided by the charge. This quantity is called the electric potential difference or, more commonly, the voltage.

What is electric current?

Current is the flow of charge. It is measured in amperes (or "amps"), named after the French mathematician and physicist André-Marie Ampère (1775–1836). One ampere is equal to one coulomb of charge passing through a wire divided by one second. The greater the voltage difference across the wire, the larger the current.

What is electrical resistance?

All objects encounter friction when moving. Electrons are no different, but we refer to the friction that electrons encounter as resistance. The electrons collide with the atoms in a wire and are deflected from their paths. For the same voltage difference, the greater the resistance, the smaller the current. Resistance causes the electric charges to lose energy. The energy goes into the thermal energy of the wire or other conductor. That is, they get hot! The thermal energy can produce heat in a toaster and heat and light in an incandescent lamp.

What factors determine the amount of resistance of a conductor?

The resistance of a wire or other conductor depends upon the following:

- The length of the conductor (the longer the wire, the more resistance).
- The cross-sectional area of the conductor (the thinner, the more resistance).
- The properties of the material (for metals, the fewer the number of free electrons, the more resistance).
- The temperature of the conductor (for metals, the warmer, the more resistance; for carbon, the cooler, the more resistance).

What is Ohm's Law?

Ohm's Law states that the current through a conductor between two points is directly proportional to the voltage across those points. This is expressed by the equation:

$$V = IR$$

V is voltage, I is intensity (or current), and R is resistance in this equation.

German physicist Georg Simon Ohm (1789–1854) developed this law that bears his name.

What are resistors?

Resistors are devices used in electrical circuits to put a definite resistance in a circuit. Normally, they are made of graphite or a thin carbon film coated on glass. Larger resistors are cylindrical and have four color bands that encode the value of the resistance. On computer boards, they are tiny, rectangular devices barely a millimeter on a side with conductive ends that are soldered to the board. If they are designed to dissipate a large amount of power, they are made of high-resistance wire.

SUPERCONDUCTORS

What is a superconductor?

Superconductors allow electrical current to travel with no resistance, and, therefore, no voltage drop across them or energy loss within them. Superconductors must be cooled below their critical temperatures to have no electrical resistance. Some elements, compounds, and alloys that are superconductors are lead, niobium nitride, and a niobium-titanium alloy. All these require liquid helium to cool them to their critical temperatures. In the 1980s, some ceramics were found to have much higher transition temperatures that could be reached using much cheaper liquid nitrogen. The first found was yttrium barium copper oxide. As of 2008, a family of materials including iron, such as lanthanum oxygen fluorine iron arsenide, was developed.

What technologies have developed as a result of superconductivity?

Superconductors are most commonly used in large electromagnets. With no resistance, once the current is started, it will continue forever without change. Therefore, the magnets dissipate no power and do not heat up. These magnets are most often used in magnetic resonance imaging (MRI) machines. An MRI allows a doctor to view the inside of the human body without using harmful radiation. They are also used in particle accelerators that reveal the fundamental structure of matter by smashing the nuclei of atoms together. The most powerful accelerator is the Large Hadron Collider (LHC) in Switzerland. Another application of superconductivity is the SQUID (Superconducting QUantum Interference Device), which is an extremely sensitive detector of magnetic fields used in geological sensors for locating underground oil.

Who discovered superconductivity?

The creation of materials without resistance was thought to be impossible, but a Dutch physicist by the name of Heike Kamerlingh Onnes (1853–1926) proved it was possible in 1911. Onnes lowered the temperature of different metals, including mercury, close to absolute zero. He then measured the electrical resistance of the materials at such low temperatures and found that mercury, at only 4.2 K (–269°C), had zero resistance to electrical current.

How can superconductivity be used in transportation?

On a monorail line made with superconducting magnets, a monorail train can travel by being held above the rail by a strong magnetic field and pushed along with almost no friction. These are often called Maglev (magnetic levitation) trains, and although they are still not profitable to run, these trains have been in operation for many years.

Who won the Nobel Prize for their work in superconductivity?

Three American physicists—John Bardeen, Leon N. Cooper, and John R. Schrieffer—explained why superconductivity occurs in metals and alloys. Their development of the BCS (Bardeen-Cooper-Schrieffer) theory for superconductivity was cited when they were awarded the Nobel Prize in Physics in 1972.

Fifteen years later, two other physicists were awarded the Nobel Prize in Physics for discovering superconductive materials that achieved zero resistance at temperatures thought to be too high for superconductors to exist. Physicists Georg Bednorz and Alex Mueller of IBM Corporation found that a ceramic substance called lanthanum barium copper oxide became a superconductor at 35 K (–238 °C)—a much higher temperature than anyone thought possible at the time.

What can superconductors do to improve power transmission?

Superconductors may be used in the near future to improve the efficiency of transporting electrical energy. With modern technology, a substantial amount of such energy is lost due to heating effects of transmission lines—enough to power millions of homes and businesses each year. Superconductors may also be used in the future to produce far more efficient motors and power generators.

ELECTRICAL SAFETY

How much electric current is dangerous?

Approximately 1 milliampere (0.001 ampere, or 1 mA) is enough to produce a tingling sensation. A current of 0.01 ampere is painful, and 0.01 to 0.02 ampere is enough to paralyze muscles, making it impossible to let go. A current of 0.06 to 0.10 ampere causes ventricular fibrillation of the heart—that is, the heart is beating in such a way that it cannot pump blood through the circulatory system. Greater than 0.2 ampere can cause the human heart to clamp down and stop beating.

How much resistance does the human body have to electric current?

On average, the human body has an electrical resistance between 50,000 and 150,000 ohms. Most of this resistance is across the skin. If the skin is wet, the resistance drops to about 1,000 ohms. If the skin is broken, then resistance across organs in the body is on the order of a few hundred ohms. In this condition, 10 volts is sufficient to cause serious or even fatal damage.

Do electric eels really use electric fields to capture their prey?

Electric eels do indeed set off electrical pulses to stun and even kill their prey. These eels have special nerve endings bundled together in their tails that can produce 30 volts in small electric eels to 600 volts in larger eels. Besides using the electrical shocks for hunting, the eels produce a constant electrical field for use in navi-

Birds can perch safely on electric wires as long as they don't come into contact with objects that have two different voltages.

gation and self-defense. Most people do not have to worry about encountering electric eels, however. This variety of eel is native only to the rivers of South America.

Why is it said that electricians work "with one hand behind their back"?

When working on high-voltage circuitry, many electricians like to place one hand behind their back because this way there is little chance for each hand to touch objects of different electrical potentials and cause a shock.

Why don't birds or squirrels on power lines get electrocuted?

Birds and squirrels can perch safely on electric wires as long as they don't come into contact with objects that have two different voltages. The difference in voltage along the wire over the distance between the animal's feet is very small. The animal would be in danger

---

How does an electric "stun gun" work?

A typical stun gun that might be used by a law-enforcement official is about the size of a flashlight and contains a number of transformers, oscillators, and a capacitor. Two electrodes a few inches apart are at the front of the stun gun. If both electrodes are touched to a conducting medium at the same time, the circuit closes, and electricity flows from one electrode to the other. Stun guns are designed so that the voltage is very high—up to 100,000 volts or more—but the current is very low (at most a few milliamperes). That way, if the conducting medium is a person, the resulting electric shock is enough to incapacitate the person but not enough to cause permanent injury or death.

only if it made contact with both a high-voltage wire and either the ground or a wire connected to ground, both of which have low voltage. If that happened, a current would run from the high-voltage wire through its body to the low-voltage wire or ground.

Is high voltage a dangerous shock hazard?

Signs around power plants and breaker boxes often state: "CAUTION: High Voltage Area." High-voltage environments can sometimes also create high current, and it is the electrical current that flows through your body that can produce serious and sometimes fatal consequences. Voltage alone, however, is not as dangerous. A Van de Graaff generator creates hundreds of thousands of volts, but it produces such a low amount of current that the sparks it emits only cause muscles to tingle.

ELECTRIC POWER

Who invented the electric light bulb?

The first person to generate light from a wire filament was the British inventor Humphry Davy (1779–1829). Davy was known for his work on electric arc lamps in the early 1800s, and his breakthrough discovery was the first electric light. His lamp used a very thin piece of platinum wire that had high resistance and emitted a soft glow. The lamp didn't last long, so it was not practical, but it did pave the way for others to make advancements in light bulb technology.

What did Thomas Edison do to contribute to electric light technology?

The American inventor Thomas Alva Edison (1847–1931) tried literally hundreds of different materials as filaments for the lamp. He found that heating a cotton thread in the air left a thin length of almost pure carbon. The carbon filament was connected to wires sealed in a glass bulb from which the air was removed. In 1878 this first practical electric lamp lasted for many hours more than any other lamp.

Why is a 100-watt light bulb brighter than a 25-watt bulb?

Generally in light bulb design, the more electric power runs through any given light bulb, the more visible light it will produce. For old-fashioned incandescent light bulbs, the electric power heats up the filament, so the higher electric levels create more heat, and a hotter object gives off more light than a cooler object of the same size. For LED light bulbs that are more common today, more electric power can cause more light-emitting diodes inside the bulb to be illuminated.

What is the difference between a kilowatt and a kilowatt-hour?

A kilowatt, 1,000 watts, is the unit used to describe power, or the rate at which the energy is being converted. Energy is the power multiplied by the time it is used, in this case hours.

Therefore, a kilowatt-hour, the product of power and time, is a unit of energy. The utility company charges you for the number of kilowatt-hours of electricity you use in a month.

For example, a 100-watt light bulb uses 100 watts (or 0.1 kilowatt) of power. If that light bulb were left on for an entire 30-day month, the energy that the bulb consumed would be 0.1 kilowatts × 24 hours × 30 days, which equals 72 kilowatt-hours. If the energy cost is $0.12 per kilowatt-hour, the bill for that one light bulb would be $8.64 for that month. Replacing the 100-watt incandescent lamp with a 23-watt compact fluorescent lamp that is equally bright would cost only $1.98 per month. An LED lamp equally bright has a power rating of only 13 watts and, therefore, would cost $1.12 to light. The compact fluorescent and LED bulbs would cost more at first, but they would save a lot of electric power—and thus lower the electric bill—in the long run.

CIRCUITS

What is needed to create a circuit?

A circuit is a circular path through which charge can flow. So, the first requirement is a complete, unbroken conducting path. Second, there must be a source of potential difference—most often a battery. The battery provides the voltage that will produce the current in the circuit. For a useful circuit, there must be a third element—a device with resistance. This may be, for example, a resistor, lamp, or motor.

In terms of energy, the energy input to the circuit is the chemical energy stored in the battery. When the circuit is complete, the chemical energy becomes electrical energy in the wires. That energy is then converted to thermal energy in the resistor or lamp or kinetic energy in the motor. The hot resistor or lamp then radiates heat and light into the environment.

How does a circuit power a home?

In a typical household, the source of potential difference is not a battery but the electric generating station operated by a utility company. It may use the chemical energy in fossil fuels such as coal, oil, or natural gas to boil water (i.e., produce thermal energy). The steam from the boiling water then turns the generator, converting the energy to rotational kinetic energy. The generator then converts this energy of rotation into the electrical energy that is transmitted to the home. A nuclear power plant uses the nuclear energy in the nucleus of the uranium atoms to heat the water and produce steam. From that point on, the nuclear and fossil fuel power plants are essentially the same.

What is an open circuit?

A circuit is a closed loop through which electric charges can flow. If the loop is opened, then it is called an open circuit, and charges no longer flow. A switch is commonly used to open and close a circuit. If a wire breaks, the connection between the wire and an-

What is a short circuit?

If a circuit consists only of a battery or another potential difference and a wire connecting the two ends of the battery, the resistance in the circuit will be almost zero, and the amount of current will be very high. This situation is called a short circuit, and it can be dangerous. The wire will become hot enough to burn you or perhaps destroy the electrical device it is in. Therefore, a circuit should always have a safe source of resistance in it to prevent a short circuit.

other part of the circuit fails, and the circuit opens. Similarly, if a lamp burns out (its filament breaks), then the circuit opens, and there is no current.

What is the danger of short circuits in a home?

If there is only a tiny amount of resistance in a circuit, the current is very large, and the wires get extremely hot. If, for example, the insulation on the wires in an appliance fails and the wires touch each other, the resistance drops and current rises. The wires, including those in the walls, can get hot enough to cause a fire. Household circuits are protected by fuses or circuit breakers. They are designed to open when the current exceeds a predetermined limit. With the circuit now "broken" or open, current stops flowing, and the wires will cool.

AC/DC

What is a direct-current (DC) circuit?

In a DC or direct-current circuit, charges travel only in one direction. The voltage source, a battery or direct-current power supply, has one positive and one negative terminal, so there is current in only one direction.

What is an alternating-current (AC) circuit?

In an AC, or alternating-current circuit, the polarity of the voltage source changes back and forth at a regular rate. In the United States, one terminal of the source changes from positive to negative and back to positive 60 times each second. Therefore, the flow of charge also alternates in direction 60 times a second as the electrons in the circuit vibrate back and forth. An alternating current is usually found in wall outlets in buildings. Most of our electrical appliances run on alternating current.

Who was Nikola Tesla?

Nikola Tesla (1856–1943) worked for Thomas Edison's company in Europe. Edison offered Tesla a large sum of money if he would invent an improved generator. When Tesla did, Edison refused to pay, saying he had been joking. Tesla quit. Tesla's AC induction motor, which he invented in 1883, and his three-phase generator became important components in the modernization of electric power transmission.

Did Thomas Edison use AC or DC for electrification?

Edison invented the first practical incandescent lamp in 1878. In 1882 he connected 59 customers in the neighborhood around his New York City laboratory to a DC generator that supplied 100 volts. The customers used the electrical power for lamps and motors. The relatively low voltage matched the resistance of the lamps and

Some people these days consider Nikola Tesla to have been a greater inventor than his contemporary Thomas Edison when it came to the study of electrical engineering.

was not believed to be very dangerous. Unfortunately, in order to carry so much power, the current had to be high at low voltage, so this heated the wires. Customers had to be within two miles of the generating system to avoid serious loss of energy in the wires.

Is AC or DC used in our homes today, and why?

In the late 1880s, Edison and Westinghouse battled over the relative merits of DC and AC power distribution systems. After the electric chair had been developed, Edison attempted give electrocution the nickname "Westinghousing" because it used AC, but he failed.

The AC system was eventually chosen because with AC power, transformers could raise the voltage to thousands of volts for transmission and then lower it for use in homes and businesses. At high voltages, the current needed to transmit large amounts of power is reduced, and so is the energy lost to thermal energy because the heating of transmission wires depends on current and resistance.

How do we use DC today?

Every electronic device that uses a battery—rechargeable or disposable—runs on DC power. That means your computers, cell phones, cameras, and flashlights are almost certainly DC devices.

How does the AC power running to our wall outlets get converted to DC power for our devices?

Every DC device that can be charged from an AC outlet needs to have the AC power converted to DC power. The large, often block-shaped power supplies in the plugs or charging cords of your DC devices usually contain an AC-to-DC converter. Some devices have their converters inside them. The basic components of a converter include a transformer, a rectifier, a capacitor, and a regulator.

ELECTRICAL OUTLETS

Why do power outlets usually have three holes?

Electrical outlets in the United States have two slots, one longer than the other, and a "D"-shaped hole. The contacts in the shorter slot are connected to a black wire. This is the "hot" connection that carries 120 volts. The contacts in the longer slot are connected to a white wire, called the neutral wire. The white wire is connected to ground in the electric distribution box. Thus, there is a potential difference of 120 volts across the two contacts.

The third hole is attached to a green wire that is at ground potential. Not all appliances use that third hole—that is, the ground. So, why do you need that extra connection to the ground potential? Because when the appliance plugged in draws current, there is current through both the black and white wires. Each wire has resistance, so there will be a voltage drop across the white wire, and it will be above ground potential at the outlet. While this voltage will be small, it could be dangerous. The green wire, which carries no current, will remain at ground. It can be connected to the metal case of the appliance, assuring that the case will remain at ground potential.

What if a tool or appliance has a three-prong plug but you have only two-slot outlets?

Do not use the appliance if you do not have the proper outlet for the device. Cutting off the grounding prong will defeat the safety feature of the separate ground wire.

It is sometimes feasible to use a three-to-two adapter in this case. On the

When installing an electrical outlet, note that the green wire (okay, you can't tell in this black-and-white photo, but it is there!) is the grounding wire, which must be attached to the grounding contact on the outlet.

adapter, there is a little green wire or plate; that is the grounding wire. Since the adapter is circumventing the ground prong, an alternate means of grounding is needed. If the screw on the outlet plate is grounded, the green wire on the adapter should be attached to it. This way, if there is an electrical short, the current can still flow through the grounding wire. If the screw is not grounded, then the adapter should not be used. An outlet tester that is available at most hardware stores can be used to make sure the screw is grounded.

What is a GFI, or ground fault interrupter?

A ground fault interrupter (GFI) outlet is now required by building codes for outlets within six feet of a sink or in any other environment where water could be close to the outlet. Normally, the currents in the black and white wires will be equal, but if the water provides an alternative current path, then the two currents will no longer be the same. The GFI detects this difference and shuts off the circuit within milliseconds. GFI units should be tested periodically to make sure the electronic circuit is still working.

Why do different countries use different voltages?

The United States was the first country to establish widespread use of electricity for the public. At the time it was implemented, a 120-volt system seemed to provide enough voltage for users, yet was not enough to cause dangerous electrical arcs in small motors. Years later, when electrical wiring was installed in European countries, technological advances had already allowed those devices to run safely at higher voltages. The advantage of a 220-volt system is that the current is only about half as much as used by the same appliance on a 120-volt system, so power losses due to resistance in the house wiring are smaller. Thus, the standard became 220 volts for Europe, while in the United States, it remained 120 volts.

Why is it dangerous to operate electrical devices near bathtubs, showers, and sinks?

Water reduces the resistance of the human body and thus makes it more susceptible to electrical shock. More important, however, for places like bathtubs, showers, and sinks, it is the plumbing that is the main hazard. Take, for example, a person who likes to watch a plugged-in TV while sitting in the bathtub. If the TV is not connected to a protected circuit and it falls into the tub, the water will come in contact with the 120-volt wires in the TV. With the metal plumbing of the tub connected to the ground, the grounding path would cause a current through the water—and potentially a very dangerous or even fatal electric shock to the person in the tub. An MP3 player or battery-powered smartphone would probably be a much safer device to use in this environment.

SERIES AND PARALLEL CIRCUITS

What is a series circuit?

A series circuit consists of electrical devices such as resistors, batteries, and switches arranged in a single line. There is only one path for the charges to flow through, and if there is a break anywhere in the circuit, the current will drop to zero.

What is a parallel circuit?

A parallel circuit allows the charges to flow through different branches. For example, the wire from the battery would be connected to one terminal of each of three bulbs. The other terminals are connected together and to the negative terminal of the battery. The charges now have three separate paths through which they can flow. If one bulb burns out or is removed from the socket, that bulb would no longer light, but the current through the other two lamps would not change. They would continue to glow.

What happens to a series circuit if more resistance is added?

If more light bulbs or other resistors are placed in a series circuit, there is more resistance in the circuit. Therefore, the current, and the brightness of each of the lamps, would be reduced.

What happens in a parallel circuit if more bulbs are added?

In a parallel circuit, the current goes through separate branches. If another branch is added with another bulb, the current has an additional path to take. But the battery (or

Are holiday lights in series or parallel circuits?

Years ago, holiday lights used large bulbs designed to work on 120 volts. These strings of lights were wired in parallel. Today, although some of those older kinds of parallel strings are still available, most holiday lights with very small bulbs are wired in series so that the tiny bulbs have only low voltages across them. In a true series circuit, if one light bulb burns out, there will be no current through the entire string. However, there will still be 120 volts across the failed bulb. Thus, the bulb has a thin wire touching the two thicker wires that deliver current to the bulb. That thin wire is covered with a thin insulating film. The film effectively insulates when the voltage difference across the bulb is small, but when the bulb fails and the full 120 volts comes across the bulb, sparks will break the insulating film and weld the thin wire to the thick wires. This short-circuits the bulb, completing the circuit through the remaining bulbs so they can remain lit.

generator) produces a constant voltage, so the current through the original bulbs does not change, and neither does their brightness.

Are series or parallel circuits used in our homes?

Each circuit, by itself, contains a series connection of a switch and lamp or appliance. The circuits themselves are connected in parallel so that they can be used independently. Thus, both series and parallel circuits exist throughout our homes and electrical devices depending on what kind of circuit is best for each purpose it is used.

MAGNETISM

When was magnetism discovered?

The discovery of rocks that attracted certain metals is lost to history. As was the case with electrostatics, Aristotle (384–322 B.C.E.) credited Thales of Miletus (625–545 B.C.E.) with the first scientific discussion of the attractive power of the rock later called lodestone. The word "magnet" comes from the region of Greece where lodestone is found.

The power of lodestone was found by other people around the world at around that same time or before. At the time of Thales's life, an Indian surgeon, Sushrata, used magnets to aid surgery. Written in the fourth century B.C.E., the Chinese *Book of the Devil Valley Master* says, "Lodestone makes iron come."

When was magnetism first used for navigation?

In the eleventh century C.E. the Chinese scientist Shen Kuo (1031–1095) wrote about the use of a magnetized needle as a compass in navigation. By the next century, the Chinese were known to use a lodestone as a shipboard compass. In modern Chinese, compasses are called "south-pointing needles."

One hundred years later, British theologian Alexander Neckham (1157–1217) described the compass and how it could be used to aid navigation. Some people thought that the Pole Star attracted the compass, while others thought that the source was a magnetic island near the North Pole. In 1269 the Frenchman Petrus Peregrinus wrote a detailed paper on the properties of magnets.

The most comprehensive and important work in this area was written by the English scientist William Gilbert (1544–1603) in 1600. Gilbert concluded that Earth was a giant magnet—which explained why magnetic needles could be used for compasses.

What are some key properties of magnets?

You've probably played with magnets since you were a child. It is likely that you found that magnets attract some materials but not others. You may have found that you can use a magnet to magnetize items like paper clips, nails, and screws. If you played with two magnets, you found that they could either attract or repel each other. Whether you played with bar-shaped metal magnets or rectangular or circular ceramic magnets, you found that the magnet exerted stronger forces at the ends or faces of the magnets. Those regions are called "poles." If you hang the magnet from a string so it can rotate freely, you'll find the magnet orienting itself north-to-south. The end facing north is called the north pole, the other the south pole. Like poles repel each other, while unlike poles attract, but either end can attract other materials.

What are magnetic dipoles and monopoles?

Magnetic poles always come in north-south pairs called dipoles (two poles). Some theories predict the existence of isolated north or south poles, called monopoles, but there have been extensive searches for monopoles over the past decades, and none has ever been found. The magnetic monopole remains a topic at the frontier of physics research.

What is a magnetic field?

Just as the gravitational field is the region around a massive object that causes an attractive force on another object with mass, a magnetic field is the region around a magnet that causes forces on magnetic materials or other magnets. A magnetic field has a distinctive shape around the poles of a magnet, which can be seen if you sprinkle iron filings on top of a magnet on a sheet of paper.

What materials do magnets attract?

Iron, nickel, and cobalt and most of their alloys are attracted to magnets. Other metals, like silver, gold, copper, tin, stainless steel, zinc, brass, and bronze, are not attracted. Nonmetals are not attracted.

Iron, nickel, and cobalt are called ferromagnetic, or exceptionally magnetic. All materials respond to magnetic fields, but most respond so weakly that the forces are hardly felt. Those that are repelled are called diamagnetic; those attracted are paramagnetic.

Why do magnets attract some materials but not others?

The ultimate cause of magnetism is electrons. When electrons are in a magnetic field, the forces they experience cause them to move in tiny circles. The circling electrons create their own magnetic fields, which give rise to diamagnetism. Electrons are tiny magnets themselves, with

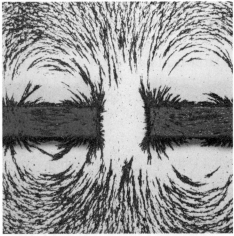

Children often discover some of the properties of magnetism by playing with bar magnets and metal shavings. In this way, you can easily discover that magnets have opposite poles and create magnetic fields.

north and south poles. In most atoms, these magnets are paired so their fields cancel. But if there are an odd number of electrons, the unpaired electron produces a paramagnet. Oxygen, for example, is paramagnetic.

In ferromagnets, the unpaired electrons in large groups of atoms interact with each other so that they point in the same direction. This group is called a domain.

When a ferromagnet is put in a magnetic field, the domains can line up with their poles facing the same direction, making the material a magnet. In most materials, when the magnetic field is removed, the domains revert to their former random directions, and the material is no longer a magnet. For certain alloys, however, the domains remain aligned, resulting in a permanent magnet.

What materials make the strongest permanent magnets?

Traditional permanent magnets were made of an alloy of aluminum, nickel, and cobalt, called alnico. Ceramic and rubber magnets use ferrites, an iron oxide material. In the 1980s, the automobile companies searched for materials to reduce the weight of motors in their cars. They found that an alloy of cobalt and samarium, a rare earth, made strong, lightweight magnets but were extremely brittle and expensive.

Today, the strongest permanent magnets are made from an alloy of iron, boron, and metals from the group of elements called the "rare earths" such as neodymium and lanthanum. They are sometimes called NIB, LIB, or rare earth magnets, and their strength is many times that of alnico magnets. They are also brittle, so they are coated with a plating of nickel and copper to reduce that quality. Their price has fallen so much that they are used to hold sunglasses to eyeglass frames, in necklace clasps, and in children's toys.

How are refrigerator magnets made?

Examine a refrigerator magnet. It is flexible, it feels like rubber, and only one surface is attracted to metals. It doesn't stick to a stainless steel door unless the stainless has been coated with steel. It's made of rubber that has been impregnated with ferrite particles and magnetized. Small pieces, each a dipole, are then pressed together under heat to bond them into one thin sheet that can be cut, folded, and bonded to other sheets.

What is the structure of a refrigerator magnet?

Which of the three arrangements shown in Figure 22 would have the properties of a refrigerator magnet as described above?

The top two wouldn't because both surfaces would act as a magnet. The top right-hand arrangement would be a very

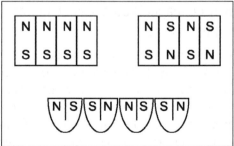

Figure 22.

weak magnet on both faces because the alternating poles would essentially cancel each other out.

In the third drawing, the sheets have been folded and then pressed together so that the poles are at only one surface, so only that surface would act like a magnet. The alternating N and S poles attract steel and stick to it. You can check this idea by taking two refrigerator magnets and holding the magnetic surfaces together, and then try sliding one over the other. You'll find that they skip as first N and S poles touch each other and attract. Then the like poles try to touch each other but repel, making the magnets skip.

What happens when a magnet is cut in two pieces?

When a magnet is cut, the atoms within the domains remain aligned. In almost every case, the cut would be between two domains, leaving aligned domains in the two halves. If you cut a domain, you would create two smaller domains, each with a north pole and a south pole. So, no matter where you cut, the result is two magnets, each with its own north and south pole. The more domains there are in a magnet, the stronger the magnet is.

EARTH'S MAGNETIC FIELD

Is Earth a giant magnet?

Our planet Earth has a core that contains liquid ferrometals—almost all of it iron and nickel. As Earth spins, those ferrometals spin as well but at a slightly different rate. This difference creates what is called a dynamo effect, generating a magnetic field around Earth, as if our planet were a giant magnet. It's not a solid magnet, though, so the magnetic field is not constant in either strength or direction. Many other planets, and even the sun and other stars, have magnetic fields as well. Details of how the dynamo effect works are still topics of modern astrophysics and geophysics research.

What are the Van Allen belts?

Charged electrons and protons from solar wind and cosmic rays entering Earth's magnetic field feel the Lorentz force, an electromagnetic force that traps them into spiral orbits about the magnetic field lines. They create doughnut-shaped regions of charged particles called the Van Allen belts. The belts are concentrated around the equator and become thinner as they approach the poles. The two belts are located at 3,200 kilometers and 16,000 kilometers above the surface of Earth.

Why are the Van Allen belts concentrated at the equator?

At the equator, the magnetic field is parallel to the ground, and the electrons and protons from the solar wind can become trapped around the field lines. At the poles, however, the magnetic field strengthens, the lines become closer together, and forces on the particles push them back toward the equator. Some of the most energetic particles

Fortunately for us, Earth generates its own magnetic field, which effectively fends off dangerous particles from the sun.

are able to penetrate the atmosphere, where they interact with oxygen and nitrogen atoms and produce the natural light shows that are called auroras.

What causes an aurora?

Disturbances on the sun can send large numbers of charged particles into space. When they reach Earth, they follow the field lines of Earth's magnetic field to the magnetic north and south poles. These streams of charged particles also disturb the Van Allen belts, causing the belts to dump particles into Earth's atmosphere. All these charged particles interact with the gases in the atmosphere, causing them to emit light. The scientific name for this phenomenon is an aurora, and it can happen not just on Earth but around any planet with a sufficiently strong magnetic field.

What are the names for the aurora on Earth?

An aurora in the Northern Hemisphere, known as the Northern Lights, is officially called the aurora borealis, while an aurora in the Southern Hemisphere, or Southern Lights, is called the aurora australis.

How is Earth's magnetic field oriented?

Because opposite poles attract, the north pole of a hanging magnet or compass must point toward a south pole. So, the south pole of Earth's magnet must be near the north geographic pole. The poles are actually far below Earth's surface, so Earth's field is not parallel to its surface.

How has Earth's magnetic field changed over hundreds or millions of years?

The direction of magnetic poles in iron in Earth's crust demonstrates that Earth's magnetic field has reversed its polarity— that is, the magnetic north pole has become the magnetic south pole and vice versa—about nine times over the past four million years. It's not known how long such reversals last, nor what happens to the field during the reversal. The locations of the magnetic poles also wander; Earth's magnetic north pole has moved as much as 800 kilometers toward the geographic North Pole since 1831.

Astronaut Jack Fischer took this photo of the aurora from space during a June 2017 mission on the International Space Station.

What is magnetic declination?

Magnetic declination is the angular difference between north as shown by a compass and the direction to the geographic North Pole, Earth's axis of rotation. Declination depends primarily on the location on Earth but, because the magnetic poles move, also on time.

How is a magnetic compass made?

A compass is a magnetized metallic pointer that can rotate about a low-friction pivot point. Sometimes, the pointer is placed in a container of liquid to dampen the movement of the pointer. The magnetic pointer aligns itself with the north-south orientation of Earth's magnetic field, and people using the compass can determine what direction they are headed by looking at the pointer.

Does a compass sometimes point downward along with pointing north?

For hundreds of years, navigators using compasses noticed that, on occasion, the compass pointer would try to point downward in addition to pointing north. This phenomenon was observed and recorded by the English compass maker and navigator Robert Norman in the sixteenth century. Using this knowledge, he balanced a compass needle on a horizontal axis, which allowed the needle to point both northward and down. He thus made the first "dip needle."

What is a dip needle, and how is it similar to a compass?

A dip needle is just like a conventional compass, but instead of being held horizontally, it is held vertically. It is a magnetic needle used for navigational purposes just like a compass, but it is used predominantly when traveling around the North and South

Poles. Instead of measuring horizontal magnetic deflection, the dip needle measures vertical magnetic inclination. When over the equator, the magnetic field of Earth is parallel to the surface of Earth. The closer pilots get to the magnetic poles, the less they rely on compasses and the more they rely on dip needles to tell them how close they are to the poles. The closer one gets to a pole, the more vertical the magnetic field becomes because it is turning into the surface of Earth. Therefore, when it is directly over the magnetic poles, the dip needle points directly downward.

ELECTROMAGNETISM

How was the connection between electricity and magnetism discovered?

The close connection between electric current and magnetic fields was discovered quite by accident. In 1820, Danish physicist Hans Christian Oersted (1777–1851) gave a lecture on the heating effects of an electric current on a wire. A compass happened to be near the wire, and he was surprised to see the compass rotate when the current was on. He had been looking for connections between electricity and magnetism for several years, but he expected that the compass would point away from the wire. Instead, he found that the compass pointed in a circle around the wire. Above the wire, it pointed perpendicular to the wire; below the wire, it also pointed in the perpendicular but in the opposite direction.

What were the implications of Hans Christian Oersted's discovery?

The fact that moving charge in a wire could create a magnetic field generated a great deal of excitement and enthusiasm in the scientific community. A week after hearing about Oersted's discovery, French physicist and mathematician André-Marie Ampère (1775–1836) gave a presentation at the French Academy of Sciences that extended Oersted's experiments and contained detailed analyses. A day later, he found that two parallel current-carrying wires would either attract or repel each other depending on the relative directions of the currents. Ampère's greatest contribution, however, was the mathematical theory he created for electricity and magnetism.

What is the Lorentz force?

British chemist and physicist Michael Faraday's (1791–1867) philosophy led him to search for connections among phenomena like electricity, magnetism, and light. In 1821 he invented what is now called a homopolar motor. One end of a wire was suspended from a support so that it could swing in any direction. The other end of the wire contacted a pool of mercury. When Faraday put current through the wire, the end in the mercury traced out a circle.

The force that Faraday had observed was formalized in 1891 by the Dutch physicist Hendrik Antoon Lorentz (1853–1928). This force, called the Lorentz force, is proportional to the current through the wire, the magnetic field, and the length of the wire. The force, which is perpendicular to both the current and the magnetic field, is strongest when the current and field are at right angles. This force is the basis of motors and many other applications.

What important electromagnetic device did Michael Faraday invent?

When Michael Faraday published his results about the Lorentz force, he failed to give credit to two other important scientists, and he was given assignments to work in other fields. Nevertheless, he continued to do experiments on the effects of magnetic fields. For example, he found that when dense glass was put in a magnetic field, the direction of polarization of light going through the glass was rotated. He spent ten years searching for ways to create a current from a magnetic field. Finally, in 1831 he tried changing the

Researching magnetic fields and electric currents, Michael Faraday and, independently, the American physics teacher Joseph Henry invented an early form of electric generator called the dynamo. Today, generators like those pictured above use the same principles to produce electrical energy from other forms of energy.

magnetic field and made the crucial discovery that an electric current is produced by a changing magnetic field. The current, a flow of charges, is produced by an electric field exerting forces on the charges. Faraday went on to invent the dynamo, an early electric generator. The American high school physics teacher Joseph Henry (1797–1878) made the discovery at almost the same time.

What contribution to electromagnetism was made by James Clerk Maxwell?

Electric charges create electric fields. Moving charges—that is, currents—create magnetic fields, and changing magnetic fields produce electric fields. In the 1860s, Scottish physicist James Clerk Maxwell (1831–1879) added a crucial additional connection: changing electric fields can produce magnetic fields.

With that idea, Maxwell recognized that these relationships meant that electric and magnetic fields could move through space. The fields move through space as transverse waves that are perpendicular to each other. Maxwell calculated the speed and found that it was equal to the speed of light. He published his results in 1864 and a textbook on electromagnetism in 1873. In 1881 Oliver Heaviside (1850–1925) wrote Maxwell's famous four equations in the form they are used today.

In 1888 Heinrich Hertz (1857–1894) transmitted electromagnetic waves across his laboratory, confirming Maxwell's theoretical work.

Why can electromagnets be so strong?

An electromagnet is a coil of current-carrying wire wound on a iron core that is at the center of an iron cup. The magnetic field created by current in the wire is strengthened by the iron core. The strength of the magnetic field produced by such electromagnets creates a large force, as described by the Lorentz Force Law, that allows people to more easily move large, metal objects, such as steel cars, from one location to another.

What is the difference between a motor and a generator?

In each device, a magnet and a coil of wire are employed to change one form of energy into another form. A motor consists of multiple loops of wire placed in a magnetic field. Either the loops or the magnet can rotate. The current through the wires in the field causes a force that results in rotation and thus mechanical energy. Motors in a home are used in fans, hair dryers, and food processors. There are over a hundred motors in a modern automobile. The starter motor is the largest and most powerful.

A generator does the opposite of a motor; it changes mechanical to electrical energy, but it still consists of multiple loops of wire in a magnetic field. Either the loops or the magnet can rotate. In an automobile, a form of a generator, called an alternator, uses some of the energy from the engine to charge the battery. Backup generators use the energy from a gasoline engine to produce enough electrical energy to keep some of the lights and appliances running in a house when the electrical power fails. Electric utilities use huge generators to provide power for a city or larger area. The generators get their energy from steam turbines. The heat required to turn water into steam can come from coal, oil, natural gas, or nuclear "burners." Wind power uses generators turned by the propeller blades.

How are earphones and electromagnetism connected?

An earbud or earphone contains a membrane made of thin plastic. In the center of the membrane is a coil of wire called the voice coil. It fits in a cylindrical slot in a permanent magnet. The center rod of the magnet is one pole, and the outside tube is the other; this results in a magnetic field perpendicular to the wire. When there is an electric current through the wire, the Lorentz force on the wire pushes the membrane in and out. The membrane exerts forces on the air molecules, producing the longitudinal waves that constitute sound.

How are magnetic materials used in computers?

Magnets are used in the compact motors that turn the disks in the CD or DVD drive and that move the laser that reads the disk to the correct position. Motors rotate the disks in a hard drive. The arm on which the read/write head is mounted has a coil of wire on it in a magnetic field. When there is a current through the wire, the force moves the arm to the correct position.

The disk itself is often made of aluminum coated with an extremely thin (10–20 nanometers) film of magnetic material that is divided into submicrometer-thick regions that are perpendicular to the surface of the disk. Each region is magnetized one way to represent a "1" and another way to represent a "0." A tiny coil in the read/write head carries the current that magnetizes the regions. The state of the magnet is read using magnetorestriction—a process where the magnetic field causes a change in resistance of a very thin wire in a coil in the read/write head.

How do airport metal detectors work?

Built into the frame of a metal detector are coils of wire that carry a current. When metal is close to the coils, the magnetic properties of the metal change the current in the coils of wire, which is detected by the electronic circuits in the detector.

Can traffic lights tell when a car is at the intersection?

Many traffic lights are triggered to change by the approach of a car. The principle is similar to that of the metal detector in that

When you walk through a metal detector with metal anywhere on your person, that metal changes the current in the coils in the frame of the detector.

there are coils of current-carrying wire just below the road where the vehicles stop at the intersection. When a large enough amount of metal passes over the coil, it induces a change in the current that creates a signal in the electronic circuits that control the traffic light.

What are Maglev trains?

Magnetically levitated (Maglev) trains are different from conventional trains in that they use electromagnetic forces to lift the cars off the track and propel them along thin, magnetic tracks. Some demonstration trains have reached speeds of 500 kilometers per hour (300 miles per hour). Although the United States has neither an

China has developed an efficient system of Maglev trains, monorail transportation using the force of magnetism to propel passengers at high speeds.

active Maglev train nor an active research program in this technology, Germany and Japan have conducted a great deal of research in the field.

What are the two main forms of Maglev transportation?

The German system uses the attractive forces between electromagnets to lift the underside of the train 15 centimeters (6 inches) above its guide rail. The coils in the train and guide rail form a linear motor—sort of like an ordinary motor that has been unrolled.

The Japanese have taken a slightly different approach toward Maglev technology. The track and train repel each other. Propulsion also uses a linear motor. Levitation works well at high speeds, but when starting and stopping, traditional wheels must be used.

Where was the first commercial Maglev train started?

The Shanghai Maglev Train (SMT) currently operates between Shanghai Pudong International Airport and Longyang Road Metro Station. With German technological assistance, it became the world's first commercial magnetic levitation line early in the twenty-first century. Its designed maximum operating speed is 430 kilometers per hour (267 miles per hour), and the entire 30-kilometer (19-mile) trip takes about seven minutes.

ATOMIC AND QUANTUM PHYSICS

What is matter?

Ancient people in many parts of the world believed that all matter was made of four elements: earth, air, water, and fire. No matter how much you divided a material, you could not separate an element into a combination of other materials. But if you kept dividing the amount of material into smaller and smaller pieces, what would you obtain?

Democritus, a Greek who lived around 410 B.C.E. and was a student of the fifth-century-B.C.E. philosopher Leucippus of Miletus, stated that all matter is made up of atoms and the void. Atoms are the smallest piece into which an element can be divided; they are uncuttable. They could be neither created nor destroyed and, thus, were eternal. The void was empty space. This viewpoint was expanded by the first century B.C.E. by the Greek Titus Lucretius Carus (95–55 B.C.E.) in his epic poem "On the Nature of Things." Aristotle (384–322 B.C.E.), on the other hand, regarded the atomist philosophy as pure speculation that could never be tested. He rejected the possibility of empty space and believed you could divide matter until it was infinitely small.

Did cultures other than the ancient Greeks develop ideas about matter?

The Greeks were not the only ones to develop a philosophy of atomism. In India, the school of philosophy known as Vaisesika, and in particular the philosopher Kanada (also called Kashyapa) in the second century B.C.E., held that earth, air, fire, and water could be divided into a finite number of indivisible particles. These ideas were also adopted by several other Indian schools of philosophy.

How did the ideas of matter from Aristotle and Democritus spread throughout history?

For almost 2,000 years, Aristotle's philosophy was taught in schools and accepted by educated people throughout Europe. In the sixteenth century, doubts about Aristotle's science increased. A number of natural philosophers—those whom today we recognize as pioneering scientists—actively opposed Aristotle's dominance of the curriculum in schools. Englishman Francis Bacon (1561–1626) developed an early version of what we today call the scientific method in opposition to Aristotle's philosophy. In 1612 Galileo Galilei (1564–1642) published his *Discourse on Floating Bodies*, in which he envisioned atoms as infinitely small particles, views that he later expanded in *The Assayer* (1623) and, more completely, in *Discourses on Two New Sciences* (1638). Nevertheless, the debate over whether atoms, much too small to see, really existed or were just a successful model of matter continued for another two centuries.

How did chemistry contribute to the physics of matter?

Chemistry is, in many ways, the science that emerged when alchemy combined with physics. Robert Boyle (1627–1691), an Irishman who wrote *The Sceptical Chymist* in 1661, is often considered the father of chemistry. He held that matter was made of atoms or groups of atoms that were constantly moving. He urged chemists to accept only those results that could be demonstrated by experiment.

How did Antoine and Marie-Anne de Lavoisier contribute to the scientific understanding of matter?

The great French chemist Antoine-Laurent de Lavoisier (1743–1794) demanded that measurements be made with precision and that scientific terms be clearly defined and carefully used. With the discovery that air contained oxygen, which is necessary to support animal life, he clearly demonstrated that air was not an element but a mixture of oxygen and nitrogen. By showing that hydrogen, discovered by Henry Cavendish (1731–1810), when mixed with oxygen formed water, Lavoisier showed that water also was not an element but a compound of two elements. Lavoisier's care with weighing both the reactants and products of reactions allowed him to make one of the earliest statements that mass is conserved, that is, neither created nor destroyed in chemical reactions.

Marie-Anne Paulze de Lavoisier (1758–1836), Antoine's wife, was a brilliant woman who collaborated with her husband in the research. She also illustrated their laboratory experiments, translated letters from foreign scientists, and published the results of their work after Antoine died.

Sadly, Antoine de Laviosier was executed during the French Revolution for being a tax collector. In supporting the guilty verdict, the judge ignorantly said, "The Republic needs neither chemists nor scientists." The First French Republic failed just 10 years later.

How did Joseph Priestley contribute to the scientific understanding of matter?

Antoine de Lavoisier met with Joseph Priestley (1733–1804) shortly after Priestley discovered oxygen in 1774. Lavoisier did extensive work on oxygen and gave it its present name. Priestley invented carbonated water and experimented with ammonia and laughing gas (nitrous oxide), but when he opposed much of the work of chemists of his generation, he was pushed to the sidelines. After writing theological books that opposed traditional Christianity in England, Priestly was forced to flee to America in 1794.

How did John Dalton explain the chemistry of atoms?

John Dalton (1766–1844) was an English chemist who, because he was a Quaker, could not obtain a position in a state-run university. For a number of years, he taught at a college in Manchester for dissenters from the Church of England. His strengths were his rich imagination and clear mental pictures and especially his astonishing physical intuition.

Also known for his work on color blindness, English chemist, meteorologist, and physicist John Dalton introduced the concepts of atomic theory into the field of chemistry.

His first interest was meteorology. He wondered how Earth's atmosphere, consisting of gases of very different densities, could have the same composition at different altitudes. His meteorological studies led him to the conclusion that atoms were physical entities and that their relative weight and number were crucial in chemical combinations. Dalton's atomic theory of chemistry was published in *A New System of Chemical Philosophy* in 1808 and 1810. Briefly, it has five parts:

1. Elements are made of tiny particles called atoms that cannot be divided into smaller particles nor can they be created or destroyed or changed into another kind of atom.

2. All atoms of an element are identical. Therefore, there are as many kinds of atoms as there are elements.

3. The atoms of an element are different from those of any other element in that they have different weights.

4. Atoms of one element can combine with atoms of another element to form a chemical compound (today called a molecule). A compound always has the same relative numbers and kinds of atoms.

5. In a chemical reaction, atoms are rearranged among the compounds; they are neither lost nor gained.

The first and last parts give an atomic basis for the conservation of mass that Antoine de Lavoisier had confirmed in many careful experiments. With the advent of atomic physics, we know today that two of Dalton's specifications were not completely correct. Atoms can be changed from one kind to another by the process of radioactive decay, and not all atoms of an element have the same weight.

Who really discovered oxygen?

Three scientists are often credited with the discovery of oxygen: Carl Wilhelm Scheele (1742–1786), Antoine-Laurent de Lavoisier, and Joseph Priestley. The Swedish scientist Scheele discovered what he called fire air in 1772. The name came from the way an ember would burst into flame when immersed in the gas. Scheele wrote a book describing his work, but it took four years to be published. In the meantime, Priestley in 1774 discovered what he called dephlogisticated air. Lavoisier, who met Priestley in 1774 and was told about Priestley's discovery, claimed to have discovered what he called vital air. Although he was sent letters from both Scheele and Priestley describing their earlier work, he never acknowledged their receipt. Lavoisier's great contribution was to make precise studies of the role of oxygen in a variety of reactions and to give it the name oxygen.

Who introduced modern chemical symbols?

Swedish chemist Jöns Jacob Berzelius (1779–1848) introduced the modern chemical notation using one or two letters to represent an element in 1813. Rather than today's notation for water, H_2O, Berzelius wrote H^2O. Berzelius also did much more; he published the first accurate list of the weights of atoms.

How did John Dalton represent atoms?

John Dalton used pictographs to represent atoms. In the first line in Figure 23 are some examples. Dalton would represent a chemical reaction (carbon plus two oxygen to yield carbon dioxide) as shown on the second line.

Are atoms indivisible?

Thanks to the invention of the electric battery in 1800 by Alessandro Volta (1745–1827), chemists had a new tool to create reactions and isolate new elements. Humphry Davy (1778–1829) was one of the most active and isolated sodium, potassium, and calcium from their salts. But it was his assistant, Michael Faraday (1791–1867), who most contributed to the discovery that the atom was not indivisible. He found that the amount of charge needed to liberate an element from a solution was proportional to the mass of the element. In modern language, the amount of charge was proportional to the number of atoms liberated. It took until 1881 for the import of this result to be realized when Hermann von Helmholtz (1821–1894) pointed out that if elements are composed of atoms, then electricity could be divided into portions that could be called atoms of electricity. George Johnstone Stoney (1826–1911) named these atoms electrons. But what were they?

How was the electron discovered?

Further advances came not from electrolysis but from studies of gases. In the 1700s

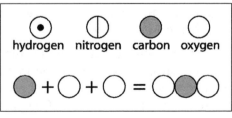

Figure 23.

and early 1800s, physicists used the vacuum pump, invented by Otto von Guericke (1602–1686) in 1690, to reduce the pressure in glass tubes fitted with electrodes to allow electricity to pass through the tubes. In 1838 Michael Faraday passed an electric current through such a tube and noticed a strange, arc-shaped light starting at the cathode (negative electrode) and ending almost at the anode (positive electrode).

What are cathode rays?

When Heinrich Geissler (1814–1879) was able to reduce the air pressure to about 1/1,000 of an atmosphere, he found that the tube was filled with a glow, like the neon lamps used today. By the 1870s, William Crookes (1832–1919) was able to reduce the pressure to 1/1,000,000 of an atmosphere. As the pressure was reduced, the glow gradually disappeared. Instead, the glass near the anode began to glow. Without air to disrupt their passage, rays of some sort were able to travel from the cathode to the anode. At the anode end, they were going so fast that they caused the glass to glow or fluoresce. Coating the glass with zinc sulfide made the glow brighter. The mysterious, invisible rays could be shown to travel in straight lines by placing metallic objects in the tube and finding that they cast sharp shadows at the anode. Because the rays came from the cathode, they were called cathode rays.

How did J. J. Thomson confirm the existence of electrons?

Joseph John (J. J.) Thomson (1856–1940) conducted three experiments with Crookes tubes that showed the nature of cathode rays. His first experiment used an electrode at one side of a tube, out of the direct path, which was connected to an electrometer that could detect electric charge. Thomson could deflect the path of the rays using a magnet and follow the path by observing the fluorescent glow on the tube's surface. He found that the electrometer showed a negative charge but only when the rays were deflected to its terminal. Thus, he showed that the rays consisted of a beam of negative particles.

Using a tube with the best possible vacuum, Thomson next explored the effect of an electric field on the rays. He added two parallel metal plates to the tube, connected a battery across the

Modern vacuum chambers like this one are used in laboratories to study atoms, molecules, nuclei, and electrons. Otto von Guericke invented the vacuum pump in 1690 to help him study electricity in a vacuum, and since then vacuums have been used to study a wide variety of natural phenomena.

plates, and found that the rays were attracted toward the positive plate and away from the negative one.

In his third, and most important experiment, done in 1897, he combined the deflection of an electric field with one by a magnetic field. In doing so, he could calculate the ratio of the mass to the charge of the particles. He found that this ratio was 1,800 times lower than that of a positively charged hydrogen ion. Thus, the particles must be either very light or very strongly charged. He later showed that they had the same charge as the hydrogen ion, so they were very light. Further experiments showed that the particles had the same properties no matter what metal was used for the cathode or whether the cathode was cold or incandescent. For his work, Thomson was awarded the Nobel Prize in 1906.

Who first measured the amount of charge an electron has?

The charge of the electron was first measured in 1909 by the American physicist Robert Andrews Millikan (1868–1953). Before Millikan's experiments, some physicists claimed that Thomson's results could imply that electrons had an average mass-to-charge ratio given by his experiments but that they could have a variety of masses and charges. But Millikan showed that all electrons had the same charge and thus the same mass.

What are the charge and mass of electrons?

The charge of an electron is negative 1.602×10^{-19} C (coulomb). Its mass is 9.11×10^{-31} kg. Both values are known to great precision today—about 85 parts per billion.

How big is an electron?

The size of an electron has never been successfully measured. Indeed, if you direct beams of very fast, high-energy particles at electrons, they are deflected by the $1/r^2$ force between two charges no matter how close they come. If the electron had a measurable radius, then when the incoming particle penetrated the electron, the deflection would be different. This suggests that electrons are not only tiny but have no fixed size.

What was J. J. Thomson's idea of the structure of the atom?

The discoverer of electrons, J. J. Thomson, pictured the atom as a swarm of electrons in a positively charged sphere. This model is called the "plum pudding" model, after a then-favorite English Christmas treat. (Americans might picture a ball of pudding filled with raisins.) This model turned out to be helpful but incorrect.

Who experimentally verified the structure of an atom?

The New Zealand-born physicist Ernest Rutherford (1871–1937) developed a method of testing Thomson's model. Rutherford's first scientific work was done in Montreal, Quebec, Canada, beginning in 1898. He and Frederick Soddy (1877–1956) conducted a study of radioactivity and radioactive materials, for which Rutherford was awarded the 1908

Nobel Prize in Chemistry. He studied the radiation emitted by thorium and uranium, which he named alpha and beta rays. He recognized that alpha rays would be ideal probes for studying materials.

Rutherford moved to the University of Manchester in Britain in 1907, where he worked with Hans Geiger (1882–1945; the inventor of the Geiger counter) on ways of detecting individual alphas. They determined that the alpha particles were doubly charged. They allowed the alphas to go through a window made of very thin mica (a mineral that could be cut into very thin slices). He noticed that the rays that penetrated the mica were deflected slightly more than they should have been by a plum-pudding atom. With Geiger and Ernest Marsden (1889–1970), he directed a beam of alphas on extremely thin foils of gold. To their great surprise, they found that a significant number of alphas were deflected into very wide angles—some even greater than 90 degrees! Rutherford commented that it was as if a 15-inch shell from a naval cannon bounced off a sheet of tissue paper.

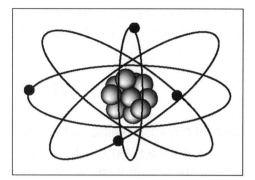

The Rutherford model, as pictured above, is used in many common symbols like that of a nuclear power plant or the flag of the International Atomic Energy Agency.

What model did Ernest Rutherford develop for the structure of the atom?

By 1911, based on his gold-foil experiments, Ernest Rutherford had developed his own model of an atom. It consisted of an extremely small, positively charged particle (later called the nucleus) surrounded by electrons. Although the paths of the electrons were not mentioned in the 1911 paper, it was later assumed that they orbited the nucleus as planets orbit the sun. The nucleus was 1/10,000 as large as the atom, but it had all the positive charge and essentially all of its mass. If the atom's mass was equal to N hydrogen masses, then the nuclear charge was about $N/2$. The electrons carry a total charge

What were some flaws with Ernest Rutherford's model of atoms?

Electrons moving in circular orbits are experiencing centripetal acceleration. All accelerating charges radiate energy. As a result, the electrons in Rutherford's atom would lose all their energy in a tiny fraction of a second. How then could atoms last billions of years? Rutherford offered no answer. In addition, as the electron spiraled into the nucleus, it would create a smear of all colors of light, but hydrogen was known to produce only specific colors, called an emission-line spectrum. More refinement was needed to create an accurate physical picture of atomic structure.

equal and opposite that of the nucleus so that the atom would be electrically neutral. Rutherford's atom model was almost entirely empty space.

THE NUCLEUS

How did physicists realize that the nucleus contained other particles?

Ernest Rutherford (1871–1937) had determined that the charge on the alpha particle, actually the helium nucleus, was two proton charges, but its mass was four proton masses. Dmitri Mendeleev (1834–1907), in building the periodic table, had arranged the elements in order of their mass and had arbitrarily assigned them sequential numbers that he called the atomic number. The English physicist Henry Moseley (1887–1915) measured the wavelengths of X-rays emitted when metals were struck by other X-rays. He was able to provide a physical basis for the atomic number and, in doing so, greatly strengthened the case for Rutherford's nuclear model.

But, as is the case for helium, if the atomic number, and thus the nuclear charge, was about half the atomic mass number, what made up the additional mass of the nucleus? The first proposal was that the missing particle was a combination of a proton and an electron, which would have the correct mass and charge. But the Heisenberg uncertainty principle showed that if an electron were confined to the size of a proton, its energy would be larger than ever observed. In addition, by the late 1920s, the angular momentum of the nitrogen nucleus, with charge 7 and mass 14, had been measured. The result could not be obtained from a combination of 14 protons and 7 electrons.

How was the neutron discovered?

In 1931 two German physicists found that when energetic alpha particles struck light elements, a very penetrating, electrically neutral radiation was produced. The next year, Marie Curie's daughter, Irène Joliot-Curie (1897–1956), and her husband, Frédérick Joliot-Curie, found that if this radiation struck paraffin, protons were ejected, suggesting that the radiation was actually a neutral particle with a mass near that of a proton. The next year, James Chadwick (1891–1974), working in Manchester, England, experimentally confirmed the suggestion. The particle was named a neutron, combining neutral with the ending of the word proton. Chadwick was awarded the Nobel Prize in Physics for his discovery in 1935. The Joliet-Curies also received the Nobel Prize in Chemistry that same year.

What are some basic properties of the neutron?

The neutron has a mass slightly larger than that of the proton. While neutrons are stable in nonradioactive nuclei, if they are free from the nucleus, they decay with a half-life of about 10 minutes. Neutrons are used extensively in creating nuclear reactions and are necessary for sustaining the atomic chain reactions that cause nuclear fission.

What is an isotope?

The number of protons in the nucleus of an atom determines what element it is, so the number of protons is fixed. But the number of neutrons can vary. Nuclei with the same number of protons but different numbers of neutrons are called isotopes. For example, carbon, with six protons, can have five, six, seven, or eight neutrons. The isotopes are respectively called carbon-11, carbon-12, carbon-13, and carbon-14. A more compact notation is ^{11}C, ^{12}C, ^{13}C, and ^{14}C. Chemical properties in general do not depend on the isotope. There are at least 3,100 isotopes of all the elements presently known.

How do the numbers of protons and neutrons in a nucleus compare?

For lighter elements, the number of protons and neutrons are approximately equal. For example, the most common isotope of helium with two protons is ^4He, so it has $4/2 = 2$ neutrons. Oxygen, with eight protons, has eight neutrons in its most common isotope, ^{16}O.

In elements heavier than calcium (20 protons, 20 neutrons) the number of neutrons is larger than the number of protons. In uranium-238 (92 protons, 146 neutrons), the ratio is almost 3 neutrons for every 2 protons.

What force holds a nucleus together?

Protons, all being charged positively, will repel each other. So, there must be an attractive force that holds the nucleus together. A strong nuclear force acts between protons and protons, protons and neutrons, and neutrons and neutrons, all with the same strength. Because the strong force acts the same on protons and neutrons, the two particles are frequently lumped together under the name "nucleon." While the repulsive electromagnetic force acts over long distances, the strong nuclear force only acts between nucleons that are in contact. Nucleons have angular momentum, or spin, and the strong force depends on the relative orientation of the spins. The force is stronger if the spins are in opposite directions.

How were the details of the strong nuclear force determined?

Examining the kinds of isotopes that exist gives clues about the nature of the strong force. The most stable nucleus yet discovered, helium-4, has two neutrons having spins in opposite directions and two protons also having spins in opposite directions. A majority of nonradioactive isotopes have even numbers of neutrons and even numbers of protons, allowing them to form pairs with opposite spins. Most of the other isotopes have either an even number of protons and an odd number of neutrons or the opposite. Isotopes with an odd number of both protons and neutrons are extremely rare. Therefore, the strength of the strong force clearly depends on the spins of the nucleons.

The mass of a nucleus is less than the sum of the masses of the protons plus the sum of masses of the neutrons. The larger the mass difference, the stronger the forces holding the nucleus together and the more energy needed to pull the nucleus apart. Studies of the mass of the nuclei can thus be used to gain further insight into the strong force. Maria Goeppert-Mayer (1906–1972) was a German-born American physicist who explained why nuclei with certain numbers of protons and/or neutrons, called "magic" numbers, were extremely stable. Her theory showed that the nuclear force depended on both the spin of the nucleon and its orbital angular momentum. The magic numbers are 2, 8, 20, 28, 50, 82, and 126. Thus, helium-4 has a magic number of protons and a magic number of neutrons, and it is called doubly magic. Oxygen-16, calcium-40 and calcium-48, and the heaviest stable nuclide, lead-208, are also doubly magic. All other atomic nuclei are subject to radioactive decay into smaller pieces to at least some extent.

RADIOACTIVE DECAY

How did a fogged photographic plate lead to the discovery of radioactivity?

In 1896 the French physicist Antoine-Henri Becquerel (1852–1908) was exploring the properties of a compound of uranium that glowed in the dark. He placed the compound in a dark drawer on top of a photographic plate that he had wrapped with heavy, black paper to keep it from becoming exposed to light. The next morning, he was surprised to find that the plate was already exposed or fogged. Becquerel presumed that some unknown rays had been emitted by the uranium compound, gone through the paper, and created the same chemical reaction in the plate that visible light would have produced. In further experiments, he found that a thin piece of metal would block the rays. Soon, it was found that compounds of both uranium and thorium emitted these strange rays, even though they did not emit visible light. In 1903, Becquerel was awarded the Nobel Prize in Physics, along with Pierre and Marie Curie, in recognition of their discovery and research on what would later be called radioactivity.

Antoine-Henri Becquerel earned the 1903 Nobel Prize for his discovery of radioactivity.

What scientific contributions did Marie and Pierre Curie make?

Marie Sklodowska Curie (1867–1934), born and raised in Poland but working in France, used an electrometer to measure the ionization of the air caused by radioactive minerals. Because the amount of radioactivity produced by a uranium compound depended only on the amount of uranium present, she concluded that the atom itself must be the source. She found that the uranium-containing mineral pitchblende was more radioactive than the uranium itself and concluded that the mineral must contain a small quantity of another element that was more radioactive than the uranium. Her husband, Pierre (1859–1906), a talented scientist and inventor in his own right, stopped his own work and joined Marie in searching for the element.

They started by grinding up 100 grams of pitchblende, but by the time they had found the element, they had processed tons of the mineral. In July 1898, they announced the discovery of an element they named polonium in honor of Poland, where she had been born. In December of the same year, they announced they had found an even more radioactive element that they named radium. It took until 1902 for them to separate one-tenth of a gram of radium chloride from a ton of pitchblende. In 1910 Marie announced that she had obtained pure metallic radium.

Pierre and Marie shared the 1903 physics Nobel Prize with Antoine Henri Becquerel for their work on radioactivity. Sadly, Pierre died in 1906 when he slipped on a wet street and was run over by a horse-drawn wagon. Marie later was also awarded the 1911 chemistry Nobel Prize for the Curies' discovery of polonium and radium. She is still the only person ever to have received a Nobel Prize in two different sciences.

What kinds of rays does radioactivity produce?

Among the scientists who explored radioactivity immediately after Becquerel's discovery was Ernest Rutherford (1871–1937), then at McGill University in Montreal, Canada. He and Frederick Soddy (1877–1956) found that uranium and thorium emitted two different kinds of rays. One could be stopped by paper, and the other required metal about a centimeter thick. Rutherford named them alpha and beta rays after the first two letters in the Greek alphabet. In 1907 he named the even more penetrating rays produced by radium gamma rays.

What are alpha rays?

In late 1907, Ernest Rutherford demonstrated that alpha rays were helium atoms with the two electrons removed. (He had not yet discovered the nucleus of the atom.) Today, physicists use the term "alpha particle" interchangeably with the helium-4 nucleus.

What are beta rays?

In 1900 Pierre (1859–1906) and Marie Sklodowska Curie (1867–1934), using an electroscope, discovered that beta rays are negative particles. Antoine-Henri Becquerel (1852–1908) used the same kind of apparatus J. J. Thomson (1856–1940) had used to measure the ratio of an electron's mass to charge and determined that beta rays are identical to electrons that are moving very fast—typically at about half the speed of light.

What are gamma rays?

Gamma rays are a highly energetic form of light—very-short-wavelength electromagnetic waves far beyond the range of visible light. Gamma rays are emitted from the nucleus along with an alpha or beta decay. When an alpha or beta decay produces a daughter nucleus, that nucleus is often in an excited state. One or more gammas can be emitted as the nucleus settles down to its lowest energy, or ground state. Gammas are like high-energy X-rays, but they are emitted from the nucleus, not the electrons, of an atom.

Is radioactivity dangerous?

In large doses, alpha, beta, and gamma rays all are hazardous to our health. On the other hand, if radioactivity is carefully used in controlled doses, it can be used for medical purposes.

Alpha particles are blocked by skin, but if a radioactive material, such as the gas radon, is inhaled, the alpha particles can cause damage to the lungs and cause cancer. Beta rays can penetrate skin and tissue and, if they strike a cell, can cause mutations to the DNA. Gamma rays can cause mutations and kill cells.

What is the half-life of a radioactive element?

It's impossible to know exactly when any particular nucleus will emit an alpha, beta, or gamma ray. It is possible, however, to know the average time between formation and decay. And we know that the number of decays will be proportional to the number of nuclei present.

Suppose we start with a large number of atoms of the same kind of radioactive element. In a given time interval, a certain percentage of these "parent" nuclei will undergo decay and emit a radioactive ray, creating "daughter" nuclei that are different kinds of nuclei altogether. After some period of time, the rate of decays will be only half of what it was originally, and there will only be half as many as parent nuclei as there were at the start. That time is called the half-life. One half-life after that, there will be only 1/4 as many decays as at the beginning. After another half-life, there will be half again as many, or 1/8 the number of initial decays. The number of remaining parent nuclei is cut in half again after each successive half-life.

Why is the half-life of a radioactive element important?

For health and safety purposes, the half-life of a radioactive substance is extremely important in its effects. For example, if a substance emits radiation for a few hours or days, it might be possible to use it as a radioactive tracer in the human body to find out how certain organs or systems are working or malfunctioning. But if a substance will continue to emit radiation for a much longer time, it could pose a danger to life in its vicinity as the radiation could cause diseases like cancer. Much of the ra-

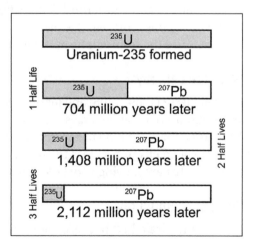

Radioactive uranium-235 decays into lead (Pb-207) very gradually. The half-life is 704 million years!

How can radioactivity be used for medical purposes?

As with many physical phenomena, if radioactivity is carefully used, it can be used for beneficial or medicinal purposes. For example, weak radioactive substances are sometimes mixed in with fluids that a person with specific illnesses might drink; over the course of several hours or days, these substances are absorbed and processed by different organs in the person's body. Medical imaging devices can then take pictures of those organs with radiation detectors to help diagnose how those organs are working. In another medicinal use of radiation, strong beams of gamma rays can be precisely aimed to kill cancerous cells in the body.

dioactive waste from nuclear power plants has a half-life of up to several thousand years; it must be stored in environmentally isolated locations to protect people from radiation poisoning and disease. Radioactive uranium-235, the fuel used in many nuclear power plants and atomic bombs, decays into lead-207 with a half-life of 704 million years!

How can you use radioactive decay to measure the ages of old things?

Many elements that undergo radioactive decay do so naturally and release radioactivity so slowly and gently that humans will not be harmed by small amounts of their radioactive by-products. Those elements can be used to measure the ages of the objects they're in, like pieces of wood or rock, by measuring the ratio of parent nuclei to daughter nuclei in the objects. This is called radiometric dating, and it is a very important scientific technique.

How is carbon-14 used in radiometric dating?

Radioactive carbon-14 (^{14}C) has a half-life of 5,730 years. It is produced in the atmosphere when cosmic ray neutrons strike nitrogen atoms, so a tiny portion of ^{14}C is mixed into Earth's environment at all times. The ^{14}C then reacts with oxygen to form carbon dioxide in the atmosphere and dissolves into the oceans. Plants take up the atmospheric CO_2 in respiration, and animals ingest carbon. Therefore, all living things exchange both ^{12}C and radioactive ^{14}C throughout their lives. When a living thing dies, the exchange stops, and the amount of ^{14}C in their bodies is no longer replaced. As time goes on, the amount of ^{14}C decreases due to radioactive decay. By measuring the ratio of ^{14}C nuclei to the number of daughter nuclei in an object, objects up to 50,000 years old—such as mummies, wooden buildings, or prehistoric clothing—can be measured in age using ^{14}C radiometric dating. Willard F. Libby (1908–1980), an American chemist, was awarded the Nobel Prize in Chemistry in 1960 for his development of carbon-14 radiometric dating in 1949.

What happens to a nucleus that undergoes alpha decay?

The number of nucleons in a radioactive decay does not change, and in an alpha decay, the number of protons of the parent nucleus must equal the number of protons in the daughter nucleus plus the number of protons in the alpha. The same is true for the number of neutrons. So, for example, when uranium-238, with 92 protons and 146 neutrons, emits an alpha, the daughter has 90 protons and 144 neutrons. It is thorium-234. The alpha is emitted with a specific energy. Generally, only nuclei with a lot of protons and neutrons decay by alpha decay.

What key natural resource is caused by alpha decay?

The decay of uranium and thorium produces all the helium gas that exists on Earth. Most of the helium is mixed with natural gas and is extracted from gas wells. It is expensive to separate and store the helium, but, given the increasing needs for this resource for everything from helium balloons to cooling superconducting magnets used in hospital MRI machines, it is an effort that must be made.

Where in your home is alpha decay going on?

A tiny amount of alpha-emitting radioactive elements, mostly americium-241, is used in home smoke detectors. The charged alpha particles leaving the element are collected on a metal plate where they produce a small electric current that keeps a circuit loop closed. When a fire is burning nearby, smoke from the fire scatters the alpha particles, reducing the current, opening the circuit, and triggering the alarm.

Believe it or not, bananas also emit alpha particles. Their radioactive decay comes from tiny amounts of radioactive potassium-40. Every banana produces about ten alpha particles per second. Don't be alarmed, though; a BED ("banana-equivalent dose") of radiation is completely harmless. You'd need the radiation of many millions of bananas all at once to receive any ill effects.

What type of force is responsible for beta decay?

The weak nuclear force is the cause of beta decay. It is 10^{-13} (one ten-trillionth) as strong as the strong force. As a result, the half-life of most beta-decaying nuclei is long.

In 1968 Abdus Salam, Sheldon Glashow, and Steven Weinberg showed that the electromagnetic force and the weak nuclear force were actually different aspects of a single force, the electroweak force. How can these two forces that are so different in strength and range be unified? It only happens at extremely high energies (corresponding to a temperature of 10^{15} K), such as what occurred in the first few moments after the Big Bang or in an accelerator. They were awarded the 1979 Nobel Prize in Physics for their work.

What happens to a nucleus that undergoes beta decay?

When a nucleus undergoes beta decay, the number of protons increases by one, and the number of neutrons decreases by one. So, for example, carbon-14 (6 protons and 8 neutrons) becomes nitrogen-14 (7 protons and 7 neutrons) with the emission of a beta particle (electron) and an antineutrino.

How did the study of beta decay lead to the discovery of a new subatomic particle?

When beta decay occurs, a nucleus emits an electron. The electron isn't in the nucleus originally; it results from the change of a neutron into a proton. Studies of the energy of the emitted electron showed that, instead of having a single energy like an alpha has, the energies of the electrons were spread from near zero to a maximum energy. Early investigators recognized that this suggested that energy was not conserved in beta decay. Austrian physicist Wolfgang Pauli (1900–1958) proposed in 1930 that a second particle was emitted along with the electron. This particle had to be neutral and have an extremely small mass. The neutron was discovered a few years later, but neutrons were too massive to be this second particle.

This particle was eventually named the neutrino, or "little neutral one." The neutrino wasn't detected experimentally until 1956 because it is so tiny and has such low energy. We know today, though, that the neutrino emitted in beta decay is actually an antineutrino and that neutrinos are an important by-product of many nuclear reactions, including the nuclear fusion processes that go on in the sun.

How are X-rays produced in atoms?

X-rays are electromagnetic waves of very short wavelength. They are emitted by atoms with many electrons, such as those high in the periodic table. The more electrons, the greater the charge of the nucleus and the higher the energy of the electrons that are close to the nucleus. Therefore, when the atom is disturbed, as it is in an X-ray tube that is bombarded by high-energy electrons, one of the electrons can be knocked out of the atom. When another electron in the atom loses energy and takes the place of the knocked-out electron, an X-ray can be emitted.

How are gamma rays produced in atomic nuclei?

Gamma rays, which are very-high-energy photons or short-wavelength electromagnetic waves, are emitted from the nucleus along with alpha decay or beta decay. When alpha decay or beta decay produces a daughter nucleus, that nucleus is often in an excited state. One or more gammas are emitted as the nucleus settles down to its lowest energy, or ground state. Gammas are like high-energy X-rays, but they are emitted from the nucleus, rather than from the electrons, of an atom.

ANTIMATTER

What is antimatter?

In 1932 Carl Anderson (1905–1991) was studying particles produced when cosmic rays struck lead sheets in a cloud chamber that was in a magnetic field. He found low-mass particles that curved the opposite direction from electrons, showing that they had positive charge. He later confirmed the existence of these particles by using a laboratory source of high-energy gamma rays. This positive electron was named the positron and was the first form of antimatter found. Anderson shared the 1936 Nobel Prize for his discovery.

Around that time, physicists began to predict the existence of an antimatter version of the proton called the antiproton. Its mass was so large compared to that of a positron, however, that laboratory technology at that time was not powerful enough to produce it. Two decades later in 1955, physicists Emilio Segrè (1905–1989) and Owen Chamberlain (1920–2006) were able to detect and discover antiprotons. They and their coworkers bombarded a copper target with very-high-energy protons, using the proton synchrotron facility at the University of California at Berkeley. For their discovery, Segrè and Chamberlain were awarded the 1959 Nobel Prize in Physics.

How can antimatter particles be produced?

When a gamma ray with sufficient energy strikes matter, it can produce an electron-positron pair. This process is called pair production. Energy is converted into particles with mass. The minimal amount of gamma-ray energy needed is given by a version of Einstein's famous mass-energy equation, $E_{gamma} = m_{electron}c^2 + m_{positron}c^2$. The uncharged gamma produces a negatively charged electron and a positively charged positron, so electric charge is conserved.

Positrons are also emitted in radioactive decay of isotopes that have a deficit of neutrons. For example, stable carbon exists as ^{12}C or ^{13}C: six protons and either six or seven neutrons. As was discussed above, ^{14}C decays by emitting an electron. One of the neutrons changes to a proton with the emission of the electron and antineutrino. On the other hand, ^{11}C, with only five neutrons, is a positron emitter. One of the protons changes to a neutron with the emission of a positron and a neutrino.

What happens when antimatter strikes matter?

When a positron strikes matter, the positron and an electron annihilate each other, producing two or three gamma rays. Similarly, when an antiproton strikes matter, it and a proton annihilate one another, producing even more gamma rays. Particles with mass are converted to energy.

How is antimatter used in medicine?

A three-dimensional image that shows biological activity in a person can be made using PET, or positron emission tomography. PET uses a short-lived positron-emitting isotope, typically ^{11}C, ^{13}N, ^{15}O, or ^{18}F. The isotope is chemically attached to a molecule that is involved in the activity that is to be studied. Fluid containing that molecule is injected into the person, and after enough time passes for the molecule to reach its target, the person is put into the PET machine. The positron that is emitted by the decaying nucleus strikes an electron and decays into two gammas that are simultaneously emitted back-to-back, that is, 180 degrees apart. Detectors record the arrival of the two gammas, and computers extrapolate them back to the location of the gamma emission. When a sufficient number of events are recorded, a three-dimensional image of the region where the biologically active molecule accumulated can be made. PET scans can be combined with other kinds of medical imaging scans to pair information on the anatomy with biological activity for diagnostic purposes.

What are particle accelerators, and how do they work?

The two highest-energy particle accelerators are the Tevatron at Fermilab near Chicago, Illinois, and the Large Hadron Collider (LHC) at CERN (the European Organization for Nuclear Research) near Geneva, Switzerland. Both machines accelerate and store protons moving in circular paths in a metallic tube, called the beam line, from which all but minute amounts of air have been evacuated. Beams of particles circulate through the tube in both a clockwise and counterclockwise direction. Long series of superconducting magnets bend the paths of the particles into the circle while other magnets focus the particles into small beams. Electric fields created in high-power vacuum tubes provide the accelerating forces on the particles.

The beams intersect at several places around the circle so that the particles can collide head-on. When collisions occur, the very high energies that result can produce subatomic particles and their by-products, which ordinarily would not be observable in the world. This is how known particles are examined and new particles are discovered. Tiny amounts of antimatter are regularly created and destroyed in particle accelerators.

Large Hadron Collider at CERN in Geneva, Switzerland.

NUCLEAR FISSION

What is the "island of stability"?

All the isotopes of elements with atomic numbers greater than that of lead are radioactive. Some have lifetimes of tens of millions of years, while others are fractions of a second. Nuclear physicists and chemists know that isotopes that have magic numbers of neutrons or protons are more stable than others. They have proposed that very heavy isotopes with a neutron number around 180 and a proton number around 110 should be more stable than those with fewer or greater neutrons and protons. A glance at the table of elements beyond uranium shows that elements around 110 have longer lifetimes than others. But researchers have not yet been able to create isotopes with enough neutrons to reach this island.

What is nuclear fission?

Radioactive nuclei—in particular, some with atomic numbers greater than that of lead—can occasionally decay in a very dramatic way: they can split into two smaller nuclei that are both larger than alpha particles. This process is called spontaneous fission. Fission can also be produced artificially by bombarding nuclei with low-energy neutrons. In fact, fission was discovered in this manner.

What did Enrico Fermi contribute to the early study of nuclear fission?

Italian physicist Enrico Fermi (1901–1954) was appointed professor at the University of Rome at the age of 24. Among many projects in which he and his group were involved, perhaps none was more important than his studies of the reactions produced when slow neutrons struck nuclei. Fermi had discovered in 1934 that slowing neutrons by passing them through paraffin greatly increased this ability to produce nuclear reactions. He did systematic studies of the results of bombarding a series of materials with slow neutrons.

How did Lise Meitner and Otto Hahn contribute to the discovery of nuclear fission?

The German chemist Otto Hahn (1879–1968) had a distinguished career that included inventing the field of radiochemistry in 1905. Using chemical techniques and measurements of half-lives to study the results of nuclear reactions, he discovered dozens of isotopes and at least one element. Three times he was nominated for the Nobel Prize. In 1907 he started a scientific collaboration with the Austrian physicist Lise Meitner (1878–1968). The teamwork between a physicist and a chemist proved to be extremely productive.

Winner of the 1938 Nobel Prize in Physics, Enrico Fermi is sometimes called the father of the atomic bomb, having been behind the construction of the world's first nuclear reactor, the Chicago Pile-1.

Hahn and Meitner, together with Hahn's young assistant Fritz Strassman (1902–1980), employed Fermi's slow neutron techniques to create nuclear reactions and, thus, more isotopes. When, in 1938, they tried bombarding uranium with neutrons, they expected to create new elements beyond uranium in the periodic table. But they kept finding the element barium in the bombarded uranium.

Starting in 1933, the Nazi regime forced people of Jewish origin out of all laboratories and universities. Meitner, who had Jewish parents who had converted to Protestantism in 1908, was protected because she was Austrian. But when Austria was incorporated into Germany, she lost that protection. In July 1938, she took the train from Berlin to the Netherlands. Thanks to the intervention of two Dutch physicists, she was allowed to leave Germany but with no possessions. She soon moved to Sweden and kept up her collaboration with Hahn by mail.

On December 17, 1938, Hahn and Strassman submitted their findings for publication but admitted that they had no explanation for the appearance of barium. Meitner and her nephew Otto Frisch (1904–1979) utilized Niels Bohr's "liquid drop" model of the nucleus and Einstein's $E = mc^2$ equation to propose that the nucleus had split into two, releasing both extra neutrons and a large amount of energy. The Meitner-Frisch paper was submitted a few days after Hahn's. Frisch returned to his laboratory in England and confirmed Hahn's result in January 1939. Hahn won the Nobel Prize in Chemistry for his work in 1944, but Meitner did not. The element meitnerium (with atomic number 109) is named in Lise Meitner's honor.

293

How did scientists influence the use of nuclear fission for military purposes?

Lise Meitner had recognized that extra neutrons could produce a chain reaction that would produce a very large amount of energy. In early 1939, physicists from many countries attempted to create such chain reactions by slowing down the released neutrons. Among these were Enrico Fermi and a Hungarian-born physicist named Leo Szilard (1898–1964). They saw signs that such a reaction had occurred in experiments conducted in Germany.

In August 1939, Szilard was worried that Nazi Germany would use fission technology to create a bomb, and he helped convince Albert Einstein to write a letter to U.S. president Franklin D. Roosevelt (1882–1945) that explained this possibility. Roosevelt subsequently directed the government to support nuclear fission research and created the Uranium Committee.

While there were several important studies during the next three years, it was British scientists who made the breakthrough finding that the rare isotope uranium-235 could be used in a weapon. The Americans were informed but ignored the results until a personal visit by one of the British team members convinced the Uranium Committee of the need for action. The United States then established a new office that could authorize large-scale engineering projects for military atomic research.

What was the Manhattan Project?

In 1942, the Manhattan Project was started to produce atomic weapons. It was named after the Manhattan Engineering District in New York City from where it was run. The

The Hanford B Reactor in Washington state, which was one of several important facilities that were part of the Manhattan Project, is now open to public tours.

chief military leader of the Manhattan Project was U.S. Army General Leslie Groves, who appointed physicist J. Robert Oppenheimer (1904–1967) as scientific director. Its first major success was at the University of Chicago where, in December 1942, Enrico Fermi's uranium reactor created the first self-sustained nuclear chain reaction.

How was uranium turned into fuel for atomic bombs?

British scientists discovered that naturally occurring uranium, a mixture of uranium-238 and only 0.7 percent uranium-235, would have to be highly enriched in uranium-235 in order for nuclear fusion to occur. This created a need for uranium enrichment plants. One method chosen had been developed in California. Uranium metal would be evaporated in a vacuum. The atoms went through a narrow slit and then into a region with a strong magnetic field. Because of their mass difference, the two isotopes followed slightly different paths. The atoms condensed on the surfaces of separate containers.

Dozens of these giant machines, called calutrons, were built in a plant in Oak Ridge, Tennessee. That location was chosen because abundant electricity was available from the nearby hydroelectric plants. Not enough copper was available to wind the coils for the magnets, so 70 million pounds of silver bullion was borrowed from the U.S. Treasury to be formed into wires for the calutrons.

Somewhat enriched uranium from the calutrons was then combined with fluorine to produce the gas UF_6. Because of the mass difference of the two isotopes, $^{235}UF_6$ would diffuse through porous membranes slightly faster (about 0.5 percent) than its more massive $^{235}UF_6$ counterpart. Thousands of separations were needed to produce weapons-grade uranium (85 to 90 percent ^{235}U). The plant at Oak Ridge that was built to accomplish this gaseous diffusion had an area of 2 million square feet, employed 12,000 workers, and cost the equivalent of more than $10 billion in today's dollars. At one time, it consumed 17 percent of all the electricity produced in the United States—more than all of New York City at that time.

What human-produced element became used as nuclear fuel for bombs?

Plutonium is not found in nature, but it is produced in reactors by bombarding uranium-238 with neutrons. Plutonium-239 (^{239}Pu) can be made to undergo fission by slow-moving neutrons, so it could be used in weapons. In December 1942, Hanford, Washington, was chosen as a site for reactors that would produce plutonium. Hanford was selected because it was isolated but also on the Columbia River, which afforded a source of cooling water.

Where were the world's first atomic bombs developed?

In September 1942, General Leslie Groves and Robert Oppenheimer chose Los Alamos, New Mexico, as the site for the top-secret laboratory at which atomic weapons would be developed. Thirty-five miles northwest of Santa Fe, it was almost totally isolated. During World War II, hastily erected housing held Nobel Prize-winning scientists, younger scientists and engineers recruited into the project, wives and children, and soldiers. A scientific research facility is still active today there: Los Alamos National Laboratories.

What nuclear fission work happened at Los Alamos?

After determining the critical mass (the minimal amount of enriched uranium needed to create a bomb), the scientists and engineers at Los Alamos designed and built the uranium-based weapon called "Little Boy." The uranium was divided into two halves and placed in a cannon-like container. An explosive charge drove the two masses together, forming a large enough mass of uranium to sustain a rapid chain reaction and explode. This weapon was never tested but immediately used in war. It contained 64 kilograms (141 pounds) of uranium, about 2.5 times the critical mass.

The second task of Los Alamos was to design and build a weapon using plutonium. Originally, they had expected to use the cannon-type method used with the uranium bomb, but it turned out that the plutonium produced in the reactors at Hanford contained too many impurities. Another design had to be developed. They arranged a subcritical plutonium mass in the shape of a sphere and used specially designed explosive charges to compress the plutonium very quickly, increasing its density above the criti-

A photo of the mushroom cloud over Nagasaki does not adequately capture the horrors of those who died from the August 9, 1945, explosion.

cal point. Scientists were uncertain that the design would work, so they decided to test the device first.

What was the first atomic bomb test explosion like?

"The Gadget" was a test version of the plutonium bomb. It was installed on the top of a 30-meter (100-foot) tower in the New Mexico desert at a location 35 miles southeast of Socorro, New Mexico, on the White Sands Proving Ground near Alamogordo. The explosion, called "Trinity," occurred on July 16, 1945. The energy yield was about 20,000 tons of TNT, more than twice what had been expected. The implosion-type bomb proved to be safer and more effective than the cannon-style "Little Boy" design. It became the standard method used to build other nuclear bombs.

What happened when atomic bombs were detonated over cities?

"Little Boy" was detonated over Hiroshima, Japan, on August 6, 1945. Less than 1 kilogram (2 pounds) of the uranium fissioned, and only 0.6 grams (0.001 pounds) of matter was converted into energy. Nevertheless, the result was the release of energy equivalent of more than 12,000 tons of TNT. More than 100,000 people were killed in the blast, in the resulting fires, and from the effects of radiation.

A plutonium bomb, named "Fat Man" for its shape, was detonated over the city of Nagasaki, Japan, on August 9, 1945. It contained 6.4 kilograms (14 pounds) of plutonium-239. About 20 percent of it fissioned, and, again, less than 1 gram was converted into energy; the explosion was even larger than the "Little Boy" explosion, producing the equivalent of more than 20,000 tons of TNT. More than 80,000 people were killed.

How many nations today have developed atomic bombs?

The Soviet Union exploded a nuclear device in 1949. China, Britain, and France developed nuclear weapons in the 1950s. India tested its first atomic bomb in 1974. Pakistan tested its first atomic bomb in 1998. North Korea first successfully detonated an atomic bomb in either 2006 or 2009. South Africa probably had an atomic bomb, but the country abandoned that weapons program. Israel is thought to have atomic bombs, but its government has never admitted to having them. Several other nations have the technology and resources to construct an atomic bomb within weeks or months; beyond that, several other nations have tried to develop nuclear weapons secretly.

How has the spread of nuclear weapons in the world been controlled?

By 1953 there had been 50 aboveground tests of nuclear weapons that created radioactive fallout, contaminating milk and animals. These effects alerted the public to the danger of such testing. The Cold War, however, created an atmosphere in which treaties could not be negotiated. In 1963 a partial test ban treaty was signed, prohibiting tests in the atmosphere, underwater, and in space. In 1968 the nuclear nonproliferation treaty was signed. Non-nuclear-weapon states were prohibited from building or acquiring nuclear weapons. Many nations have signed the treaty, although some major states did not on the basis that the treaty makes no effort to curb development by states that already have such weapons. In 1996 the Comprehensive Nuclear-Test-Ban treaty was adopted by more than two-thirds of the members of the United Nations general assembly. The United States signed the treaty but rejected its ratification in 1999. Nevertheless, 337 facilities around the world monitor compliance with the treaty. They send data to a center in Vienna, Austria, for analysis and distribution to the states that have signed the treaty.

How is nuclear fission used peacefully?

Nuclear reactors produce electric power. The energy from uranium fission heats water that circulates through the reactor. The heated water produces steam that turns turbines connected to generators. Many safeguards are built into every nuclear reactor to prevent a catastrophic explosion or release of any dangerous radioactive by-products of the nuclear reaction. They do not always work, regrettably.

As is the case with all electric power plants, only about one-third of the energy produced by the reactor is converted into electrical energy. The remaining energy heats local rivers, lakes, or the atmosphere. In the United States today, there are about 100 reactors that provide about 20 percent of the electricity used by the country.

One major disadvantage of nuclear power is the cost of the plant and the extremely long time scale associated with obtaining approval and constructing the facility. Costs are difficult to calculate precisely, but nuclear power is currently one of most expensive methods of generating electricity; wind, solar, oil, and natural gas are all cheaper. As a result of these uncertainties, factors other than costs are increasingly important.

Another major disadvantage of nuclear power is the production of nuclear waste, which poses long-term dangers to people due to its intense radioactivity. The problem of long-term storage of nuclear wastes still needs to be solved. While recycling nuclear fuel is an attractive option, the plutonium in used fuel rods raises issues of nuclear weapon proliferation. To date, no safe and effective long-term storage plans have been fully approved. Underground storage in salt deposits is one possible storage method for the future.

What are the pros and cons of nuclear power?

Electric power in the United States today is produced primarily by power plants that use hydrocarbon fuels like coal, oil, and natural gas. These fossil fuels are limited in quantity and produce a great deal of the carbon dioxide that contributes to global warming and climate change. Nuclear power can reduce our reliance on such fuels, and nuclear power plants do not produce greenhouse-gas emissions like carbon dioxide. However, uranium is also a limited resource, and mining it and storing nuclear waste can be dangerous.

Are nuclear power plants safe?

Although almost every nuclear power plant to date has had no safety problems, the results of a major accident at nuclear power plant can be catastrophic. There was a close call in the United States in 1979, when a water-cooling failure at the Three Mile Island nuclear power plant near Harrisburg, Pennsylvania, caused a partial meltdown of the reactor core. Had the core fully melted down, the more than 600,000 people that lived within a 20-mile radius of the plant would have been in serious danger and would have had to flee their homes.

In 1986, the graphite-moderated nuclear reactor at Chernobyl, Ukraine (at that time part of the Soviet Union), had a huge power spike while tests were being run. The spike, caused by a combination of poor reactor design and terrible human decision-making, led to massive explosions and fire. Radioactive material totaling 400 times more than the amount released by the atomic bomb detonated over Hiroshima was spread over a huge area of the Soviet Union and Europe. Thousands of people are estimated to have died as a result of the accident, and the area near the now-defunct power plant is still uninhabited.

In March 2011, the largest-ever earthquake recorded off the coast of Japan and the tsunami it caused destroyed the Fukushima Daiichi nuclear power plant. A series of failures and poor decisions led to its explosion, which scattered radioactive materials across a wide area. As in Chernobyl, large parts of the region near the destroyed plant remain uninhabitable.

Will more nuclear power plants be built?

A few years ago, concerns over the buildup of greenhouse gases and the environmental problems caused by coal mining led to renewed interest in building new nuclear power plants. Advocates of construction point out that if a standardized plant could be designed, then licensing delays could be reduced and design costs minimized. In addition, several new types of plants have been suggested and have undergone small-scale testing; they may be simpler and safer than traditional designs.

After the Fukushima Daiichi nuclear power plant disaster in 2011, however, many governments seriously reconsidered their plans to build more nuclear plants because of the tremendous safety risks they presented. The government of Germany even decided not only to stop constructing new nuclear plants but to decommission all of its existing nuclear plants. Today, the cost of generating electricity with solar and wind technologies has become increasingly cheaper; if this trend continues, within a few years, it could be much more economical and much less risky for the United States to do the same thing as Germany did and move away from nuclear power generation altogether.

NUCLEAR FUSION

What is nuclear fusion?

Nuclear fusion is the opposite of fission. Two nuclei join, or fuse together, forming a more massive nucleus. This process is called nucleosynthesis, and if the mass of the resultant nucleus is less than that of the reacting nuclei, energy is released. The reacting nuclei are both positively charged, so there is a large repulsive force between them. To overcome this force, the reacting nuclei must have very high energy. The most common fusion reaction is the formation of helium by fusing hydrogen nuclei. Nuclear fusion was first observed in the laboratory in 1932 by Australian physicist Mark Oliphant (1901–2000).

One example of a nuclear fusion reaction involves two isotopes of hydrogen: ^2H, or deuterium, and ^3H, or tritium. They fuse to produce ^4He, releasing a neutron. The energy released is more than a million times greater than that released when an electron combines with a proton to produce a hydrogen atom.

What is the closest operational nuclear fusion reactor?

Humans have not yet successfully constructed a controlled fusion reactor. Throughout the universe, however, such controlled fusion reactors are abundant—in the hearts of stars. Our sun, the star in our solar system, is 150 million kilometers (93 million miles) away from Earth, and its nuclear fusion provides the heat and power that warms our planet and sustains life here.

How did we learn about nuclear fusion in the sun?

In stars like the sun, the principal reaction is called the proton-proton chain. This reaction was first described in 1939 by German-American physicist Hans Bethe (1906–2005). In the first step, two protons fuse into a deuterium (^2H) nucleus. The deuterium has a proton and a neutron, so the second proton changes into a neutron, releasing a positron and a neutrino. The positron annihilates with an electron, producing two gamma rays. In the second stage, the deuterium fuses with another proton to produce helium-3 (^3He) plus a gamma ray. In the sun, the third stage is primarily a fusion between two He^3 nuclei producing ^4He (an alpha particle) plus two protons. So, the net reaction is an input of four protons and an output of one ^4He plus six gammas and two neutrinos, as well as a lot of energy due to the loss of 0.7 percent of the mass of the protons. To accomplish these reactions, the protons must be moving with a large amount of kinetic energy, the equivalent of a temperature of 15,000,000 K.

How does a nuclear fusion bomb work?

Even before the nuclear fission bomb was completed, some physicists at Los Alamos

A nuclear power plant near Antwerp, Belgium, is one of 450 currently in operation around the world.

started work on what they called the "super," a weapon based on fusion. After World War II ended, there were heated discussions, both scientific and political, about the wisdom of developing a new, even more destructive weapon. Due to tensions in the Cold War, both the United States and the Soviet Union embarked on programs to create these weapons, informally called hydrogen bombs or H-bombs because they fused hydrogen. The United States tested such a weapon in 1952; the Soviet Union did so in 1955. Britain, China, and France are also known to have tested fusion bombs since then.

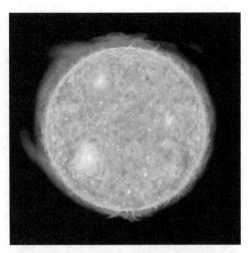

Nuclear fusion occurs naturally in stars like our sun, but reproducing a fusion reaction in the laboratory has proven to be very difficult because it takes a lot of energy to fuse nuclei together.

Some aspects of fusion weapons are known to the public, while others are still kept secret. What is known is that the bomb starts with a uranium or plutonium implosion bomb with deuterium and tritium gases inside its core. The neutrons released by the fusion increase the fission of plutonium or uranium, boosting the efficiency of the fission device. The energy released is transferred to a lithium-dihydride fusion fuel in a process that remains classified. That undergoes fusion, and the energy released causes fission in a surrounding layer of natural uranium.

The most powerful H-bomb ever built was detonated by the Soviet Union on October 30, 1961. It released an amount of energy equivalent to 50 million tons of TNT. More recently, significant effort on weapons development has been made to decrease the size of the weapons' payloads and increase their accuracy so that they can be precisely deployed using smaller delivery systems.

Can nuclear fusion be used for peaceful purposes?

The energy released in a fusion reaction could be captured and converted into electrical energy. If a controlled fusion reactor were successful, it could use hydrogen from ocean water as fuel, effectively ending any fuel shortages for generating energy for human use.

The difficulty in creating a fusion reactor is confining the reactants, typically deuterium and tritium nuclei, at temperatures needed for fusion. There are two approaches: magnetic and inertial confinement. Most effort has gone into magnetic confinement, where the positively charged nuclei are trapped in evacuated regions containing strong magnetic fields that keep the nuclei from colliding with the metallic walls of the equipment. The swarms of charged nuclei, a plasma, must then be raised

301

A model of a fusion reactor on exhibit at the 2013 International Fusion Energy Days in Monaco. Should fusion reactors become feasible, we could use water as fuel and replace nuclear and fossil-fuel burning power plants altogether.

to very high temperatures and held at that temperature long enough for fusion to take place. So, there are three variables: number of nuclei, their energy (or temperature), and the confinement time. The product of these three variables determines whether or not fusion will occur.

What is the ITER project?

The ITER reactor project is an international endeavor to pursue nuclear fusion as a source of energy by building a tokamak, or magnetic fusion device. More than 30 nations are participating, including the United States, Russia, the European Union, India,

Japan, China, and South Korea. The plan is to spend 10 years in construction and 20 years in operation. No electrical energy would be generated—that would require a more advanced reactor. Still, using 0.5 grams of deuterium-tritium fuel, the reactor is designed to produce 500 megawatts of energy over 1,000 seconds. For the first time in a fusion reactor, more energy would be produced than needed to run the reactor. In May 2010, contracts were signed to construct the first buildings in the project, which are sited in the south of France. Fusion reactions are scheduled to begin in the near future.

How might lasers help produce nuclear fusion?

One method of producing fusion is called inertial confinement. According to theoretical models, tiny glass, plastic, or metallic spheres containing about 10 milligrams of deuterium-tritium fuel mixture at high pressure are bombarded on all sides by extremely powerful lasers. The lasers vaporize the sphere, driving a strong, inward shock wave that compresses the fuel enough to create fusion.

The National Ignition Facility is a nuclear fusion laboratory located at Lawrence Livermore Laboratory in California. It was completed in 2009 and houses the most powerful laser in the world. Its design goal is to create a huge flash (2 million joules) of ultraviolet energy that hits a tiny target from 192 separate beamlets. All of the beamlets must strike the target within a few picoseconds. In January 2010, the laser beams delivered 700,000 joules to a test sphere, heating the gas inside to 3.3 million K. It takes hours for the lasers to cool enough to deliver another pulse. The eventual goal was to deliver one pulse every five hours.

When might controlled fusion reactors start generating electricity for human use?

Nobody knows for sure. There are many technical challenges to solve to make power plants that operate in this way, even if facilities like ITER and the National Ignition Facility are successful in producing controlled fusion. For example, before deuterium-tritium fusion fuel can be used at a laser-based power plant, neutron shields have to be built to protect the lasers. There are also no accepted engineering plans to convert the energy released to electrical energy. One proposed method is to surround the bead with a blanket of liquefied lithium that would be further heated by the energy released, but this has yet to be tested.

Although the challenges are daunting, controlled nuclear fusion has so much potential to benefit humanity that continued scientific research on the topic is very worthwhile. Like many questions at the edge of the scientific frontier, the answers always seem to lie just beyond our grasp. For scientists working on controlled fusion, it has been a running joke that no matter when this question has been asked since the 1960s, the answer has always been, "about 40 years from now."

QUANTUM PHYSICS

What is quantum physics?

Beginning around 1900, physicists began to understand that the rules that govern motion, matter, and energy in the world around us do not work the same way at the size scales of molecules, atoms, and subatomic particles. Rather than being continuous properties, concepts like the amounts of energy and momentum that such a tiny system could contain are quantized—that is, limited to certain quantities for each system. Quantum physics is the broad field within physics dedicated to the study of the laws of nature at the microscopic scale and below, including how those laws can affect us at both the tiniest and largest scales of the universe.

Are electrons waves or particles?

The electrons in an atom are not confined to one region of space; they are spread out. They act more like waves than particles. Louis de Broglie (1892–1987) proposed in his 1924 doctoral thesis that electrons behave like waves with a wavelength given by $A = h/mv$, where h is Planck's constant, and m and v the mass and velocity of the electron. The thesis was forwarded to Albert Einstein, who enthusiastically endorsed the idea and recommended that the thesis be approved. De Broglie was awarded the Nobel Prize in 1929 for this work. The de Broglie wavelength is associated with any particle, although for an object the size of a baseball, it is much smaller than the diameter of an atomic nucleus.

What is the importance of the wavelength computed by Louis de Broglie?

The de Broglie wavelength of a particle determines its wavelike properties. Just as light photons interfere with themselves in a two-slit experiment, so do particles. The interference of electrons, atoms, and even molecules as large as C_{60} (buckminsterfullerene) has been observed, and the measurements fit de Broglie's wavelength perfectly. In other words, just about any particle can be considered to be both a particle and a wave at the same time.

Is light a wave or a particle?

Light (and other forms of electromagnetic radiation) has the properties of both a particle and a wave. As a wave, it is described by its wavelength, frequency, amplitude, and polarization. It has the ability to diffract and interfere. As a particle, it has energy, momentum, and angular momentum. It has the ability to be emitted and absorbed and to scatter off other particles, transferring energy and momentum. In some experiments, it acts like a wave in part of the experiment and a particle in other parts.

For example, if you put a beam of light through a pair of narrow, closely spaced slits—the so-called Young two-slit experiment—you get an interference pattern with alternating stripes of light, showing light where the interference is constructive and darkness where the interference is destructive. If you now greatly reduce the intensity of the light and use a detector that can detect individual photons, you will have regions where a large number of photons arrive and others where none arrive. The regions are exactly where the dark and light stripes were.

If the intensity is so low that there is only one photon in the apparatus at a time, how can the dark and light regions be understood?

Particles can't split, with half going through one slit and half through the other so the two halves interfere. If you try to modify the experiment so you can tell through which slit the photon came, you destroy the interference pattern.

Physicists have grappled with this mystery of the wave-particle duality of light for more than a century. It is certainly not easy to understand or describe exactly how light can seem to behave so strangely.

Does light interact with matter as a wave or a particle?

A wave carries energy continuously over time; the more energy there is in the wave, the faster the energy is transferred. A particle, on the other hand, delivers its energy all at once. When an atom either absorbs or emits light, the transfer is almost instantaneous. Therefore, light interacts with an atom like a particle.

The idea that light comes in packets of energy was first stated by Albert Einstein (1879–1955) in 1905. He called the packet a light quantum. The quantum was given the name "photon" in 1926. The photon has no mass or charge, but it does carry angu-

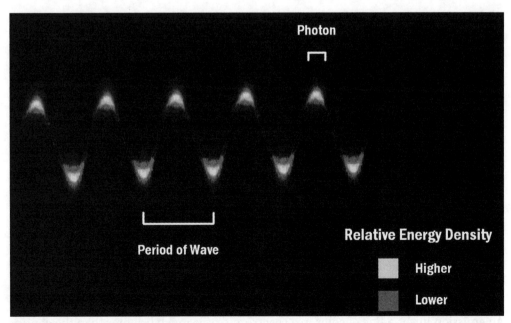

This illustration puts forth a good effort to convey how light waves can also behave like particles, with energy packets (photons) moving along a wave pattern.

How does the photon relate to the way human eyes work?

The biochemist George Wald (1906–1997) discovered in 1958 that vision is best explained in terms of photons. The molecule $C_{20}H_{28}O$ is called retinal (or retinaldehyde) and is a component of both rod cells and cone cells in the human retina. Retinal can exist in two forms, one straight and one bent. When a photon strikes the molecule, it changes its shape from bent to straight. The shape change creates an impulse in the nerves of the retina. The energy of the photon—and thus the wavelength of the light—that causes the transition depends on the other molecules, called opsins, in which the retinal molecule is embedded.

lar momentum. It always moves at the speed of light, 299,792,458 meters per second, represented by the lowercase letter c.

Each photon carries an amount of energy $E = hf$, where f is its frequency. Therefore, a photon of blue light has more energy than one of red light. The energy carried by a beam of light depends on both the frequency and the number of photons per second in a vacuum, leaving the source.

ATOMIC SPECTROSCOPY

In what ways do atoms emit and absorb light?

The German physicist Gustav Kirchhoff (1824–1887), who also made important discoveries about electric circuits, published three laws that described how materials emit and absorb light, even before the atomic theory of matter was confirmed:

- A hot solid or a hot, high-density (opaque) gas produces a continuous spectrum.
- A hot, low-density (transparent) gas produces an emission-line spectrum.
- A continuous spectrum source viewed through a cool, low-density (transparent) gas produces an absorption-line spectrum.

What is a continuous spectrum?

Isaac Newton (1643–1727) showed in his famous work *Opticks* that when white light is passed through a prism, it is split into a spectrum of all colors from violet through red. When there are no gaps between the colors, even if the brightness of the colors varies, the spectrum is called continuous.

What is an emission-line spectrum?

An emission line is a line of color in a spectrum that is much brighter than the continuous light on either side. Emission lines typically occur at only one wavelength. A single substance can produce many emission lines. In the visible light portion of the electromagnetic spectrum, for example, the element sodium produces two yellowish emission lines. Hydrogen produces four emission lines of visible light. The emission spectrum of iron can have hundreds or even thousands of lines.

What is an absorption-line spectrum?

An absorption-line spectrum occurs when a low-density gas absorbs distinct colors, leaving dark gaps in the otherwise continuous spectrum. For example, the spectrum that the German physicist Joseph von Fraunhofer (1787–1826) took of the sun in 1814 showed 574 dark lines. In 1859 they were explained as being absorption lines from the cooler gases in the sun's atmosphere.

Fraunhofer made the best optical glass of any glassmaker of his era. He made great improvements to the achromatic lens, which refracts light of all colors the same amount. Regrettably, like most glassmakers of his era, he died young, very possibly from the poisonous effects of the materials used to make the glass.

What happens when atoms emit and absorb light?

Figure 24 roughly illustrates the emission of light when the electron goes from a higher-energy to a lower-energy orbit and the absorption of light when the electron's energy is increased.

How do hydrogen atoms emit and absorb light?

In 1885 Johann Balmer (1825–1898) found a formula that accurately calculated the wavelengths of the visible light emission lines of the hydrogen spectrum. It was purely empirical; that is, there was no physics-based explanation of it. In 1888 Johannes (Janne) Rydberg (1854–1919) generalized Balmer's results to allow calculation of hydrogen emission in the ultraviolet and infrared as well. Rydberg's formula could be expressed

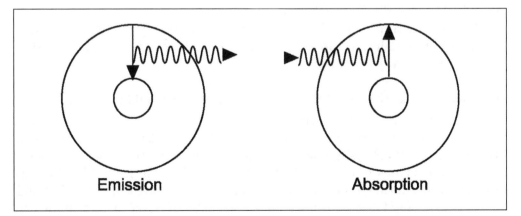

Emission Absorption

Figure 24.

as a sequence produced when light of different energy was emitted from hydrogen gas with frequencies or wavelengths that corresponded with the colors of the lines. The cause of the sequence, however, was unknown at the time.

QUANTUM MECHANICS

How does quantum mechanics explain emission lines and absorption lines?

In 1911 a young Danish physicist named Niels Bohr (1885–1962), who had recently earned his Ph.D. at the University of Copenhagen, joined Ernest Rutherford (1871–1937) at Cambridge University. He quickly began work on the Rutherford model. He published his results in 1913, basing them on three postulates that became known as key tenets of a new branch of physics called quantum mechanics.

1. Electrons only move in certain allowed orbits at discrete radii and with specific energies. That is, their radii and energies are "quantized." When in these orbits, their radii and energies are constant. The atoms do not emit or absorb radiation.

2. Electrons gain or lose energy when they jump from one allowed orbit to another. Then they emit or absorb light with a frequency given by the equation $hf = E_2 - E_1$, where E_2 and E_1 are the energies of the electrons in the allowed orbits. The constant h is called Planck's constant, 6.6×10^{-34} J/Hz (joules per hertz).

3. When the electron is very far away from the nucleus, classical physics must give the same answer as the new quantum physics. This property was called the correspondence principle.

Bohr in later years changed the third postulate from the correspondence principle to the requirement that the angular momentum of the electron be "quantized," or proportional to an integer called the quantum number. The results didn't change, but the derivation of them was more straightforward. This method is presented today in almost every textbook on quantum mechanics.

Niels Bohr's idea of quantized energies explained very precisely what Balmer and Rydberg had experimentally measured. The Bohr atom was thus a major advance in establishing the structure of atoms.

What were some of Bohr's accomplishments as a scientist and as a citizen?

Niels Bohr returned to the University of Copenhagen as a professor of physics in 1921. With the help of the government and the Carlsberg Beer Foundation, he established the Institute of Theoretical Physics. Bohr's institute attracted all the major theoretical physicists from around the world for short visits or extended appointments. When the Germans occupied Denmark, Bohr made a daring escape, first to Sweden, then to England, then to the United States. He worked on the Manhattan Project to develop the first atomic bomb, believing it was necessary to defeat Nazi Germany in World War II. After the war, however, he tried to get U.S. president Harry Truman (1884–1972) and British prime minister Winston Churchill (1874–1965) to agree to share the secrets of the bomb with all the world's countries, including the Soviet Union, to promote world peace. They both rejected Bohr's proposal. But after the Soviets developed their own atomic bomb in 1949, Bohr's ideas helped found the United Nations' International Atomic Energy Agency. Until his death in 1962, he worked to reduce the threat of a nuclear war. In the centenary of his birth, Denmark issued a postal stamp showing Bohr and his wife, Margarethe.

What were the limits of the model of the Bohr atom?

Although the Bohr atom was a major advance, it was still simplistic in many important ways. Today's model of the atom has advanced substantially from Bohr's

Winner of the 1922 Nobel Prize in Physics, Danish physicist Niels Bohr was the first to propose the idea of quantized energies.

1913 model. Bohr's model, for example, could explain only the spectra of hydrogen and helium from which an electron was removed. With some modifications, it could also explain the spectra of the alkalis like lithium.

From 1913 through 1926, physicists tried to extend the model, with some successes, but the lack of a physics-based explanation of the postulates led to research aimed at a model that was not based on classical ideas like Bohr's. One of the first steps was taken by the young German physicist Werner Heisenberg.

What is the Heisenberg Uncertainty Principle?

Werner Heisenberg (1901–1976), together with Max Born (1882–1970) and Pascal Jordan (1902–1980), tried a totally different approach using mathematical matrices to explain how atoms behave. As part of their work, Heisenberg developed a principle that demonstrates that in the atomic world, our knowledge is limited. In words, the uncertainty of a particle's position times the uncertainty in its momentum is never less than a factor of about one-twelfth of Planck's constant. If it has a precise location, then its momentum, and thus its speed (measured at the same time), must be imprecise. Planck's constant is extremely small, so the uncertainty principle is important only for objects the size of atoms or smaller. The position and momentum of a baseball, for example, can both be known at the same time precisely enough that an outfielder can easily catch a soft fly ball.

What does the Heisenberg Uncertainty Principle imply about atomic structure?

The uncertainty principle shows why Bohr's original electron orbits cannot exist. If you know the radius of the circle precisely, then it must have some velocity along the radius, which smears out its orbit. The uncertainty principle also exists in a form linking energy and time. In this form, it says that if an electron is in a state that lasts for only a short time, then its energy is not precisely defined. Electron orbits must thus be quantized as Bohr explained but yet also somehow imprecise at the same time.

How did probability become part of the model of atomic structure?

In 1926 the Austrian physicist Erwin Schrödinger (1887–1961) published an equation for a wave function that describes the probability of finding an electron at a particular position. It agrees with Bohr's model in that the most probable radius for an electron is the one given by the Bohr model, and the energy of the electron is the same as Bohr calculated, but its results are fundamentally different.

The solution of Schrödinger's equation can be shown as a probability cloud that shows the most probable locations for the electron. The electron's position can never be precisely predicted, but it can be predicted where the electron is likely to be.

The $n = 1$ state of the hydrogen atom is small and spherical. There are two $n = 2$ states. The s state has angular momentum = 0 and another spherically symmetric cloud. The p state, with angular momentum = 1, has two most probable locations, the top and bottom. The $n = 3$ state has three possible values for the angular momentum, called 0, 1, and 2 respectively. The d state, with angular momentum = 2, has four angular regions with high probability.

What is the nature of modern quantum mechanics?

The early work in quantum physics by physicists like Max Planck, Albert Einstein, Niels Bohr, Werner Heisenberg, Erwin Schrödinger, and many others helped establish the modern field of quantum mechanics—the study of atoms by themselves and their behavior in molecules, liquids, and solids. The atoms obey the Schrödinger equation, but for any atom more complicated than hydrogen, the equation gets extremely complicated and can be solved only with computers. Using computers to solve equations, run complicated models, and simulate experiments is an important part of the study of physics today; computational physics has equal status with theoretical and experimental physics.

What is the importance of quantum mechanics in our everyday lives?

Some of the recent accomplishments of quantum mechanics are in the areas of condensed matter physics, or the study of solids, and atomic physics. All of the integrated circuits used in smartphones, computers, televisions, cameras, and automobiles are designed using quantum mechanics. They are all based on diodes and transistors whose properties are explained by quantum mechanics. Quantum mechanics also guides materials scientists to select appropriate materials for their construction. One of the most exciting new materials is graphene, a film of carbon where the atoms are in hexagon-shaped arrays, but the film is only one atom thick.

What is a Bose-Einstein condensate?

In the 1920s, Albert Einstein and the Indian physicist Satyendra Nath Bose (1894–1974) predicted that if an atomic gas were cooled enough, the atoms could form a new state of matter that would exhibit quantum effects on a macroscopic scale. The first experimental confirmation of this prediction occurred in 1995, when physicists Eric Cornell (1961–) and Carl Wieman (1951–) at the University of Colorado cooled a gas of rubidium atoms to about 1/6 of a millionth of a Kelvin. They and Wolfgang Ketterle (1957–) at MIT shared the 2001 Nobel Prize for their work. While this new state of matter, called a Bose-Einstein condensate, has no current applications, physicists are using it to improve their knowledge of how atoms interact at very low temperatures and to explore possible future applications to atomic clocks and computers.

How was the laser invented based on quantum mechanics?

One of the most useful applications of quantum mechanics in modern life is the laser—a device that uses stimulated emission of light predicted by Albert Einstein using quantum theory. In 1953 Charles Townes (1915–2015) of Columbia University (and later MIT) developed and patented the first application that he called a "maser"—an acronym for microwave amplification by stimulated emission of radiation. It used a beam of ammonia molecules and later led to the hydrogen maser, now used as an extremely accurate clock. In 1958 Townes and Arthur Schawlow (1921–1999) described how molecules

Why did Einstein say that "God does not play dice with the universe"?

As quantum mechanics developed in the 1920s and 1930s, its statistical nature became more evident. Einstein was convinced that there were variables that could not be seen but that controlled the outcomes of these probability-based uncertainties. Einstein had many debates with his colleagues Niels Bohr and Max Born about the statistical and probabilistic interpretation of quantum mechanics. In a 1926 letter to Max Born, Einstein wrote, "Quantum mechanics is certainly imposing. But an inner voice tells me that it is not yet the real thing. The theory says a lot but does not really bring us any closer to the secret of the 'old one.' I, at any rate, am convinced that He does not throw dice."

In 1935 Einstein, Boris Podolsky (1896–1966), and Nathan Rosen (1909–1995) wrote a famous paper that explored what they considered to be the incomplete description of the world provided by quantum mechanics. In the paper, they proposed an experiment that has now been performed—and confirmed the predictions of quantum mechanics.

could be used to extend the maser concept to optical frequencies. After their paper was published, a number of physicists at university and industrial labs rapidly tried to apply these ideas to working devices.

In May 1960, Theodore Maiman (1927–2007) at Hughes Aircraft Company demonstrated an "optical maser" that used a ruby crystal with a flash lamp (similar to the camera flash lamp) to put the chromium atoms in the ruby into their excited states. Maiman was involved in a court fight over the validity of his patent for the laser with Gordon Gould (1920–2005), who worked with Townes at Columbia. In 1973 Gould was awarded the patent rights. Maiman won number of awards but never the Nobel Prize.

What are some modern uses for lasers?

While the laser was first described as a "solution looking for a problem," over the past 50 years, lasers have become a multibillion-dollar industry. Lasers have been constructed using gases, as in the familiar helium-neon (HeNe) laser; the carbon-dioxide laser used to cut fabrics and metals; the argon-ion laser used in surgery; the ultraviolet excimer laser used for eye surgery; and the free-electron laser used in basic physics research. Lasers made of tiny, semiconducting crystals are used in CD and DVD players and laser pointers as well as optical fiber communications equipment that brings television, video, and internet signals into your home. Lasers have revolutionized research in physics, chemistry, and biology. Powerful lasers are being used to study controlled nuclear fusion.

What is a laser tweezer?

When a laser is sent through a microscope, it creates a tiny spot of very intense light in the material on the slide. A tiny, plastic sphere will interact with the light in such a way

Lasers are now commonly used in industry, surgery, and communications technology, among other uses. Theodore Maiman demonstrated the first laser in 1960, but Gordon Gould was awarded the patent.

that it is pulled into the center of the light. The light can be moved around the slide, dragging the sphere with it. That is the essence of a laser tweezer. The sphere can be chemically attached to the end of a long molecule, such as DNA. When the other end of the DNA is similarly fastened to the surface of the slide, the sphere can be used to stretch the DNA, straightening it out, and measuring properties such as the force needed to stretch it. Proteins, enzymes, and other polymers can be used in place of the DNA and the forces they exert similarly measured. In addition, the tweezers can be used to sort cells, moving them to specific locations on the slide.

THE STANDARD MODEL OF MATTER

What is the Standard Model of matter?

The fundamental description of the most basic subatomic particles that comprise the building blocks of matter is called the Standard Model. It was developed in the 1970s to organize all the information known at the time about subatomic particles. A model is something like a scientific theory in that it is intended to explain observations. It has practical use but is not as complete as a theory. The Standard Model works very well, but there are still many questions about it at the frontier of scientific research.

What evidence is there that protons and neutrons are made of smaller particles?

Ernest Rutherford (1871–1937) scattered alpha particles off gold atoms and discovered that the atom was not filled with a uniform positive material. In the same way, scattering of protons at high energies off hydrogen nuclei showed that the proton is not composed of uniform positive material. Rather, it is composed of three much smaller charged particles called quarks. The neutron is also composed of three quarks.

How are the results of collisions in particle accelerators detected?

A collision between two protons or a proton and antiproton can produce a large number of particles of different kinds. The task of a detector is to find the direction in which each particle is going; find its charge, momentum, and energy; and identify it.

Detectors at particle accelerators are huge devices. The CDF (Collider Detector at Fermilab) weighs 5,000 tons and is 12 meters (39 feet) in each dimension. It is operated by a collaboration of about 600 physicists from 60 universities and organizations. ATLAS (A Toroidal LHC ApparatuS) at the Large Hadron Collider is even larger. It weighs 7,000 tons and is 44 meters (144 feet) long and 25 meters (82 feet) in diameter. Some 2,000 physicists are involved in building and operating this experiment.

In both CRF and ATLAS, position detectors locate the particles close to where they are produced in the beam line. Somewhat farther away from the beam line, a strong magnet deflects the paths of charged particles. Outside the magnet are additional position detectors and finally detectors that measure the energy of the particles. The curved paths that the charged particles take in the magnet are used to find their charge and momentum.

Each collision can produce billions of pieces of data, only a small number of which are interesting. Electronic circuits and fast computers use data from the position detectors to decide whether the set of data are interesting enough to be recorded. ATLAS produces 100 megabytes of data each second. The entire Large Hadron Collider produces 4 gigabytes of data—about the disk space of a high-definition, full-length feature film—each second, which is a rate 30 times faster than the highest internet streaming speed currently available.

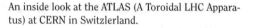

An inside look at the ATLAS (A Toroidal LHC Apparatus) at CERN in Switzlerland.

How high are the energies of particles and collisions in accelerators?

In particle accelerators, energies are measured in electron volts (eV). One eV is about the energy that an infrared light photon contains. Modern accelerators give particles energies of many GeV (billions of electron volts) or TeV (trillions of electron volts). The Tevatron in Chicago, Illinois, creates collisions between protons and antiprotons, each giving an energy of almost 1,000 GeV, resulting in an energy of 2 TeV when the two beams collide. In the Large Hadron Collider at CERN in Europe, both beams are protons. The design energy is 7 TeV in each beam for a total collision energy of 14 TeV.

What are quarks?

The American physicist Murray Gell-Mann (1929–2019) first proposed that there exist subatomic particles that make up protons and neutrons. He thought at first that there were three kinds of such particles, so he lightheartedly borrowed the name from the phrase "three quarks for Muster Mark" from *Finnegan's Wake* by James Joyce and called them quarks. The name stuck. Gell-Mann was awarded the 1969 Nobel Prize in Physics for his work.

Today, we know of six quarks in the Standard Model of matter, known as "flavors." They come in three pairs, known as "generations," all with unusual names: up and down, top and bottom, charm and strange.

What are some properties of quarks?

Quarks are fractionally charged; their charges are either positive 2/3 (up, top, and charm quarks) or negative 1/3 (down, bottom, and strange quarks) the charge of a proton. A proton is composed of two up quarks and one down quark, while the neutron is composed of one up and two down quarks. The charges of the quarks add up correctly to +1 and 0.

Quarks have half-integral spin, like electrons and neutrinos, so there cannot be two identical quarks that have the same quantum properties in the same location. Particles with that property are called fermions.

Quarks have an additional property called color charge that can be red, green, or blue. A proton or neutron must have quarks with one of each of the colors so that they add to white, just like ordinary colors do.

Why do quarks have such strangely named properties like flavor and color?

It does seem odd at first to describe subatomic particles with properties like flavor and color that have nothing to do with those properties of ordinary matter that we can taste or see. On the other hand, they help physicists understand these particles more intuitively, and using these somewhat whimsical terms, as Murray Gell-Mann did with the name quark, shows a sense of fun with science—an approach that invites other people into the discussion of deep concepts rather than creating barriers to keep others out.

Are there antiquarks?

Yes. Just as there are antiprotons and anti-electrons (commonly called positrons), there are anti-up quarks with charge –2/3 and antidown quarks with charge +1/3. A quark and an antiquark of the same flavor (up, down, top, bottom, charm, or strange) that come into contact will annihilate each other, resulting in energy release.

An antiproton is composed of two anti-up quarks and one antidown quark. Antiquarks have color charges that are the complementary colors to red, green, and blue: cyan (antired), magenta (antigreen), and yellow (antiblue).

What holds quarks together in a proton or neutron?

Eight different massless force carriers called gluons exert the strong nuclear force that holds quarks together. Gluons are members of the group of particles called bosons; in the Standard Model, there are four types of bosons—gluons, photos, W and Z particles, and gravitons—that carry force from one particle to another.

What are some properties of bosons and fermions?

Bosons have integral spin, while fermions like quarks, electrons, and neutrinos have half-integral spins. Bosons can be created or destroyed, while fermions can be created only if an antifermion is also created. A single fermion cannot be destroyed unless an antifermion is simultaneously destroyed.

Must quarks always be parts of larger particles?

There have been many searches for so-called free quarks, but they have never been observed. The theory that describes the interactions between quarks and gluons is called quantum chromodynamics (QCD for short), and according to QCD quarks can never break free because the more that the gluons are stretched, the stronger the force they exert. In this way, they act like springs whose force also increases the more they are stretched.

How do electrons and neutrinos interact with quarks and gluons?

Electrons and neutrinos are a kind of fermion called leptons, meaning "lightweight ones." They do not participate in the strong nuclear force interactions, so they do not interact directly with quarks and gluons. They are, however, involved in the weak nuclear force interaction through two additional force carrier particles, the W and Z bosons—and those two bosons can indeed interact with quarks.

How is beta decay explained using quarks?

Beta decay involves neutrons, protons, electrons, and neutrinos and can be described in two steps using quarks. First, the down quark in a proton changes to an up quark with the emission of a W⁻ boson. Then, almost instantaneously, the W⁻ changes into an electron and anti-neutrino. Physicists diagram the two beta decay processes as shown in Figure 25.

When a nucleus undergoes beta decay, the number of neutrons goes down by one, the number of protons goes up by one, and an electron and antineutrino are emitted. In terms of quarks and leptons, one of the down quarks in the neutron is changed to an up quark in the proton, and a lepton (electron) and antilepton (antineutrino) are emitted. The change from down to up quark is a flavor change, and that can occur only in beta decay. When a positron is emitted in inverse beta decay (see Figure 25), an up quark

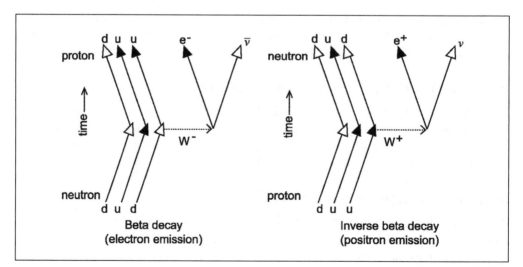

Beta decay
(electron emission)

Inverse beta decay
(positron emission)

in the proton is changed to a down quark in the neutrino, and a lepton (neutrino) and antilepton (positron) are emitted.

What other kinds of particles contain quarks?

In the 1950s, new particles with masses in between those of the electron and the proton were found. They were given the name mesotrons, in which "meso" meant intermediate. This was soon shortened to meson.

In the Standard Model of matter, mesons are composed of a quark and an antiquark. The most commonly produced meson in particle accelerators is the pi-meson or pion. Pions can be positively or negatively charged or uncharged.

It is also possible to create particles with different combinations of three up (u) and down (d) quarks. These particles have higher masses than protons and neutrons and can only be created by colliding protons and neutrons together with very high energies. Examples are the combinations uuu with charge +2 and ddd with charge –1. These particles decay via the strong interaction. The uuu decays when its extra energy creates a down-antidown pair. The down quark replaces one up quark in the uuu particle, resulting in a proton, and the antidown quark combines with the extra up quark to form a pion.

What are the three generations of particles in the Standard Model?

The first generation of fundamental particles in the Standard Model contains the up and down quarks and two leptons, the electron and electron neutrino. This generation of particles comprises the building blocks of atomic nuclei—protons and neutrons—and pions.

In the second generation of particles, the quarks are called charm and strange, and the leptons are called muons and the muon neutrino. The strange quark was named after particles that contain these quarks. The particles decayed in very different ways and were therefore called strange particles. The charm quark was found in a particle that had a very long, or "charmed," life. The muon was first called the mu-meson before its true identity was discovered.

In the third generation, the quarks are called top and bottom, and the leptons are called tau and the tau neutrino. These particles are rarely if ever observed on Earth outside of particle accelerators.

What role do the second- and third-generation quarks and leptons play in the physical universe?

The second- and third-generation quarks and leptons require large amounts of energy to be created. Particle accelerators like the Tevatron and the LHC have been built to create and study them. They can also be created when very-high-energy cosmic rays strike Earth's atmosphere. Some astrophysicists think there is evidence that there are unusual stars in the universe that contain the strange quark.

These other-generation particles most likely also existed at times less than a minute—or maybe even a fraction of a second—after the Big Bang, when temperatures   **317** in the universe were extremely high. Studying high-energy collisions using particle accelerators is thus one way of studying what the universe was like when it was very, very young, even though the universe is currently 13.8 billion years old.

What are the masses of quarks and leptons?

Because quarks can't be separated, their masses can be found only by using one or more theories to calculate them. The masses of the charged leptons can be measured to high precision. According to the Standard Model, neutrinos do not have mass, but recent experimental results show that they do have a tiny amount of mass.

The masses and mass limits of the particles in the Standard Model are given in the table below. The mass units are given as an amount of energy (in eV, MeV, or GeV) divided by the speed of light squared, based on Einstein's famous equation $E = mc^2$.

| | Quarks | | | | Antiquarks | | | |
Generation	Charge +2/3	Charge −1/3	Leptons		Charge −2/3	Charge +1/3	Antileptons	
1st	up	down	electron	electron neutrino	Anti-up	Anti-down	positron	Anti-electron neutrino
	u	d	e	ν	$\bar{u}$	$\bar{d}$	e^+	$\bar{\nu}$
2nd	charm	strange	muon	muon neutrino	Anti-charm	Anti-strange	Anti-muon	Anti-muon neutrino
	c	s	μ	ν_μ	$\bar{c}$	$\bar{s}$	$\bar{\mu}$	$\bar{\nu}_\mu$
3rd	top	bottom	tau	tau neutrino	Anti-top	Anti-bottom	Anti-tau	Antitau neutrino
	t	b	τ	ν_τ	$\bar{t}$	$\bar{b}$	$\bar{\tau}$	$\bar{\nu}_\tau$

How are neutrinos detected?

Neutrinos hardly interact at all with matter. Some 100 trillion neutrinos pass through your body every second, yet the chance that a neutrino ever interacts in your body over your lifetime is only one in four. For this reason, neutrino detectors must be huge and can expect to detect only an extremely small fraction of the neutrinos striking them. Most detectors are large cavities filled with extremely pure water or mineral oil. Neutrino interactions result in flashes of light that are seen using sensitive phototubes. These detectors have identified neutrinos from the sun, from cosmic rays, from a supernova, and from reactors and particle detectors. Some experiments have used beams of neutrinos that are aimed through Earth by a reactor or accelerator at a distant detector.

Where were neutrinos from the sun first detected?

The first effective neutrino detector was set up in 1967 deep underground in the Homestake Gold Mine near Lead, South Dakota. There, American astrophysicists Ray Davis Jr.

(1914–2006) and John Bahcall (1934–2005) set up a tank filled with 100,000 gallons of nearly pure perchlorate (used as dry-cleaning fluid), and they monitored the liquid for very rare neutrino interaction events. In 2002, Davis was awarded the Nobel Prize in Physics.

How is Antarctica used to detect neutrinos?

Near the South Pole, thousands of sensitive photodetectors have been embedded on long chains, in a huge, prism-shaped pattern, deep into the Antarctic ice. The ice provides the atoms needed to interact with neutrinos that pass through our planet, and it is transparent enough for the photodetectors to see the tiny flash of light that is produced when a neutrino interaction occurs. This remarkable experimental facility is called the IceCube Neutrino Observatory.

The Astronomy with a Neutrino Telescope and Abyss environmental RESearch (ANTARES) neutrino detector is located at the bottom of the Mediterranean Sea. Operational since 2008, it uses photomultiplier tubes inside a three-dimensional array of detector modules (illustrated above) to analyze neutrinos filtering through 1.5 miles (2.5 kilometers) of ocean water.

What is the evidence that neutrinos have mass?

The number of neutrinos detected from the sun is only a fraction of what should be detected if our understanding of the nuclear reactions in the sun is correct. A proposed solution to this problem is that neutrinos change from one flavor to another as they travel from the sun to Earth. This phenomenon is called neutrino oscillation. It can occur only if the neutrinos contain at least a small amount of mass. Most experiments that support neutrino oscillation measure the lack of neutrinos of one flavor that they interpret as an oscillation into an undetected flavor. Recently, an experiment has detected tau neutrinos in a muon-neutrino beam from an accelerator at CERN and a detector in the Gran Sasso tunnel in Italy, 732 kilometers away. This detected flavor change is evidence that supports neutrino oscillation.

Are there more than three generations of quarks and leptons?

Scientists have conducted many experiments to search for additional generations of the fundamental particles. All of the experiments have had negative results so far, and theoretical considerations of the Standard Model suggest that there are no more generations. However, they may indeed exist but require such high energies that we cannot detect them with current technology.

What is the origin of the masses of quarks and leptons?

The British physicist Peter Higgs (1929–) proposed a quantum mechanical mechanism that gives rise to the masses of the quarks, leptons, and the two weak-interaction bosons, **319** the W and Z. The mechanism successfully predicted the mass of the Z boson and requires the existence of an elusive mass-carrying particle, which is known today as the Higgs boson. In 2013, Higgs and his Belgian colleague François Englert (1932–) were awarded the Nobel Prize in Physics for their work on the Higgs mechanism.

How was the Higgs boson discovered?

Higgs bosons, compared with many other kinds of subatomic particles, are very hard to detect, in part because they are thought to exist for only extremely short periods of time. To confirm the existence of these elusive particles, scientists used the Large Hadron Collider and measured the results of thousands upon thousands of acceler-

Peter Higgs, shown here in 2013 when he earned the Nobel Prize in Physics for his research on the existence and nature of the Higgs boson, was proven correct in his theories thanks to experiments at the CERN supercollider in Switzerland.

ated particle collisions, examining those huge sets of data for any subtle signs that Higgs bosons were ever-so-briefly present in them.

On July 4, 2012, scientists based at the European Organization for Nuclear Research (CERN) announced the discovery of a particle that could be the Higgs boson. Using the Large Hadron Collider, two international teams of physicists found convincing evidence of new subatomic particles that match theoretical predictions of the characteristics of Higgs bosons. The two teams used different equipment and different methods to achieve the same results, greatly increasing the odds that the result is correct.

Why doesn't the total mass of three quarks add up to the mass of the proton?

The combined mass of two up quarks and one down quark sum to 8.81 MeV/c2, and gluons have mass zero, but a proton's mass is 938 MeV/c2. How can that be? The remainder of the mass—the vast majority of the proton's mass—is actually due to the high energies of the quarks and gluons within the proton.

What are the force carrier bosons?

The photon is called the force carrier for the electromagnetic interaction. That means that, for example, the attractive force between two oppositely charged objects can be described mathematically not by an electric field but by an exchange of (invisible) photons. The photon has neither mass nor charge, but it does have angular momentum.

Similarly, the gluon carries the color force, which, in combination with quarks, creates the strong nuclear force. The gluon also has neither mass nor charge. The weak nuclear force is carried by the massive W+, W–, and Z_0 particles. Gravity is thought to be carried by a particle called the graviton—but it has never been detected, and its mass may well be zero.

How does gravity fit into the Standard Model?

Actually, gravity does not fit well at all into the Standard Model. Gravity, explained by Albert Einstein's General Theory of Relativity, is extremely successful on the macroscopic scale. But all attempts to create a theory of quantum gravity have failed. Attempts to fit gravity into the Standard Model, which is fundamentally a theory of quantum physics, also have not yet been successful.

Could there be more than four forces?

For decades, physicists have explored the possibility that there are more than the four known fundamental forces in nature. Usually, a fifth fundamental force is proposed when experiments or observations produce strange results that cannot seem to be easily explained with the four known forces alone. In 1986, for example, researchers reported an unusual experimental result that could have been caused by a fifth force, but later experiments did not confirm that result. More recently, a research team based in Hungary has been studying an experimental anomaly that may be evidence for a previously undiscovered boson; if the existence of this particle is confirmed, it might mean that there would exist a corresponding fifth force. Such confirmation, however, has not yet happened.

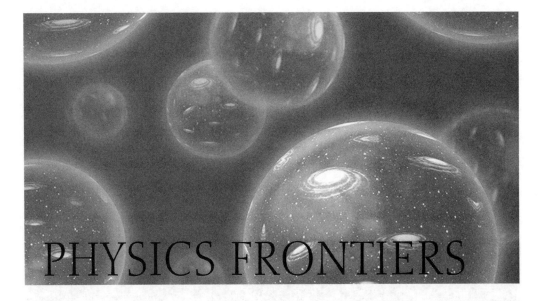

PHYSICS FRONTIERS

DARK MATTER

What kinds of matter are there in the universe?

Protons, neutrons, and electrons are particles that make up almost all the matter on Earth. In the entire universe, these particles are known as baryonic matter or luminous matter because they interact with light and can be detected by the electromagnetic waves they produce. Amazingly, though, baryonic matter comprises only about 5 percent of the contents of the universe. About 70 percent of the contents of the universe are a kind of mysterious "dark energy," and the remaining 25 percent or so is matter that is not composed of protons, neutrons, and electrons. This matter is called nonbaryonic dark matter.

How did we first learn about dark matter in the universe?

The first hints that there is more to the universe than visible stars and galaxies came in 1933, when the Swiss-American astronomer Fred Zwicky (1898–1974) studied the motion of galaxies in the Coma cluster. From his measurements, he estimated that to account for their rapid motion, there must be 400 times as much mass as could be accounted for using light emitted from the galaxies in the cluster.

In the 1970s the American astronomer Vera Rubin (1938–2016) studied the rotation of galaxies and found that at least half of the matter in galaxies like the Milky Way must be invisible, further confirming the existence of large amounts of dark matter.

Dark matter is also in evidence in gravitational lensing, where the gravitational interaction of mass with light causes the light from very distant galaxies to be bent by large concentrations of mass closer by. The result is a distortion of shape and position in the images of the distant galaxies as if they are being seen through a lens.

How have we measured just how much dark matter there is?

Gravitational lensing studies are very useful in determining the amount of dark matter present in any particular part of the universe. In one set of colliding clusters of galaxies, for example, the center of mass of the system as determined by gravitational lensing is separated from the center of optical and X-ray brightness. This separation stems from the effect of the collision of the clusters, and it is evidence that most of the matter in the clusters is not visible and not producing any electromagnetic radiation.

A very strong piece of evidence about the amount and distribution of dark matter in the universe comes from the cosmic microwave background—the earliest electro-

This image, taken with the Hubble Telescope's Advanced Camera for Surveys in 2002, shows distant galaxies, trillions of stars, and lensing (you can see distorted images around the periphery of the photo) that are the result of dark matter distorting light.

magnetic radiation still visible from the Big Bang, the event that started time and began the expansion of the universe. Studies of the patterns of this radiation show that in order for the universe to look the way it does today, there must be several times more non-baryonic dark matter than baryonic (luminous) matter in the universe.

What is dark matter made of?

Many candidate particles have been proposed as possible constituents of dark matter, but there is no agreement at this point. Some ideas over the years have included neutrinos with mass, supersymmetric particles, weakly interacting massive particles (WIMPs), neutralinos, and other types of particles that have never been definitively detected. The composition of dark matter is one of the most important frontiers of research in physics today.

DARK ENERGY

What is dark energy?

The kinds of energy that we experience on Earth, like kinetic energy and potential energy, exist because of the interaction of objects and particles in the universe. There appears, however, to be another kind of energy in the universe that exists because of space itself. Within the past few years, it has been shown that this kind of "dark energy" comprises more than 70 percent of the entire contents of the universe.

What is the mathematical origin of dark energy in physics?

The theoretical idea of dark energy arose about 100 years ago when Albert Einstein (1879–1955), Willem de Sitter (1872–1934), Alexander Friedmann (1888–1925), Georges Henri Lemaître (1894–1966), and others were researching the nature and geometry of the universe. Einstein introduced a mathematical term into his equations to keep a balance between cosmic expansion and gravitational attraction. This term became known as the "cosmological constant" and seemed to represent an unseen (or dark) energy that emanated from space itself.

After Edwin Hubble (1889–1953) and other astronomers showed that the universe was indeed expanding, the cosmological constant no longer appeared to be necessary, so it was not seriously considered again for decades. Then, starting in the 1990s, a series of discoveries suggested that the "dark energy" represented by the cosmological constant does indeed exist.

What was the first strong evidence that dark energy exists?

In the 1990s, a team led by Wendy Freedman (1957–), Robert C. Kennicutt Jr., and Jeremy Mould (1949–) used the Hubble Space Telescope to measure the properties of Cepheid variable stars in a number of galaxies in the local universe. These observations measured the distances to the Cepheids with high precision and were used to determine very accurately the expansion rate of the universe that was first observed by Edwin Hubble more than half a century earlier.

The Cepheid measurements showed that the universe is expanding at such a high rate that it would cause the current age of the universe to be too short. The most likely way to resolve this paradox was if the cosmological constant first proposed by Albert Einstein were not zero—and that dark energy existed after all.

What discovery confirmed that dark energy exists?

After the Cepheid observations from the Hubble Space Telescope were published in 2001, two groups of scientists further

A Cepheid variable star like this one—RS Puppis in the Milky Way—pulsates predictably in size and in brightness, which is useful for determining distances on an intergalactic level.

studied the expansion history of the universe using exploding stars known as Type Ia supernovae. They both showed conclusively that for about the first half of the history of the universe, the attractive forces of gravity on matter—both baryonic matter and nonbaryonic dark matter—slowed the expansion of the universe. But beginning about five billion years ago, the universe began to expand at an increasing rate. For their work on this topic, three astrophysicists—Saul Perlmutter, Adam Riess, and Brian P. Schmidt—were awarded the 2011 Nobel Prize in Physics.

What causes dark energy to exist?

Though astronomers have definitely measured the presence of this dark energy, we still have no idea what causes this energy to exist nor do we have a clue what this energy is made of. The quest to understand the cosmological constant in general, and dark energy in particular, addresses one of the great unsolved mysteries in physics today.

Could dark energy be a result of quantum physics?

In the formulation of quantum mechanics that includes the Standard Model of matter, it is possible to show that there exists a "vacuum energy" that could push outward on space, exactly the way dark energy apparently does today. For example, pairs of "virtual particles" could continually appear and disappear throughout the universe. These pairs exist only for a very tiny period of time—a very small fraction of a second—and then annihilate in a burst of energy. Added together, all the energy from all the virtual particle pairs could be pushing the expansion of space faster and faster.

The problem with quantum vacuum energy is that according to the calculations, if it exists at all, it would be at least one trillion trillion trillion trillion trillion times (10^{60}), or even that amount squared (10^{120}), more plentiful than the dark energy that has been measured in the universe to date. So, if quantum vacuum energy does exist, something must also be going on to reduce its effect on space by a huge amount—and what that could be is completely unknown.

QUANTUM ENTANGLEMENT

What is quantum entanglement?

Consider what happens when a positron and electron annihilate. Two gammas (high-energy photons) are produced that go off in opposite directions, 180 degrees apart. They can be detected many meters from their source and have opposite spins, but which gamma has which spin is a random choice. That is, for each gamma, there is a 50-50 chance that it will be in a particular direction. Suppose you find that the spin of one gamma is pointing up. The moment you detect the spin of that gamma, the spin of the other gamma must be pointing down. The result of one measurement determines the results of the other. The two detectors would measure the spins of their gammas at the same time, so there is no way that one gamma could communicate with the other gamma.

Physically speaking, the spins of the two photons are "entangled," and the spin state of each photon is the superposition of the two possible spin directions. When the spin is measured, the quantum mechanical uncertainty ends, and a definitive result comes out. Albert Einstein (1879–1955) called results like this "spooky action at a distance." Others have called it quantum weirdness.

Similar results can be obtained with atoms or ions, in which case the photons are light quanta and may be transported through space (or the air) or by optical fibers. For example, if an atom is excited by the absorption of a photon, it can emit two photons that are entangled the same way the gammas are in the example above.

What is quantum teleportation?

First and foremost, quantum teleportation is not "Beam me up, Scotty!"—where a person disappears from one place and reappears at another. Rather, it involves information transfer between photons or atoms, which could be instantaneous. Quantum teleportation involves an atom whose state is to be communicated from a sender to a receiver—let's call them Alice and Bob, respectively—and a source of two entangled photons.

How does quantum teleportation use entanglement?

Alice first sets or measures the state of the atom. This is the information she will send to Bob. At the same time, she measures the polarization of one of the two photons. This polarization information is the encryption key that will allow Bob to decode the information. Because the two photons are entangled, the polarization of the transmitted photon is fixed. Bob now has the key that allows him to decode the information Alice sent. In practice, of course, the encryption key isn't the polarization of a single photon but many.

The longest distance information has been teleported to date is 143 kilometers (89 miles) through the air during an experiment conducted in the Canary Islands. Through optical fiber, the longest distance to date that information has been teleported is 102 kilometers (63 miles).

What is the state of quantum computing today?

Quantum computers are by nature parallel computing machines that can work on many problems at once. At this time, computers with up to only about 100 qubits have been demonstrated, so there is a long way to go before such a device is practical. Still, progress is being made. A way has been developed, for example, of using the kind of lasers used in DVD and Blu-ray players to create entangled photons, potentially greatly reducing the size, complexity, and cost of a quantum computer.

Will quantum teleportation be able to be extended to longer distances with error-free results?

At the present, the encryption codes for secure telecommunications cannot be broken. It is unknown if that will continue to be true or if quantum computers will ever become practical enough to break them. For now, scientists think that quantum computers could be used someday to factor very large numbers, a task used to create encryption keys to keep electronic messages secure. An important question will be to find out what other problems that are intractable for traditional computers could someday be solved with quantum computers.

What is entangled time?

In the same way that quantum entanglement causes one particle far away from another to be tied together in the transfer of information, it can also tie together two particles

Does entangled time mean the present can affect the past?

When we are talking about particles entangled in time, our standard way of talking about time as a single, one-directional arrow is not quite adequate. In quantum mechanics, time can theoretically run both "forward" and "backward," so the sources of cause and effect can get confused. Time still runs as it should, even though it might seem to us that information in the present could actually affect an event in the past.

that do not exist at the same time. This makes sense because time is a dimension like length, width, and height, but it behaves differently because we can only move through time in one direction and not two.

In 2013, physicists at the Hebrew University of Jerusalem published the results of a complex experiment where they (1) entangled two photons, which we'll call A and B; (2) destroyed photon A; (3) created another entangled pair, which we'll call C and D; and then (4) entangled B and C, untangling D in the process. When they then measured D, they found D to be as if it had been entangled with A—even though A no longer existed. This result showed that quantum entanglement seems to work through time as well as through space.

NANOPHYSICS

What is nanotechnology?

An important frontier of physics today is the study and development of structures and materials at tiny sizes of just a few billionths (10^{-9}) of a meter. A billionth of a meter is called a nanometer, so this size scale is known as the "nanoscale." The physics at this scale is called nanophysics, and the structures and devices that are being developed to work at this scale are collectively known as nanotechnology.

What is a buckyball?

Carbon is a versatile element that can be combined with many other substances or shaped into structures that have remark-

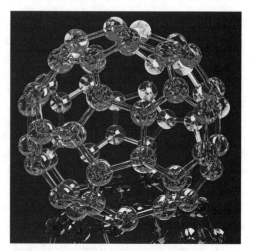

An illustration of a carbon-60 buckyball, or fullerene.

able properties. One such structure is a molecule consisting of 60 carbon atoms put together in the shape of a soccer ball. It was named buckminsterfullerene and nicknamed a "buckyball." Today, many varieties of these huge, hollow carbon molecules have been discovered, and they are called fullerenes or buckyballs.

Who are buckyballs named after?

In 1923, the German engineer Walther Bauersfeld (1879–1959) designed a planetarium with a domed roof that was supported using numerous small beams arranged in triangles. This was the first geodesic dome derived from a three-dimensional shape called an icosahedron, and Bauersfeld is widely honored among the planetarium community today for his invention.

In the 1940s, the American inventor and architect R. Buckminster Fuller (1895–1983) began experimenting with geodesic domes as building structures. In 1949, Fuller obtained the U.S. patents for the geodesic dome and began designing buildings that proved to be strong yet lightweight and aesthetically pleasing. The dome became extremely popular and is still widely used today in architecture and design. The hexagonal shapes that are used to design these domes resemble the shape of buckminsterfullerene, C_{60}.

Buckminster Fuller invented a wide variety of devices and designs in his lifetime. He was said to have declared that he "belonged to the universe" and that his inventions were "the property of all humanity." He optimistically believed that someday, thanks to the work and generosity of people, human society will become a utopia of high education, prosperity, and environmental beauty.

Who discovered the first buckyball?

The British chemist Harold Kroto (1939–2016) was using microwave spectroscopy to study the astrochemistry of red giant stars. He noticed strange signals from the stars, and he hypothesized that the signals were produced by large crystalline carbon structures. This was unexpected because at the time, only six forms of crystalline carbon were known, and none were so complicated.

In 1985, Kroto worked with his American colleagues Robert Curl and Richard Smalley and their graduate students J. R. Heath and S. C. O'Brien in a series of experiments to see if they could produce such carbon structures by simulating conditions similar to the environments of red giant stars in a laboratory. They succeeded in producing very strong and stable structures of carbon with 60 atoms (C_{60}) with the shape of a truncated icosahedron cage—the same shape traced out by the fabric of a standard soccer ball. This was buckminsterfullerene—the first buckyball.

For their discovery, Kroto, Curl, and Smalley were awarded the 1996 Nobel Prize in Chemistry.

Why are buckyballs important in nanotechnology?

The size—or, more precisely, the diameter of the physical sphere of influence—of a buckyball is about one nanometer (10^{-9} m). Buckyballs are strong, and stable, too, able to hold their shape at high temperatures and pressures. One application already envisioned for buckyballs is for use in energy technology to increase greatly the efficiency of solar cells.

Just as interesting, a buckyball is just about the right size for objects to interact with biological structures like cells or DNA molecules. Medical physicists are studying the possibility of using buckyballs as mobile "containers" for smaller atoms and molecules, perhaps to deliver medicines precisely to organs or parts of the body that are otherwise not easily accessible through injections or surgery.

Does technology exist to work at nanoscale?

Buckyballs are about one nanometer across—about 1/100,000 the size of a single grain of fine sand. Machines have not yet been made that are as small as buckyballs, but machines have now been built that can carve sculptures or write your name onto such a grain of sand in just a few minutes. A thermometer has been designed, using the thermal properties of DNA molecules, less than 10 nanometers in size. Components of machines have now been constructed that are just a few nanometers across. A major frontier of physics and engineering is the ability to put all of these pieces together into working nanoscale machines.

What is a nanotube?

A nanotube can be constructed by connecting atoms of carbon, silicon, or other elements together into tube-shaped structures that are at most about 100 nanometers in diameter. Unlike buckyballs, they are open on each end and can theoretically be made into long, hollow structures. Nanotubes have many potential applications in fields such as electronics, manufacturing, and medicine.

What is graphene?

Graphene is a carbon structure that is one atom thick and can be made into large, flat sheets. Graphene sheets were first constructed, isolated, and analyzed by taking successively thinner layers of graphite—a crystalline structure of carbon used in pencils—using regular adhesive tape until a sheet of graphite just one atom thick remained. The physicists who did that, Andre Geim (1958–) and Konstantin Novoselov (1974–) at the University of Manchester in England, received the 2010 Nobel Prize in Physics for their work.

Graphene has surprising properties as a material, including being 100 times stronger than steel for its thickness, nearly transparent, able to conduct heat and electricity very well, and having unusual magnetic properties. Graphene is being investigated as a way to construct extremely tiny and powerful integrated circuits. It may also be used as a sensor to detect single atoms or molecules of selected materials or as a transparent conductor for touch-screen computer displays. In short, scientists think graphene will have very important uses, especially in nanotechnology, in the years to come.

What is a quantum dot?

A quantum dot is a tiny piece of semiconductor material (such as silicon) only a few nanometers in diameter. Due to their small size, quantum dots behave very differently from larger pieces of material. They are sometimes called artificial atoms.

Quantum dots are a very important kind of nanoparticle at the current frontier of nanotechnology. One particularly interesting property they have is that they glow with different colors depending on their size when exposed to ultraviolet light. They show great promise for future use in microelectronics and nanoelectronics, lasers, solar cells, quantum computing, medical imaging, and much more.

STRING THEORY

What are the most fundamental particles in the universe?

The Standard Model of matter, which is the basis of particle physics today, describes subatomic particles called bosons and fermions. Fermions are further subdivided into quarks and leptons, and each fermion has an antimatter counterpart. This system has effectively described all of the matter that has ever been detected on Earth in the history of science.

Even so, there are enough uncertainties about the fundamental nature of matter that make physicists think the Standard Model might not be the complete story. For example, the dark matter that accounts for about 85 percent of the mass in the universe has never been detected on Earth, and it may not fit into the Standard Model. Thus, several ideas have been advanced and studied that could ultimately supersede the Standard Model. If one or more of these ideas are correct, they imply that there exist particles even more fundamental than bosons, quarks, and leptons.

What is supersymmetry?

Supersymmetry is a model of particle physics that supposes that the basic particles of the Standard Model have partners called superparticles, or "sparticles" for short. Each boson in the standard model has a supersymmetric fermion sparticle, and each fermion has a supersymmetric boson sparticle. They are called, respectively, sleptons, squarks, gluinos, and photinos.

In a sypersymmetric model, many of the advanced behaviors of subatomic particles that are not well explained by the Standard Model are caused by the interaction of particles and sparticles.

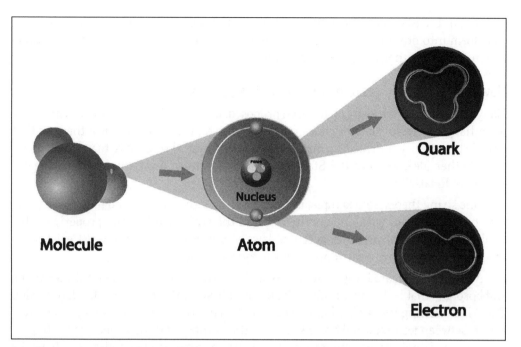

According to string theory, when you go deep enough into subatomic particles, what you find is that matter is made up of tiny, vibrating strings of energy.

Has supersymmetry been experimentally tested?

If current models of supersymmetry are correct, then sparticles must be much more massive than the Standard Model particles—that is, hundreds of billions of electron volts. Particles at those mass levels are very hard to detect experimentally, even with the largest particle accelerators. At this time, there is no direct evidence of their existence, despite quite a bit of searching by physicists using the Large Hadron Collider. It is still possible that sparticles will be detected with larger accelerators of the future.

What is string theory?

String theory is not yet a confirmed scientific theory but rather the name of a model of the structure of matter and energy in the universe that supersedes the Standard Model. Several variations of the model are called "string theory."

In string theory, it is assumed that space-time actually contains many more dimensions than the familiar four-dimensional space-time (three spatial dimensions and one time dimension) that we experience. A string is a tiny, closed loop (like a quantum rubber band) that extends into these many dimensions. In some versions of string theory, the loop can break into an open length (like a piece of yarn), but in others, it cannot. Strings vibrate at characteristic frequencies, like guitar strings. Each vibration frequency in each multidimensional string corresponds to an elementary particle that exists in our four-dimensional space-time, like a photon or gluon or other particle that is found in the Standard Model. When these particles interact, for example to build or transform into other particles, what happens would be the result of the strings' interactions in these other dimensions invisible to us.

How many dimensions are predicted by string theory?

Many different string theories have been proposed, each with its own characteristics. Any potentially successful scientific idea must be consistent with the physical theories that already have been confirmed, so mathematical and physical rules have to work when applying other ideas such as the Standard Model, quantum mechanics, or the General Theory of Relativity.

One string theory, for example, is called bosonic string theory, and it considers only how bosons interact with one another. For all the strings to interact properly in this model without violating other accepted laws of physics, each string must exist in 26 space-time dimensions—far more than the four we experience.

A more complicated way to structure a string theory is to consider both fermions and bosons—that is, to include all of the kinds of basic particles in the Standard Model. This kind of string theory is what most current string theorists study today. The interaction between fermions and bosons reduces the number of dimensions needed to produce a consistent string theory by 16; so, these current string theories have 10 dimensions—which is still more than four.

How can so many other dimensions of space exist?

One hypothetical model that explains how so many dimensions can exist in our universe is the idea of compactification. Picture a big oil or natural gas pipeline stretching across a vast plain: when you stand next to it, it clearly has the three dimensions of length, depth, and height. By moving away a few feet, it increasingly seems like the pipeline has only length and height; moving still farther away, it may seem like it only has one dimension, length. In a sense, two of the pipeline's three dimensions have now been "compactified." They are still there, but they are too small to be observed.

Why might string theory help explain how the universe works?

One great advantage of string theory is that the graviton, the particle that carries the gravitational force, is naturally accommodated. Right now, the Standard Model does not explain the graviton very well, in part because the theory of quantum mechanics and the General Theory of Relativity do not overlap. Knitting the two theories together into a single theory of "quantum gravity" would be a great triumph of theoretical physics; Albert Einstein tried to create a theory of this kind over the last 20 years of his life, but he was unsuccessful, and no other scientists have yet succeeded either.

The same concept might apply to dimensions beyond those of observable space and time. If each space dimension we see actually consists of three separate dimensions, including compactified ones, then there are actually 9 dimensions of space and 1 dimension of time, totaling 10 dimensions for a string. This idea has been around for more than half a century. However, it might be impossible to confirm the existence of these extra dimensions because we cannot observe the very tiny size scales necessary for that amount of compactification.

THEORIES OF EVERYTHING

What are theories of everything (TOEs)?

Some scientists think that all the physical laws of the universe might be described by a single, elegant theory—a so-called theory of everything (TOE). One well-known scientist who tried to develop a TOE was none other than Albert Einstein. He failed to create such a theory, but he laid the groundwork for others who have been working away at it ever since.

What are grand unified theories (GUTs)?

A grand unified theory (GUT) is almost a TOE—it is essentially a theory of everything except gravity, which is not effectively explained by quantum mechanics or the Standard Model. In 1974, American physicists Howard Georgi (1947–) and Sheldon Glashow (1932–) and European physicists Harald Fritzch (1943–) and Peter Minkowski (1941–) published the first GUTs.

Over the decades, many of these models have shown important insights and promising progress, but none of them has yet been scientifically confirmed. One experimental test of GUTs, which is the measurement of the lifetime of a proton, has shown that the simplest models of GUTs cannot be correct. They have not yet, however, ruled out the most complicated versions.

Could a string theory be the theory of everything?

String theory can bridge the gap between quantum mechanics and gravity—something a GUT does not do—so it has the potential to be a successful theory of everything. This is because string theories can contain certain mathematical properties known as symmetries that make their effects felt at larger distances. Furthermore, if supersymmetry is correct and particle-sparticle pairs of fundamental particles exist, then a 10-dimensional string theory can also take massive particles into account properly.

To be a complete theory of quantum gravity, though, the size of a string must be about the Planck length: 10^{-35} meter, which is about one-billionth of one-trillionth the diameter of a proton. There is, not surprisingly, no experiment that could currently test string theory.

What is M-theory?

By the mid-1990s, five formulations of 10-dimensional string theory had been discovered. All of them appeared to be consistent and promising, but it was not clear why they were different or whether one was correct over the others. Then, in 1995, the American mathematical physicist Edward Witten (1951–) developed an idea that all these 10-dimensional string theories are related through their connection to an 11-dimensional supersymmetric structure. This idea is known as M-theory, mainly because it implies the existence of multidimensional structures called membranes.

Theoretical physicist Edward Witten is an expert in string theory, quantum gravity, and supersymmetric quantum field theories. He developed M-theory, which helped explain other multidimensional theories being discussed in the 1990s.

Having this additional 11th dimension helps greatly to unify string theories toward a theory of everything. However, the mathematical analysis is very complex and still not fully worked out. The "M" in M-theory, according to Witten, could stand for membrane, mystery, or even magic because we won't know for sure how it works until a more fundamental formulation of the theory is developed.

What is a 'brane?

A membrane, or 'brane for short, is a multidimensional structure that perhaps exists in something like a supersymmetric bulk described above. Membranes can move around in the bulk—like a huge jellyfish floating and swirling around in a vast ocean—and interact with (that is, "bump into") one another, possibly causing vast releases or exchanges of energy. M-theories are a kind of cosmological model—and possibly a theory of everything—that implies the existence of 'branes. These models propose the existence of different kinds of 'branes, which are given names like m-'branes, n-'branes, or p-'branes.

According to membrane theory, where might our universe be located?

Hypothetically, it is possible to think of two five-dimensional membranes that contact one another in one or more dimensions. The multidimensional intersection of those 'branes could be, for example, a point, a line, a surface, or even a four-dimensional space-time. One idea, then, is that our universe, which is a four-dimensional space-time, exists because two five-dimensional membranes intersected, launching the expansion of space and the beginning of time. That moment of intersection would thus be the Big Bang, and the universe would then be at the intersection of two 'branes.

What is the Randall-Sundrum model of gravity?

The American physicist Lisa Randall (1962–) and the Indian-American physicist Raman Sundrum (1964–) developed a model within M-theory that could explain the origin of gravity in our universe. If our universe is a four-dimensional 'brane embedded within a five-dimensional bulk, a second 'brane with strong gravity could interact with our universe through that fifth dimension. Then, if that fifth dimension is strongly warped, our space-time could grow tremendously in size, but our gravity would become tremendously weak in response. The result would help explain both our expanding universe and the weakness of gravity compared to the other fundamental forces.

How can the existence of 'branes be experimentally tested?

Physicists have proposed some experiments to test the existence of membranes, but the technology required to conduct them successfully does not yet exist. For example, if the Randall-Sundrum model is correct, then it may be possible that when a very massive star explodes in a supernova, a tiny fraction of a percent of its energy might escape from our universe and "leak" onto another 'brane. But supernovae are rare—one occurs in our Milky Way galaxy only every century or so—and our modern telescopes and instruments cannot possibly measure the total energy released by a supernova with nearly enough precision.

What is superfluid, and what does it have to do with quantum gravity and dark energy?

Theories of everything are still far from confirmed, so new ideas are being proposed all the time that may provide answers to unsolved problems in very unexpected ways. For example, in 2014 a team of European physicists proposed that the vacuum of space could be a kind of quantum superfluid.

On Earth, superfluids are produced when certain substances like liquid helium are cooled down to such a low temperature that they flow with almost no resistance; they will even slowly flow up the edges of their containers, seemingly defying gravity, and spill out drop by drop. If gravity behaved like a certain kind of quantum mechanical phenomenon, it could propagate through the superfluid vacuum and push very gently on space itself—causing the accelerating expansion of the universe that scientists currently think is caused by dark energy. A physical model like this is very creative but is very hard to test, and it will require extraordinary experimental proof to confirm.

THE BIRTH OF THE UNIVERSE

What is cosmology?

Cosmology is the scientific study of the universe itself, including its origins, its structure, and its final fate. Astronomy and physics intersect at cosmology; studying the universe requires observations of the universe and its contents far beyond Earth using advanced telescopes, and it also requires experiments about the most basic and energetic pieces of matter and energy that can be examined in laboratories with equipment such as powerful particle accelerators.

Has the universe always been the same, or has it changed over time?

The universe has changed over time, and dramatically so. In the 1920s, the American astronomer Edwin Hubble (1889–1953) discovered that the universe is expanding in all directions at a rapid rate. By measuring that expansion rate precisely, as well as the geometric structure of the universe, astronomers have been able to determine that the universe began its expansion 13.8 billion years ago, when it was a tiny, almost volumeless point.

What were the conditions of space and time very early in the universe?

The universe is expanding, which means it is larger now than it ever has been before. Therefore, in the past, all of the matter and energy in the universe had to fit in a smaller volume. The earlier it was in the history of the universe, the smaller the volume that was available, so the universe was hotter and denser as well. Very early in the universe, when the cosmos was less than 100,000 years old, the temperature of the universe was more than 10,000 degrees Celsius; when it was a few minutes old, the temperature was more than 10,000,000 degrees; and when it was one-millionth of a second old, it was more than one trillion (10^{12}) degrees.

What is the Big Bang?

The moment the universe began to expand in size is called the Big Bang. The moment that time began to exist—that is, time equals zero for the universe—is the origin point of the cosmos. At that moment, the universe may have been a singularity, infinitely dense with zero volume.

Is the Big Bang a fact or a theory?

It is a theory. Scientifically speaking, that makes it more powerful than a fact. Facts are single pieces of information, while theories incorporate many, many facts into a conceptual model, which is then confirmed by a process of prediction, observation, and experiment. In science, individual facts can be weak and often turn out to be wrong, whereas theories are not easily disproved and are strongly supported by evidence.

The Big Bang theory has solid scientific evidence to support it, and its fundamental concepts have been scientifically proven to be correct. However, like all major theories in science, there are many details yet unproven and many questions still unanswered. The many important unknowns will continue to lead scientists to search for answers and make new discoveries as they try to understand the cosmos.

According to the Big Bang theory, a singularity of space and time suddenly began to expand 13.8 billion years ago, resulting in the universe we have today.

How do we know the Big Bang happened?

There are several strong pieces of astrophysical evidence that confirm the Big Bang definitely happened. First, Edwin Hubble (1889–1953) showed that the universe is expanding, and the logical conclusion from that expansion is that the universe had to begin its expansion at some time in the past. Second, the distribution of matter in the universe today—about three-fourths hydrogen, one-fourth helium, and only a tiny amount of other kinds of atomic nuclei—is consistent with a universe that was once very hot but only stayed that way for a short period of time. Third, the universe has a cosmic background temperature of just under 3 K (–454 degrees Fahrenheit, or –270 degrees Celsius), which is the temperature that should exist throughout the vast emptiness of space if a hot Big Bang occurred. For their discovery of the cosmic background radiation, the American astronomers Arno Penzias (1933–) and Robert Wilson (1936–) were awarded the 1978 Nobel Prize in Physics.

What do we think caused the Big Bang?

The Big Bang theory does not explain why the Big Bang actually happened. One well-established hypothesis is that the universe began in a "quantum foam"—a formless void where bubbles of matter, far smaller than atoms, were fluctuating in and out of existence on timescales far shorter than a trillionth of a trillionth of a trillionth of a second. In our universe today, such quantum fluctuations are thought to occur, but they happen so quickly that they never affect what happens in the cosmos. But if, 13.8 billion years ago, one particular fluctuation appeared but did not disappear, suddenly ballooning outward into a gigantic, explosive expansion, then it is possible that something like today's universe could have been the eventual result.

In another, more recently proposed hypothesis that is related to M-theory, the universe is a four-dimensional space-time that exists at the intersection of two five-dimensional membranes. Picture two soap bubbles that come in contact with one another and stick together: The "skin" where the bubbles intersect is a two-dimensional result of the interaction between two three-dimensional structures. If the 'brane hypothesis is correct, then the Big Bang event marked the moment the two 'brane made contact. Neither of these models has any kind of experimental or observational confirmation yet.

What are the Planck time and the Planck length?

The origin point of the universe is probably a singularity, where (and when) the currently understood laws of physics cannot describe what is going on. This means

German physicist Max Planck (1858-1947) discovered energy quanta, which are the basis for quantum theory.

that the behavior of the universe can only be traced back to a time after the Big Bang, when the laws of physics first begin to apply. By combining the fundamental properties of the theory of quantum mechanics and the General Theory of Relativity, physicists have deduced that the earliest time when the behavior of the universe can be traced is about 10^{-43} of a second after the Big Bang. That is a ten-millionth of a trillionth of a trillionth of a trillionth of a second! This earliest, all-but-unknowable period of cosmic history is called the Planck time, after the German physicist Max Planck (1858–1947), the pioneer of quantum theory.

The size of the universe at the Planck time was approximately the distance light can travel within that time interval. That means that the diameter of the universe was 10^{-35} of a meter, or about a billionth of a trillionth of a trillionth of an inch. This length is known as the Planck length.

What happened to force during the Big Bang?

Although this has not yet been experimentally confirmed, most physicists think that at the origin of the universe, there was only one kind of force. If certain theories of quantum gravity are correct, then at the Planck time, the force of gravity split off from the original unified force. Indeed, it may have been this process of splitting that powered the Big Bang in the first place. We do not, however, know of any way to study the universe before the Planck time, so this hypothesis remains unconfirmed.

What happened to force right after the Big Bang?

Mere moments after the Planck time—about a hundred billionth of a trillionth of a trillionth (10^{-35}) of a second—the nongravitational force split again into the strong nuclear force and the electroweak force. Then, another fraction of an instant later—one trillionth (10^{-12}) of a second—the electroweak force split into the electromagnetic force and the weak nuclear force. There have been four fundamental forces in the universe ever since.

What caused the forces to split?

Exactly what caused the unified force to break apart in the early universe remains a mystery. If M-theory or a supersymmetric string theory is correct, however, there is a mathematical way to describe what might have happened; it is known as spontaneous symmetry breaking. As an analogy, imagine throwing an ice cube very hard against a brick wall; some of the ice may split into bits, some of the ice may melt and turn into water, and some may even get superheated and turn into vapor. In each transformation, the original structure of the ice cube—that is, its "symmetry"—has been broken in some way, leading to different amounts of energy release and states of matter. On a superenergetic, cosmic scale, something like this could have happened, in stages, to the unified force.

341

What is the hyperinflation of the universe?

The hyperinflation of the universe—or, simply, inflation—is a vast, super-high-speed expansion of space in the period of time from 10^{-35} of a second to 10^{-32} of a second after the Planck time. Although there is not yet solid scientific evidence to confirm inflation happened, it appears likely to have occurred; and if it did, it would explain the geometry and remarkable uniformity observed in the universe today.

THE MULTIVERSE

What was there before the Big Bang?

It does not make scientific sense to talk about time "before" the Big Bang. That is because time itself did not exist until the Big Bang occurred. Just as there is nothing "north" of the North Pole because Earth does not extend any farther north, there was nothing "before" the first instant of time.

That said, if you can imagine the existence of more than one universe—which might be true, for example, if M-theory or string theory is correct—then it is possible that other universes, with other dimensions of space and time, could have existed while our universe did not exist.

Is there a multiverse?

The existence of a multiverse has been explored by both scientists and science fiction writers for a very long time. Could our universe simply be one among many—perhaps even an infinite number—of other universes, each coexisting but separated from one another? Current M-theory or string theories do not rule out this possibility; indeed, in some cases, they may even require it to be the case. The biggest question is whether we would be able to make contact with those other universes or any of their contents or whether we are somehow so isolated from them that it will never matter to us whether or not other universes exist.

How might inflation have created a multiverse?

The inflationary model of the universe provides an explanation for why our observable universe is so flat and so smooth in every direction. Roughly speaking, something caused the universe to expand so fast and so suddenly that we can no longer observe any part of the universe beyond the part that inflated, no matter how long cosmic expansion goes from now on. In the past few years, astrophysicists have suggested that hyper-inflationary events might occur within an existing universe at different times and places, and could even do so an infinite number of times. This is called "eternal inflation."

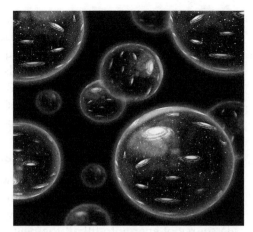

The idea that multiple universes could exist is not just a convenient plot device for comic books but a frontier of theoretical physics research that could help explain how reality works.

If this is the case, an infinite number of isolated "observable universes"—one of them our own—could be coexisting right now within a larger "multiverse."

What scientific problem does eternal inflation present?

One very important philosophical underpinning of science is that the laws of nature and the properties of the universe have an underlying and ultimately discoverable cause. If the idea of eternal inflation is correct, then every single possible set of laws and properties can occur in any given new, recently inflated universal volume purely at random. In the opinion of some scientists, that would imply that inflation is, paradoxically, unscientific.

What is the many-worlds interpretation (MWI) of quantum mechanics?

Quantum mechanical processes—such as radioactive decay or the motion of a particle—are inherently uncertain. Typically, one outcome is most likely at any given moment, but another outcome could happen instead, and there would be no way to know for certain beforehand that it would happen that way. According to the many-worlds interpretation (MWI) of quantum mechanics, both outcomes occur, and each outcome leads to a different timeline of what happens next. As a result, what was once a single timeline and a single universe becomes two. This happens with every quantum mechanical event, so a huge number of new universes are created at every moment in time and coexist throughout all of time—hence the term "many worlds."

How was the many worlds interpretation developed and promoted?

In the middle of the twentieth century, the Austrian-Irish physicist Erwin Schrödinger (1887–1961) proposed a thought experiment known today as Schrödinger's cat. Let's say that an uncertain quantum mechanical process determines whether a cat hidden in a closed box is alive or not. Before the box is opened, could the cat be both alive and not alive at the same time? In 1957, the American physicist Hugh Everett (1930–1982) proposed his "relative state" formulation of quantum mechanics that addressed this question, known today as the many-worlds interpretation (MWI). Another American physicist, Bryce Seligman DeWitt (1923–2004), promoted Everett's formulation for many years afterward, helping to establish the MWI as a plausible and mathematically feasible interpretation of quantum mechanics.

Does the many-worlds interpretation cause any scientific problems?

As with any scientific theory, the many-worlds interpretation of quantum mechanics brings up challenging questions. A common one has to do with energy: if the total amount of energy in a universe is conserved, as it is in ours, then there must be an outside source of energy to fill all the new universes that are continually being created. What is this outside source of energy, and where is it located?

Does the many-worlds interpretation mean my cat could live forever?

If a series of quantum mechanical processes determines at any given time whether a creature (such as the Schrödinger's cat) will live or die, there could always be a timeline and universe in the MWI where the outcome of those processes cause the cat to continue living. By following that universe each time another set of life-resulting quantum processes occur, it is theoretically possible that the cat could continue to stay alive in at least one universe indefinitely. Such an outcome is, however, highly improbable, and the odds of any of us sharing that particular universe with the cat would be exceedingly slim.

Does the many-worlds interpretation imply that there is an infinite multiverse?

The effects of the MWI are dependent on the many parallel universes moving through the dimension of time. The Big Bang sets a starting time, so all these MWI universes are likely to be following the same laws of physics and parameters of the universe that have held true since the Big Bang. In addition, our universe appears to be able to move only forward in time. That means that the number of universes that have been created by quantum mechanical processes since the Big Bang depends on the number of particles that undergo such processes and the amount of time that such processes have had available since the Big Bang to occur. If both of these quantities are finite so far, then the number of MWI universes is huge but not yet infinite.

PHYSICS OF THE MIND

What role does physics play in the human brain?

The brain is the center of all the higher-level information processing in the human body,

and it plays the major role in producing the response of a person and the person's systems to external stimuli. The human brain is made of atoms and molecules, and therefore, everything that happens in the brain follows the laws of physics and occurs as a result of physical processes.

What is the electrical behavior of the human brain?

Each cubic centimeter of a typical person's brain contains about 4,000 km (2,500 miles) of nerve cell "wiring" in a complex network. A person is typically born with about 100 billion brain cells. These cells, called neurons, transmit electrical signals that are around 0.001 volt, communicating a vast amount of information. The brain is very efficient; the total amount of power used by a human brain is about 12 watts. By comparison, a typical desktop computer uses up to 20 times that much power.

What are the physical characteristics of a brain cell?

Neurons have three basic parts: the cell body or soma; a long, wirelike projection called the axon; and small nerve endings called dendrites. Dendrites can be located on both the soma and the axon. Depending on the kind of neuron, the soma can be at one end of the neuron or in the middle, and the axon can have one or more open ends that can make connections with the ends of other neurons.

The brain and the spinal cord together comprise the central nervous system (CNS). There are neurons located throughout your body, not just in your brain; in fact, there are thousands of kinds of neurons, organized into three general categories: sensory neurons that carry electrical signals inward toward the CNS, motor neurons that carry signals outward from the CNS, and interneurons that carry signals within the CNS. Together, they make a complex electrical circuit system that transfers information and electric power to and from every organ and every part of you.

What are the physics of intelligent thought?

The human brain contains a number of substructures, the largest of which is the cerebrum. The outer layer of the cerebrum is called the cerebral cortex. The cortex is highly convoluted, with grooves called sulci and folds called gyri all over its surface; the more grooves and folds there are in the cortex, the more likely it is that complex thoughts and ideas can be formed and contemplated.

What is intelligence? What is "the mind"? What is "consciousness"? Is there such a thing as a soul? There may be some questions that defy scientific explanation.

The part of the cerebral cortex toward the front and top of your skull is called the prefrontal cortex, and that is where scientists think that a person's attention, short-term memory, and "executive function"—complex thought, judgment, decision making, and social awareness—are centered and coordinated. The electrical networks in this part of the brain are extremely complex, and biophysicists are still trying to understand how much of this part of the brain works to create intelligent thinking.

What is intelligence?

Intelligence is far more than just being able to get high scores on tests—it is not at all a simple concept. There are many categories of activity that could be measured as intelligence such as linguistic (language), logical (math), kinesthetic (body), intrapersonal (mind and emotion), interpersonal (social), spatial, and musical. Intelligence can also be organized as being analytical (the ability to examine information and reach conclusions), creative (the ability to generate original and interesting ideas), and practical (the ability to fit into and thrive in an environment). There is substantial overlap in every kind of intelligence, and different subregions of the brain tend to be electrically active depending on the kinds of intelligence that are being used at any given time.

What makes us intelligent?

Although the laws of physics do not necessarily predict the development of intelligence, it is scientifically true that, from very complex systems, unexpected properties can emerge and produce new systems and phenomena. For example, when trillions upon trillions of subatomic particles interact following the laws of physics, they begin forming mixtures, molecules, compounds, and chemicals. Eventually, they start behaving by a new set of rules that govern the behavior of these more complex things. That's how chemistry emerged from physics. Similarly, it can be argued that biology emerged from chemistry, that ecology emerged from biology, and so on.

The human brain, with its billions upon billions of neurons, originally evolved to give early humans the ability to survive and reproduce as a species millions of years ago. It seems scientifically feasible that the great complexity of the brain somehow produced an emergent property that led to humanity's ability to learn and to communicate to one another what was learned. Intelligence could have emerged in this way, at an early stage of human evolution, and become enhanced over time through language and other complex forms of communication. Ultimately, it led to our ability to write, think, reason, create, invent, construct, and study and contemplate our world and universe with physics.

What is artificial intelligence?

Artificial intelligence, or AI, is the term used to describe the activity of machines that mimic intelligent human behavior like learning and problem solving. Normally, computer software is designed in such a way that a machine or computer running the software can go through a list of possible responses to various inputs, process the input, and select the correct course of action based on various decision-making processes. Machines that use artificial intelligence are often called robots.

What is the Turing test?

The British mathematician and philosopher Alan Turing (1912–1954) is considered the father of modern computer science. In 1950, Turing proposed a way to test whether a machine exhibits human-like intelligence. A person would type questions into a computer, and responses would come back on the computer screen. The AI would pass the Turing test if the person could not tell whether it was another person or the AI answering the questions.

Turing predicted that by the year 2000, an AI would be able to pass his test about one-third of the time. This has not yet happened, however. The latest predictions are that Turing test-capable AI will be ready by about 2030.

British mathematician Alan Turing was a key figure in developing the first computers. After he saved countless lives by breaking a German code in World War II, his own government chemically castrated him for being gay, which eventually led to his suicide.

Are artificial intelligence systems able to think on their own?

AI has advanced significantly since it began as a subfield of computer science in the 1950s. In what is sometimes called the "AI effect," many tasks that used to be goals of AI systems are now routine technological capabilities. This includes the ability to read letters and numbers, understand basic human speech, and play games like chess at a high level. On the mechanical side, robots can now do complex tasks like fetch newspapers, clean hotel rooms, and carry on simple conversations with emotional-sounding tonal variations.

Properly programmed AI systems are able to learn how to achieve specific tasks—a process often called machine learning. Some AI systems are structured in a way similar to the way scientists think the human brain and its neurons are wired; these are called neural networks. AI has a long way to go, however, before we might equate their learning and processing abilities with human thought. It is likely to be years, for example, before an AI will be able to pass the Turing test consistently.

Heck, are we humans able to think on our own?

As it turns out, there is a scientific controversy among brain scientists about whether we as intelligent humans actually think freely. While it is true that the human brain is   347 a very complex system, each of the trillions of particles in the brain must behave according to the laws of physics. If a force is exerted on a particle, it must move a certain way; if energy is added or removed from a particle, it must heat up or cool down accordingly. Taken all together, every input to the human brain must produce a predictable corresponding output. That suggests all human behavior consists of reactions and responses rather than products of active decision making. We might, in other words, be no more creative or innovative than any computer; we just think we are because our brains are so complex that we can convince ourselves we are free thinkers!

Do the laws of physics imply that free will is an illusion?

Our brains definitely sometimes cause our bodies to react before we can think about it. Human reflexes, for example, can make you kick or duck before you can think about it because the electrical pulses through your nerves can move at more than 400 km per hour (250 mph), while your prefrontal cortex takes at least a part of a second to process sensory input. Some brain experiments have even shown that a person will sometimes answer questions before the person's prefrontal cortex—the thought center of the brain—is electrically activated. So, rather than acting with free will in such cases, our brains actually respond without consulting us first and then we justify our decisions after the fact.

Might the laws of physics allow for the existence of free will?

Claiming that those experiments confirm that we humans have no free will might be taking it a little too far. Our brains, after all, are very complex, and all the physics and biology research we humans have conducted on the brain have only uncovered a small part of its mysteries. If, for example, intelligence is an emergent property, then so is consciousness—and the physics of consciousness remains almost completely unexplored.

It is certainly possible that a well-planned decision, made after a period of consideration, would provide enough time to allow the conscious part of the brain to influence our actions freely. It is also true that at the cellular, molecular, and electronic level, the neurophysical networks of our brains are subject to quantum mechanical uncertainties. So, perhaps not every input will produce an exactly predetermined output, and that unpredictability may form a core component of what we humans consider to be free will.

BLACK HOLES AND
THE END OF TIME

What is a black hole?

A black hole is a region of space where the gravity is so strong that an object's escape velocity equals or exceeds the speed of light. The idea was first proposed in the 1700s, when the English scientist John Michell (1724–1793) hypothesized that Newton's law of universal gravitation allowed for the possibility of stars that were so small and massive that particles of light could not escape. Thus, the star would be dark.

More than a century later, after the confirmation of Albert Einstein's General Theory of Relativity, physicists realized that there could be locations in the universe where space was so severely curved that it would actually be "ripped" or "pinched off." Anything that fell into that location would not be able to leave. This general relativistic idea of an inescapable spot in space—a hole where not even light could leave—led physicists to coin the term "black hole."

If you saw a black hole closely enough, you would notice a lensing effect, as light around and near the event horizon would be warped.

How are black holes created?

Two categories of black holes are known to exist, and a third kind has been hypothesized but not yet detected. One kind, known loosely as a stellar black hole or low-mass black hole, is found wherever the core of a very massive star (usually 20 or more times the mass of the sun) has collapsed. The other known kind, called a supermassive black hole, is found at the centers of galaxies and is millions or even billions of times more massive than the sun.

The third kind of black hole, called a primordial black hole, is found at random locations in space. It is hypothesized that these black holes were created at the beginning of cosmic expansion as little "imperfections" in the fabric of space-time. However, no such black hole has yet been confirmed to exist.

What would happen if you fell into a black hole?

That depends on the size of the black hole. If you fell into a low-mass or stellar black hole, it has a very high density, so very strong tidal forces would cause substantial destruction to your body. The front of the body would be accelerated so much more intensely than the rear that the atoms and molecules would be pulled apart from one another. You would then flow into the black hole as a stream of subatomic particles.

If you fell into a supermassive black hole, it would have a much lower density, so there would not be such strong tidal forces to deal with. In that case, the relativistic effects of being near the event horizon, or boundary, of a black hole would become apparent. As you fell closer and closer to the event horizon, your speed would approach the speed of light, and the passage of time for you would get slower and slower. Eventually, you would essentially be frozen in place, and the event horizon will grow outward to meet you. As it did so, your body would change from matter into energy, following the formula $E = mc^2$, and disappear into the black hole.

What happens when two black holes collide?

Black holes almost never hit one another head-on. Rather, they spiral together, coming closer and closer until their event horizons come in contact. At that moment, the two black holes merge into one, and a huge amount of energy is released into the universe. Some of that energy goes into space itself, creating gravitational waves that propagate into the universe at the speed of light.

The first detection of such a gravitational wave was made on September 14, 2015, when astrophysicists using the LIGO (Laser Interferometer Gravitational-Wave Observatory) facility detected a tiny change in the fabric of space-time rippling through Earth at almost exactly the speed of light. The wave was caused by the collision of two black holes about 1.3 billion light-years (12 billion trillion kilometers) away from Earth, causing an explosion more powerful than any that had ever before been detected.

Many more such detections have been made with LIGO since 2015, ushering in a new era of astrophysical study that is not dependent on electromagnetic waves. For their work on black holes and on the LIGO project, the American physicists Kip Thorne, Barry Barish, and Rainer Weiss were awarded the 2017 Nobel Prize in Physics.

What is Hawking radiation?

According to British physicist Stephen Hawking (1948–2018), energy can slowly leak out of a black hole. This leakage, called Hawking radiation, occurs because the event horizon (boundary) of a black hole is not a perfectly smooth surface but "shimmers" at a subatomic level due to quantum mechanical effects. At these quantum mechanical scales, space can be thought of as being filled with so-called virtual particles, which themselves cannot be detected but can be observed by their effects on other objects. Virtual particles come in two "halves," and if a virtual particle is produced just inside the event horizon, there is a tiny chance that one "half" might fall deeper into the black hole, while the other "half" would tunnel through the shimmering event horizon and leak back into the universe.

Hawking radiation is a very, very slow process. A black hole with the mass of the sun, for example, would take many tril-

The late, great physicist Stephen Hawking hypothesized that radiation could escape a black hole because of imperfections along its surface. This is now called Hawking radiation in his honor.

lions of years—far longer than the current age of the universe—before its Hawking radiation had any significant effect on its size or mass. Given enough time, though, the energy that leaks through a black hole's event horizon becomes appreciable. Since matter and energy can directly convert from one to another, the black hole's mass will decrease a corresponding amount.

Could a giant black hole swallow the universe?

No, a giant black hole will not consume our universe. Remember, black holes are deep gravitational structures, not cosmic "vacuum cleaners," and they do not "suck" things in. Picture a manhole in a construction zone on a busy city sidewalk: if you fall in, you may not come back out again, but if you avoid the hole and its environment, you will not be in danger of falling in. A black hole works the same way. No matter how massive a black hole is, there is always a limit to its gravitational influence, and matter that lies beyond that limit will not be affected by the black hole at all.

What is the holographic model of a black hole?

On Earth, a hologram is created when a picture of a three-dimensional object is encoded, or drawn onto, a two-dimensional surface such as a piece of glass. When light shines onto that surface, someone looking at it sees the original three-dimensional object, as if all the 3-D information were now projected into 2-D.

In the holographic model of a black hole, a complete description of anything that falls into a black hole is preserved on the event horizon, perhaps in the form of subtle quantum fluctuations. If this model is correct, then even if the object is destroyed, it would be possible to reconstruct the object using that description. So, for example, if an apple fell into the black hole, the quantum mechanical information about each particle in the apple would be saved at the event horizon even though the apple is converted into energy; by "reading" the event horizon's information carefully enough, each particle could be restored with an equal amount of energy, and the apple (or at least applesauce) could be reconstituted with great effort.

What is the holographic principle of the universe?

In the same way that a black hole could follow a holographic model, physicists have produced models of quantum gravity that, if correct, would store complete descriptions of the contents of the observable universe on the cosmic horizon, the boundary beyond which we cannot get any information. In this model, our three-dimensional existence in space-time is still physically valid because the average temperature of the universe is very low—only about 3 degrees above absolute zero (–270 degrees Celsius).

How will the universe end?

Even though the cosmic horizon is unlike an event horizon of a black hole, they are similar in that someone inside the horizon cannot see what is beyond it. So, although

351

Is the universe a giant black hole?

There are holographic models of the universe based on string theory that appear to have some merit in describing the universe as a whole. One example of how holography works well in fundamental physics is the Anti-de Sitter/conformal field theory correspondence, or AdS/CFT correspondence for short, which may be a mathematically feasible way to connect gravity with the other three fundamental forces even though they are so much stronger. The holographic principle for the entire universe, on the other hand, is not yet mathematically solved, in part because the cosmic horizon is a very different kind of surface from the event horizon of a black hole. The cosmic horizon is a continuously growing observational boundary, whereas the event horizon of a black hole is a quantum mechanical and gravitational boundary.

we cannot say how the entire universe will end, we can predict that within the cosmic horizon, the amount that dark energy is pushing the expansion of the universe is stronger than the amount that dark matter and baryonic matter together is pulling on the universe to slow down that expansion.

What this means is that the observable universe will continue to expand faster and faster without limit. Many billions of years from now, the expansion rate will be so great that anything not held together by gravity will fly so far away from everything else that no more large structures will be assembled. Then, if the most current models of the universe are correct:

- About one thousand trillion (10^{15}) years later, all the stars in the universe will have burned out.

- About one trillion trillion trillion (10^{36}) years after that, all the protons and neutrons in the universe will have decayed.

- About one googol (10^{100}) years after that, all the black holes in the universe will have evaporated due to Hawking radiation.

All that will remain after that are fermions and bosons being carried farther and farther apart by the expansion of space. As the age of the universe approaches an infinitely long time, the temperature of the universe will approach absolute zero, and the density of the universe will approach zero.

Is there any way to change how the universe will end?

When the universe reaches the end of time, all the asymmetric wrinkles and dimples caused by massive or hot objects will disappear. The result will be a kind of complete, perfect universe, and it will stay that way forever. Considering, on the other hand, how the dramatically the universe has started off—from the Big Bang 13.8 billion years ago to the formation of quarks, leptons, bosons, stars, galaxies, black holes, planets, people, and so much more—this cold and dark finale to the universe may seem a bit quiet.

There's no need to be concerned. Physicists continue to study this final frontier of science, and a number of ideas have been proposed that may cause this end of the universe to become the start of something new. In one model, based on the General Theory of Relativity, matter decays completely, leaving only energy in the universe. This then becomes the starting condition, like the Planck time, for a new universe to be born, with new and possibly different scales of physics. In another model, based on M-theory, the universe is a four-dimensional intersection of two constantly moving five-dimensional membranes. Every time the membranes pull apart and come back together again, an existing universe ends—and a new universe begins in its place, with its own Big Bang.

When will we know if the end of the universe will be the beginning of a new universe?

Are either of these models—or any other cyclical model of the universe—correct? It will depend on many still unconfirmed scientific ideas, like whether M-Theory, string theory, supersymmetry, or quantum gravity prove to be correct, and what dark matter and dark energy really are. Much more scientific research needs to be conducted, theoretically, computationally, and experimentally, before we know for sure. Maybe someday, *you* will be the person who discovers the solutions! After all, the questions we will have to answer are the ones we have yet to ask.

Further Reading

History of Physics

Cropper, William H. *Great Physicists: The Life and Times of Leading Physicists from Galileo to Hawking.* Oxford: Oxford University Press, 2001. ISBN 019-517324-4.

Hakim, Joy. *The Story of Science: Aristotle Leads the Way.* Washington, DC: Smithsonian Books, 2004. ISBN 1-58834-160-7.

Hakim, Joy. *The Story of Science: Newton at the Center.* Washington, DC: Smithsonian Books, 2004. ISBN 1-58834-161-5.

Hakim, Joy. *The Story of Science: Einstein Adds a New Dimension.* Washington, DC: Smithsonian Books, 2007. ISBN 1-58834-162-4.

Holton, Gerald, and Brush, Stephen G. *Physics, the Human Adventure: From Copernicus to Einstein and Beyond.* Piscataway, NJ: Rutgers University Press, 2005. ISBN 0-8175-2908-5.

Physics of Sports

Adair, Robert. *The Physics of Baseball.* New York: Harper-Collins, 2002. ISBN 0-06-08436-7.

Armenti, Angelo, editor. *The Physics of Sports.* College Park, MD: American Institute of Physics, 1992. ISBN 0-88318-946-1.

Fontanella, John J. *The Physics of Basketball.* Baltimore: The Johns Hopkins University Press, 2006. ISBN 0-8018-8513-2.

Gay, Timothy. *The Physics of Football: Discover the Science of Bone-Crunching Hits, Soaring Field Goals, and Awe-Inspiring Passes.* New York: HarperCollins, 2005. ISBN 10-06-0862634-7.

Haché, Alain. *The Physics of Hockey.* Baltimore: The Johns Hopkins University Press, 2002. ISBN 0-8018-7071-2.

Jorgensen, Theodore P. *The Physics of Golf.* New York: Springer Science, 1999. ISBN 0-387-98691-X.

Leslie-Pelecky, Diandra. *The Physics of NASCAR: How to Make Steel + Gas + Rubber = Speed.* New York: Dutton Penguin Group, 2008. ISBN 978-0-525-95053.

Lorenz, Ralph D. *Spinning Flight.* New York: Springer Science, 2006. ISBN 978-0387-30779-4.

Physics of Light, Music, and the Arts

Falk, David, Dieter Brill, and David Stork. *Seeing the Light: Optics in Nature, Photography, Color, Vision, and Holography.* New York: Harper & Row, 1986. ISBN 0-476-60385-6.

Fletcher, Neville, and Thomas Rossing. *Physics of Musical Instruments.* New York: Springer Science, 1998. ISBN 378-0387-98374-5.

Laws, Kenneth. *Physics and the Art of Dance: Understanding Movement.* Oxford: Oxford University Press, 2002. ISBN 0-19-514482-1.

Rogers, Tom. *Insultingly Stupid Movie Physics: Hollywood's Best Mistakes, Goofs, and Flat-Out Destructions of the Basic Laws of the Universe.* Naperville, IL: Sourcebooks, 2007. ISBN 978-1-4022-1033-4.

Rossing, Thomas. *Light Science: Physics and the Visual Arts.* New York: Springer Verlag, 1999. ISBN 387-98827-0.

Sundburg, Johan. *The Science of the Singing Voice.* DeKalb, IL: Northern Illinois University Press, 1987. ISBN 087-58012-0-X.

General Physics with a Creative Approach

Nitta, Hideo. *The Manga Guide to Physics.* San Francisco: No Starch Press, 2009. ISBN 978-1-59327-196-1.

Walker, Jearl. *The Flying Circus of Physics.* New York: John Wiley and Sons, 2007. ISBN 978-0-471-76273-7.

How Things Work

Bloomfield, Louis A. *How Things Work: The Physics of Everyday Life.* New York: John Wiley, 2010. ISBN 978-0-470-22399-4. (See also *howeverythingworks.org*)

Macaulay, David. *The New Way Things Work.* New York: Houghton Mifflin, 1998. ISBN 0-395-93847-3.

Physics Frontiers

Carroll, Sean. *From Eternity to Here: The Quest for the Ultimate Theory of Time.* New York: Dutton/Penguin, 2010. ISBN 978-05259-5133-9.

Gates Jr., S. James, and Cathie Pelletier. *Proving Einstein Right: The Daring Expeditions That Changed How We Look at the Universe*. New York: PublicAffairs, 2019. ISBN-13: 9781541762251.

Greene, Brian. *The Elegant Universe: Superstrings, Hidden Dimensions, and the Quest for the Ultimate Theory*. 2nd Edition. New York: W. W. Norton, 2010. ISBN-13: 9780393338102.

Hawking, Stephen. *Brief Answers to the Big Questions*. London: Hodder & Stoughton, 2018. ISBN-13: 9781529345421.

Kaku, Misho. *Physics of the Impossible: A Scientific Exploration into the World of Phasers, Force Fields, Teleportation, and Time Travel.* New York: Doubleday, 2008. ISBN 978-0-307-27882-1.

Levin, Janna. *Black Hole Blues and Other Songs from Outer Space*. London: Bodley Head, 2016. ISBN-13: 9781847921963.

Liu, Charles, Karen Masters, and Sevil Salur. *30-Second Universe*, Brighton: Ivy Press, 2019. ISBN-13: 9781782408505.

Randall, Lisa. *Dark Matter and the Dinosaurs: The Astounding Interconnectedness of the Universe*. New York: Ecco, 2015. ISBN-13: 9780062328472.

Randall, Lisa. *Higgs Discovery: The Power of Empty Space*. New York: Ecco, 2013. ISBN-13: 9780062300478.

Online Resources (all accessed on May 4, 2020)

AASNova: *Research Highlights from the Journals of the American Astronomical Society*. aasnova.org.

Center for History of Physics. aip.org/history-programs/physics-history.

Cosmology: The Study of the Universe. map.gsfc.nasa.gov/universe/.

The Particle Adventure: The Fundamentals of Matter and Energy. Sponsored by the Lawrence Berkeley National Laboratory. particleadventure.org.

Physics—Spotlighting Exceptional Research. Produced by the American Physical Society. physics.aps.org.

Physics to Go: Explore Physics on Your Own. Produced by the American Association of Physics Teachers and the American Physical Society. compadre.org/informal.

Physics Today. Produced by the American Institute of Physics. physicstoday.org.

Physics World. Produced by the Institute of Physics. physicsworld.com.

Symbols

Symbol	Meaning
a	acceleration
A	ampere
AC	alternating current
AM	amplitude modulation
b	bottom quark
B	magnetic field strength
Btu	British thermal unit
c	centi (10^{-2})
C	capacitor
c	charm quark
C	coulomb
c	speed of light
°C	degrees celsius
cal	calorie
Cal	food calorie (kcal)
cd	candela
d	deci (10^{-1})
d	distance
d	down quark
da	deka (10^{1})
dB	decibel
DC	direct current
E	electric field
E	energy
e	electron
EM	electromagnetic

Symbol	Meaning
emf	electro-motive force
eV	electron volt
f	femto (10^{-15})
F	force
f	frequency
°F	degrees fahrenheit
FM	frequency modulation
G	giga (10^9)
g	gram
g	gravitational field strength
G	universal gravitational constant
h	hecto (10^2)
h	Planck's constant
Hz	hertz (cycles per second)
i	ac electric current
I	dc electric current
I	impulse
I	moment of inertia
J	joule
k	generic symbol for constant
K	kelvins
k	kilo (10^3)
KE	kinetic energy
kg	kilogram
kWh	kilowatt hour
l	length
lum	lumen
M, m	mass
M	mega (10^6)
m	meter
m	milli (10^{-3})
MA	mechanical advantage
n	nano (10^{-9})
n	index of refraction
n	neutron
N	newton
N	normal force
n	generic integer number
p	momentum
P	peta (10^{15})
p	pico (10^{-12})

Symbol	Meaning
P	power
P	pressure
Pa	pascal
PE	potential energy
psi	pounds per square inch
q	charge
R	resistance
S	entropy
s	second
s	strange quark
T	temperature
T	tera (10^{12})
t	time
t	top quark
u	up quark
V	potential difference (volt)
v	velocity
V	volt
W	charged boson (weak force)
x	direction in Cartesian coordinate system
X, x	generic unkmown
y	direction in Cartesian coordinate system
z	direction in Cartesian coordinate system
Z	neutral boson (weak force)

Greek	Letter	Meaning
α	alpha	alpha particle
α	alpha	angular acceleration
β	beta	beta particle
γ	gamma	gamma ray
γ	gamma	special relativity factor
Δ	Delta	change in, or uncertainty in
θ	theta	angle
μ	mu	micro (10^{-6})
υ	nu	frequency
π	pi	pi (3.1415926…)
τ	tau	torque
Ω	Omega	ohm
ω	omega	angular velocity

Glossary

Absolute zero—Temperature at which molecular motion is at a minimum.

Acceleration (*a*)—Change in velocity divided by the time over which the change occurs (vector).

Accuracy—How correct or how close to the accepted result or standard a measurement or calculation has been.

Acoustics—Study of the ways musical instruments produce sounds, the design of concert halls, using ultrasound images.

Active noice cancellation (ANC)—Device that creates a waveform that is the opposite of the noise so that the noise is cancelled.

Aerodynamics—An aspect of fluid dynamics dealing with the movement of air.

Air drag—Friction caused by an object moving through air. Drag depends on velocity, area, shape, and the density of the air.

Alloy—A metal that is a mix of two or more different metals.

Alpha decay—Type of radioactive decay where nucleus emits an alpha particle. Result is nucleus with atomic number reduced by two and mass number reduced by four.

Alpha particles—Helium atoms with two electrons removed emitted with high energy by some radioactive nuclei.

Alternating-current (AC)—Polarity of the voltage source and thus current changes back and forth at a regular rate.

Ampere (A)—Unit of measure of electric current. Equal to one coulomb of charge passing through a wire divided by one second.

Amplitude—The distance on a wave from the midpoint to the point of maximum displacement (crest or compression).

Amplitude modulation (AM)—EM wave amplitude changes to represent transmitted information.

Aneroid barometer—Device to measure gas pressure in which the elastic top of an extremely low-pressure drum is bent by the pressure.

Angular momentum—Equal to the product of its moment of inertia and its angular velocity.

Angular velocity (ω)—Equivalent of velocity for rotational motion.

Anode—Positive electrode or terminal on, e.g. a battery, electrolytic cell, or cathode ray tube.

Antenna—Used to transmit or receive electromagnetic radio waves.

Antinode—Locations in a standing wave where there is the largest amplitude caused by constructive interference.

Apparent weightlessness—Occurs when an object is in free fall. Force on object in direction opposite gravity is zero.

Archimedes' Principle—An object immersed in a fluid will experience a buoyant force equal to the weight of the displaced fluid.

Arrow of time—The forward direction of time is the one in which entropy increases or remains the same.

Astrophysics—The study of the universe and its contents, including planets, stars, galaxies, space, time, and the universe itself.

Atmospheric physics—Study of the atmosphere of Earth and other planets, especially effects of global warming and climate change.

Atomic and molecular physics—Study of single atoms and molecules that are made up of these atoms.

Atomic mass number—Number of protons and neutrons in nucleus.

Atomic number—Number of protons in the nucleus and thus the nuclear charge.

Atoms—Smallest piece into which an element can be divided and retain its properties. Uncuttable; can be neither created nor destroyed.

Audion—A vacuum-tube amplifier invented in 1906 by Lee DeForest.

Barometer—Device to measure gas pressure.

Bernoulli effect—One of three causes of lift. Due to difference in air pressure on upper and lower surfaces of wing.

Beta decay—Type of radioactive decay where nucleus emits a beta particle (electron) from the change of a neutron into a proton. Antineutrino also emitted. Result is nucleus with same atomic mass number but atomic number increased by one.

Beta particles—Electrons emitted at high energy by some radioactive nuclei.

Biophysics—Study of the physical interactions of biological molecules.

Black—The absence or the absorption of all light.

Block and tackle—Simple machine that is a combination of fixed and movable pulleys.

Bohr model—A nuclear model but with electrons moving in only certain allowed orbits at discrete radii and with specific energies. When in these orbits their radii and energies are constant. The atoms do not emit or absorb radiation. Electrons gain or lose energy when they jump from one allowed orbit to another. Then they emit or absorb light with a frequency given by $hf = E_2 - E_1$ where E_2 and E_1 are the energies of the electrons in the allowed orbits.

Boiling point—Temperature at which pressure of the water vapor equals the atmospheric pressure.

Bose-Einstein Condensate—Quantum effects on a macroscopic scale exhibited by extremely cold gas of atoms.

Boson—Particles like photons and gluons with integral spin.

Brass—An alloy that is typically 80% to 90% copper with zinc.

British Thermal Unit (Btu)—Unit of energy often used to measure heat in homes and industries.

Brittle materials—Usually break when under tension rather than compression.

Bronze—An alloy of copper and tin.

Buoyant force—The reduction in weight equal to the net upward force of a fluid.

Calorie (cal)—A unit of energy that is used both to measure both thermal energy and heat.

Candela (cd) —Unit of measurement of luminous intensity.

Capacitor—Device that consists of two conducting plates with opposite charge separated by an insulator to store electric charge.

Carbon dating—Method of dating objects by measuring relative amounts of radioactive and stable carbon in them.

Carnot efficiency—Largest efficiency that a heat machine can obtain. $e = (T_{hot} - T_{cold})/T_{hot}$.

Cathode—Negative electrode or terminal on, e.g. a battery, electrolytic cell, or cathode ray tube.

Cathode rays—Electron beam in an evacuated glass tube.

Celsius scale—Temperature scale on which the freezing point of water is $0°$ C and the boiling point $100°$ C.

Center of gravity—Depends on how weight of extended object is distributed. The object's motion is the same as that of a point object in which the gravitational force that acts on its center of gravity.

Center of mass—Because the force of gravity is proportional to the mass, the center of mass is at the same location as the center of gravity.

centi—10^{-2}

Centrifugal force—A pseudo-force that exists only in rotating reference frames. Apparent force outward from center of rotation.

Centripetal acceleration—Acceleration due to a centripetal force toward the center of a circle.

Centripetal force—Force that keeps an object moving in a circle. Must be exerted by an external agent.

Charge of electron—$e = -1.602 \times 10^{-19}$ C.

Charging by contact—Results when a neutral object touches a charged object.

Charging by induction—Results when a conductor is first polarized by a charged object that doesn't touch it. Then one polarity of charge is removed from conductor.

Chemical energy—Potential energy of a system due to its chemical composition. A chemical change can cause this energy to be changed into or out of other forms. Animal bodies and batteries are two common objects that have chemical energy.

Chemical physics—Study of the physical causes of chemical reactions between atoms and molecules and how light can be used to understand and cause these reactions.

Chromatic aberration—Defect of a lens that introduces colors into images.

Circuit—Circular path through which charge can flow. Needs a source of potential difference, wires and a device with resistance.

Circuit, open—Circuit that is open so no charges flow.

Code division multiple access (CDMA)—Method of sending cell phone calls in which three calls are packed together with six more calls at two other frequencies.

Coefficient of friction μ—Friction force divided by normal force.

Coefficient of kinetic friction μ_k—Coefficient when there is relative motion between the two surfaces. Smaller than static friction.

Coefficient of static friction μ_s—Coefficient when there is no relative motion between the two surfaces.

Color blindness—An inability to see some colors due to an inherited condition.

Color mixing, additive—Combining blue, green, and red light to form other colors.

Color mixing, subtractive—Combining dyes, pigments, or other objects that absorb and reflect light.

Colorimetry—Measurement of the intensity of particular wavelengths of light.

Colors, complementary—Pair of one secondary and one primary color that form what is close to white light.

Colors, primary—Blue, green, and red of light; cyan, magenta, and yellow of pigments.

Colors, secondary—Cyan, magenta, and yellow of light; blue, green, and red of pigments.

Compass—Magnetized pointed metallic pointer that can rotate about a low-friction pivot point.

Composite materials—Modern composite materials use carbon fibers with very high tensile strength but low compressive strength to increase the tensile strength of brittle materials.

Compression—Region of higher pressure in a longitudinal wave.

Compressive force—Force pushing an object together.

Condensation—Change from vapor to liquid on a cold object.

Condensed matter physics—Studies the physical and electrical properties of solid materials.

Conduction, heat—Occurs when two objects are in contact, like when you put your hand in hot water.

Conductor, electrical—A material that allows electrons to move easily through it.

Cone—Cone-shaped nerve cells on the retina that can distinguish fine details in images. Located predominantly around the center of the retina called the fovea; responsible for color vision.

Conservation of energy—The energy of a system doesn't change; energy is neither created nor destroyed. The energy put into a system equals the energy change in the system plus the energy leaving the system. True as long as no objects are added to or removed from a system, and as long as there are no interactions between the system and the rest of the world.

Conservation of mass—Mass is neither created nor destroyed in chemical reactions.

Conservation of momentum—The momentum of the system is constant when external forces are zero.

Constant in Coulomb's law—$k = 9.0 \times 10^9$ Nm^2/C^2.

Convection—The motion of a fluid, usually air or water, due to a difference in temperature. An efficient means of heat transfer.

Coordinate system—A set of axes used to locate an object. Usually defined at the reference location. Cartesian coordinates are three directions at right angles to each other, usually the x, y, and z directions.

Coriolis force—A pseudo-force that exists only in rotating reference frames. Apparent force in direction of rotation.

Cornea—A transparent membrane that contains the fluid in the eye.

Cosmology—The study of the origin, history, properties, structure, and future of the universe as a whole.

Coulomb—Equal to the charge of 6.24×10^{18} electrons (negative) or protons (positive).

Coulomb's law—Describes the strength of the electrical force between two charged objects. $F = k \, (q_1 \, q_2/r^2)$.

Crest—The highest point of a transverse wave.

Critical angle—Angle at which refracted light angle is 90°.

Critical mass—Minimal mass of enriched uranium or plutonium needed to create a bomb.

Current, electrical—The flow of electric charge.

Dark energy—Said to be cause of accelerating expansion of universe. Comprises 72% of mass and energy of universe. Its nature is totally unknown.

Dark matter—Matter in universe that does not interact with electromagnetic radiation but does interact with gravity. Comprises 23% of mass of universe. Its composition is unknown.

de Broglie wavelength—Wavelength associated with moving particle $\lambda = h/mv$.

deci (d)—10^{-1}

Decibel (dB)—Unit of relative intensity of sound.

deka (da)—10^{1}

Diamagnetic—Caused by circling electrons that create their own magnetic field. Material is repelled by magnetic fields.

Difference tone—Frequencies produced as a result of two different frequencies mixing with each other.

Diffraction—Light or other wave bends around the edge of the hole, through a narrow slit, or around a small object.

Dip needle—Like a conventional compass, but the needle rotates in vertical plane.

Direct-current (DC)—Charges travel only in one direction.

Dispersion—Light of different colors are refracted into different directions caused by wavelength dependence of refractive index.

Displacement—Distance and direction of an object from reference location (vector).

Distance—The separation between object's position and the reference location (scalar).

Domain—Group of atoms in a ferromagnet whose spins are aligned in the same direction.

Doppler effect—Change in frequency of a wave that results from an object's changing position relative to an observer.

Drag—Force that opposes the motion of an object through a fluid.

Earth's mass—$m_{Earth} = 5.9736 \times 10^{24}$ kg.

Efficiency, energy—Useful energy output divided by energy input.

Elastic collision—In a collision the kinetic energy of colliding objects is the same before and after the collision.

Elastic deformation—When the deforming force is no longer applied object returns to its original shape (see plastic deformation).

Elastic potential energy—The energy in the squeezed spring or stretched band.

Electric field—Region that surrounds a charged object. Another charged object placed in that field will experience a force.

Electricity, resinous—Produced when plastic, amber, or sealing wax is rubbed.

Electricity, vitreous—Produced when glass or precious stones are rubbed.

Electromagnet—Coil of current-carrying wire wound on a iron core that produces a magnetic field that depends on the current.

Electromagnetic spectrum—The wide range of electromagnetic (EM) waves from low to high frequency.

Electromagnetic wave (EM)—Two coupled transverse waves, one an oscillating electric field, the other a corresponding magnetic field perpendicular to it and perpendicular to motion of wave.

Electromagnetism—Studies how electric and magnetic forces interact with matter.

Electromotive force (*emf*)—Potential difference resulting from separation of charges.

Electron volt (eV)—Energy particle with charge of an electron or proton obtains by being accelerated by a 1 volt potential difference.

Electroscope—Device used to measure electrical charge on an object.

Electrostatics—Study of the causes of the attractive and repulsive forces, called static electricity, that result when objects made of two different materials are rubbed together.

Elementary particles and fields—Study of the interactions and properties of the particles from which all matter is built.

Energy transfer—Requires a source, whose energy is reduced, a means of transferring the energy, and an energy receiver, whose energy is increased.

Entanglement—An effect in an atom or nucleus when two or three photons are emitted. The spin of one photon depends on that of the other, or is entangled with it. Measuring one spin tells you what the direction of the other is.

Entropy—A measure of the dispersal of energy. The greater the energy is dispersed the larger the entropy. As a consequence of the second law of thermodynamics is that the entropy of the system and the environment can never decrease.

Evaporation—Change from liquid to gas at a temperature below the boiling point.

Extremely high frequency (EHF) waves—Frequency 30 Ghz–300 Ghz.

Extremely low frequency (ELF) waves—Frequency less than 30 kHz.

Fahrenheit scale—Temperature scale on which the freezing point of water is 32° F and the boiling point 212° F.

Faraday cage—Cage or metal grating that can shield electrical charge by gathering charges on outer shell of the cage.

Faraday's law—An electric field is produced by a changing magnetic field. The field can cause a current in a wire.

Farsightedness—Also called hyperopia. Allows only objects far from the eye to be focused on retina. Images from close objects are behind the retina so they are blurry.

femto (f)—10^{-15}

Fermion—Particles like quarks, electrons, and neutrinos with half-integral spins.

Ferromagnetic—Describes materials like iron, nickel, and cobalt, in which unpaired electrons in large groups of atoms interact with each other so that they point in the same direction. Strongly attracted by magnetic fields.

First law of thermodynamics—A restatement of the conservation of energy: net heat input equals net work plus change in thermal energy.

First-class lever—Simple machine with pivot at one end or bar, the input force applied at the other end, the effort force between the two. *MA* greater than one.

Fission—Form of radioactive decay where nucleus splits in two. Extra neutrons and a large amount of energy is released.

Flavor—Generation of quark (up/down, strange/charm, top/bottom) or lepton (electron/electron neutrino, muon/muon neutrino, tau/tau neutrino).

Fluid—A liquid or gas that can flow and assume the shape of its container.

Fluid dynamics—Study of fluids in motion.

Fluid statics—Study of fluids in a state of rest.

Focal length—Measures the strength of a lens or mirror. Distance from lens or mirror to focal point.

Focal point—Location where light rays from a very distant object are reflected from a mirror or refracted by a lens to come together.

Force (F)—A push or a pull on an object by an external agent (vector).

Fourier theorem—Any repetitive waveform can be constructed from a series of waves of specific frequencies, a fundamental and higher harmonics.

Frequency (f)—How many cycles of an oscillation occur per second.

Frequency modulation (FM)—EM wave frequency changes to represent transmitted information.

Friction force—A force due to an interaction between an object and the surface on which it rests. Always in the direction opposite the motion. Depends on the force pushing the surfaces together. In most cases is larger if the surface is rougher. Generally doesn't increase if the speed of motion increases. $F_{friction} = \mu N$.

Fuse—Protects household electrical circuits are protected by fuses or circuit breakers. Opens when current exceeds a predetermined limit.

Fusion—Joining of two nuclei to form a more massive nucleus.

Galvanic cell—Combination of two different metals separated by a conducting solution that produces a potential difference and thus a continuous flow of charge.

Gamma ray—High energy photons emitted by some radioactive nuclei.

Gas—State of matter in which the atoms are about ten times further apart than liquids or solids and forces between them are very weak. A fluid that assumes the shape of its container.

Gear—Toothed wheels that transmit torque between two shafts. The smaller gear in a pair is called a pinion and the larger one a gear.

General relativity—Study of the descriptions and explanations of the causes and effects of gravity.

Generation (nG), cell phone—1G analog voice phone calls. 2G digital signals with more simultaneous users; 3G able to receive television-like video, video images; 4G and 5G are very high-speed networks.

Generator—Converts mechanical to electrical energy. An electric field is created at the ends of multiple loops of wire on the armature rotating in a magnetic field. The field causes a current in an attached circuit.

Geometrical optics—Deals with the path that light takes when it encounters mirrors and lenses and the uses of these devices.

Geophysics—Study of the forces and energy found within Earth including tectonic plates, earthquakes, volcanic activity, and oceanography.

giga (G)—10^9.

Gigahertz (Ghz)—One thousand million or one billion hertz.

Global Positioning System (GPS)—Consists of many satellites in 12-hour orbits that broadcast their location and the time the signal was sent; a control system that keeps the satellites in their correct orbits, sends correction signals for their clocks as well as updates to their navigation systems; and a receiver.

Gluon—Eight different massless particles that carry the color force that holds quarks together.

Gold alloys—24-karat gold is pure gold. 18-karet gold has 18 parts gold and 6 parts other metals. 10-karet gold has 10 parts gold and 14 parts other metals.

Gravitational field energy—Energy stored in the gravitational field when an object moves away from the (larger) object that is attracting it.

Gravitational field strength—Ratio of gravitational force to mass of attracted object ($g = 9.8 \, N/kg$).

Gravitational mass—Mass defined by the effect of gravity on object. Gravitational force divided by gravitational field strength ($m = F_{gravitation}/g$).

Graviton—Particle that carries the gravitational force.

Greenhouse gas—Gases transparent to light and short-wavelength infrared radiation, but are opaque to the long-wavelength infrared emitted by warm objects. Examples are carbon dioxide, methane, and water vapor.

Ground fault interrupter (GFI)—Detects difference in current into and out of circuit to protect against accidental grounding, typically through water.

Half life—Time when there are half as many decays in a second as there were originally.

Harmonic—Mode of vibration that is a whole-number multiple of the fundamental mode. The first harmonic is the fundamental frequency. The second harmonic is twice its frequency, etc.

Heat—Energy transfer that results from a difference in temperature is called heat. Heat always flows from the hotter (energy source) to the cooler object (energy receiver). Heat is transferred by conduction, convection, and radiation.

hecto (h)—10^2.

Hertz (Hz)—Unit of measurement of frequency. Equals one cycle per second.

Higgs particle—Hypothesized particle associated with the mechanism that gives rise to the masses of the quarks, leptons, and the two weak-interaction bosons, the W and Z.

High frequency (HF) wave—Frequency 3 Mhz–30 Mhz.

Horsepower (hp)—Unit of power. 1 hp = 746 W = 0.746 kW.

Hue—Related to the wavelength of a color.

Hydraulics—Use of a liquid in a device, usually a machine such as a pump, lift, or shock absorber.

Image, light—Point at which light rays converge.

Image, real—Light rays do converge. Can be projected on a screen.

Image, virtual—Light rays do not converge, but your eye believes that rays came from a single point. Cannot be projected on a screen.

Impedance—Opposition to wave motion exerted by a medium.

Impedance-matching device—Placed between two media to create a smooth impedance transition to maximize energy transfer by the wave through the interface. A transformer.

Impulse—The product of force and the time the force is applied.

Inclined plane—Simple machine used to help lift an object.

Induced drag—A consequence of the lift generated by the wing. Induced drag causes vortices.

Inelastic collision—In a collision the kinetic energy of colliding objects is smaller after the collision than before it.

Inertial confinement—Method of producing nuclear fusion by using lasers to compress tiny amounts of a deuterium-tritium fuel mixture.

Inertial mass—Mass defined as the net force divided by its resultant acceleration ($m = F_{net}/a$).

Inertial reference frame—A reference frame in which there is no acceleration.

Infrasonic—Frequencies below human hearing, 20 Hz.

Inner ear—Consists of the cochlea and three semicircular canals that sense the body's motions and give rise to a sense of balance.

Instantaneous speed (v)—Limiting value of distance divided by time interval when the time interval is reduced to zero.

Insulator, electrical—A material in which electrons are strongly bound to their nuclei and thus cannot move through the material.

Interference, constructive—Two waves both either positive or negative that add producing a larger wave amplitude.

Interference, destructive—Two waves, one positive, the other negative that produce a smaller wave amplitude, even zero.

Interfernce, light—Light waves from two closely-spaced slits or two surfaces of a thin film that produces regions of bright and dark regions.

Isotrope—Nuclei with the same number of protons but different numbers of neutrons.

ITER project—International effort to produce electricity from nuclear fusion.

Jelly doughnut (JD)—Informal energy unit. Energy in a medium-sized jelly-filled doughnut. 1 JD = 239 kcal.

Kelvin temperature scale—Based on absolute zero being 0 K. A kelvin (K) is equal to a degree on the Celsius scale. 0°C is equal to 273.15 kelvins.

Kepler's first law—Planetary orbit is an ellipse with the sun at one focus.

Kepler's second law—In equal times planets sweep out equal areas.

Kepler's third law—The square of the period of the planet is proportional to the cube of the radius of the orbit.

kilo (k)—10^3

Kilocalorie (kcal or Cal)—The food calorie used to measure energy content in foods.

Kilogram (kg)—The mass of a platinum-iridium cylinder that is permanently kept near Paris.

Kilohertz (kHz)—One thousand hertz.

Kilowatt (kW)—1,000 watts or 1,000 joules per second.

Kilowatt-hour (kWh)—Unit of energy used to measure electrical energy.

Kinetic energy (*K* or *KE*)—energy of motion ($K = 1/2mv^2$).

Laminar flow—Flow of slow-moving fluid or object moving slowly through a fluid. Each thin film of water moves slightly faster than the one closer to the surface.

Large Hadron Collider (LHC)—Very high synchrotron at CERN near Geneva, Switzerland.

Laser—Light source using stimulated emission of radiation invented in 1960. Creates tiny spot of very intense, monochromatic light.

Latent heat—Energy needed to change phase of matter.

Latent heat of fusion—Energy involved in change from solid to liquid phase.

Latent heat of vaporization—Energy involved in change from liquid to gas.

Law—Summary of many observations that describes phenomena.

Law of reflection—Angle of incident light is equal to angle of reflected light.

Lens, converging—Has at least one convex side. Its shape causes the entering light rays to converge, that is, come closer together.

Lens, diverging—Has at least one concave side. The shape of the lens causes the entering light rays to spread apart when they leave the lens.

Lepton—Light-weight particles like electrons and neutrinos.

Lepton generation—Each generation has a charged and uncharged particle. In first electron and neutrino, in second muon and muon neutrino, in third tau and tau neutrino.

Lever—Simple machine consisting of a rigid bar that pivots. Three different classes.

Leyden jar—Device consisting of an insulating jar with conductors on inner and outer surfaces. Stores charges. Modern version is a capacitor.

Lift—Upward force on an airplane wing caused by the air moving past it.

Light—An electromagnetic wave to which human eyes respond.

Light detector—Converts light energy to another form of energy.

Light-year—Distance that light travels in one year. 9.4605×10^{12} km or about 6 trillion miles.

Limits of human vision—Lower and upper boundaries of light. 4×10^{14} Hz, or 700 nm and 7.9×10^{14} Hz, or 400 nm.

Liquid—State of matter in which the forces keep the atoms close together, but they're free to move. A liquid is a fluid that flows freely and assumes shape of its container.

Load—An engineering term for force.

Longitudinal wave—Wave and its energy move in the same direction as direction of oscillations.

Lorentz force law—Force on a current-carrying wire in a magnetic field. Proportional to the current, the magnetic field, and the length of the wire. Force is perpendicular to both the current and the magnetic field and is strongest when the current and field are at right angles.

Loudness—Measure of how a listener responds to intensity of sound. Measured in sones.

Low frequency (LF) waves—Frequency 30 kHz–300 kHz.

Lumen (lum)—Unit of measurement of luminous power.

Luminous intensity—Eye response to intensity of light electromagnetic waves.

Luminous power—Eye response to power of light electromagnetic waves.

Lunar eclipse—Occurs when Earth casts its shadow on the moon. Earth must be directly between the sun and the moon.

Mach number—Ratio of a velocity to the speed of sound.

MAGLEV—Magnetically levitated trains that use electromagnetic forces to lift the cars off the track and propel them along thin magnetic tracks.

Magnetic declination—Angular difference between north as shown by a compass and the direction to the geographic north pole, Earth's axis of rotation.

377

Magnetic field—Region around a magnet that causes forces on magnetic materials or other magnets.

Magnetic pole—Region of magnetic strength at ends of magnets. Always come in North-South pairs.

Magnus force—Force on a spinning ball that causes its path to curve.

Mass (*m*)—A property of the object, the net force exerted on object divided by its acceleration (scalar).

Mass energy—Potential energy the result of the mass of an object first postulated by Einstein.

Mathematical physics—Study of how physical processes can best be described by mathematics.

Maxwell's equations—Four equations that describe the nature of electric and magnetic fields.

Mechanical advantage (*MA*)—Output force divided by the input force ($F_{output} / F_{input} = MA$).

Mechanics—Study of the effect of forces on the motion and energy of physical objects including fluids and granular particles.

Medical physics—Study of how physical processes can be used to produce images of the inside of humans, as well as the use of radiation and high-energy particles in treating diseases such as cancer.

Medium frequency (MF) waves—Frequency 300 kHz–3 Mhz.

mega (M)—10^6

Megahertz (Mhz)—One million hertz.

Meson—Intermediate mass particles made up of a quark and antiquark.

Meter (m)—The distance light travels in 1/299,792,458 of a second.

micro (μ)—10^{-6}.

Microwave oven—Uses 2.4 GHz EM waves to cook food.

Microwaves—Electromagnetic waves with frequencies above about 3 Ghz.

Middle ear—Between outer and inner ear. Includes the ear drum, the hammer, anvil, and stirrup, the three smallest bones in the human body, and the oval window on the inner ear.

milli (m)—10^{-3}.

Mirage—Occurs when there is a temperature difference between the air directly above the surface, which is hot and thus less dense, and the cooler, denser, air a few meters above the surface.

Mirror, concave—Curved inward so that the incident rays of light are reflected and can be brought together. Can form real or virtual images.

Mirror, convex—Curved outward. The reflected light spreads out rather than converging at a point.

Mirror, plane—Flat mirror. Forms virtual images.

Moment of inertia (I)—Equivalent of mass for rotational motion. Depends on mass and how far it is from axis of rotation. The further the mass is from the axis, the larger the moment of inertia.

Momentum—The product of mass and velocity $p = mv$ (vector).

Monopole—Isolated North or South poles. Predicted by some theories, but never found.

Motor—Converts electrical to mechanical energy. Current through multiple loops of wires on the armature in a magnetic field causes a force that results in rotation.

nano (n)—10^{-9}.

Natural oscillation frequency—A property of an object that can vibrate. Depends on the mass and force that restores object to equilibrium position.

Nearsightedness—Also called myopia. Allows only objects relatively near the eye to be focused on retina.. Images from distant objects are in front of the retina, so they are blurry.

Neutrino oscillation—Change of neutrinos from one flavor to another that is a proposed solution to deficit of neutrinos emitted by sun.

Neutron—Uncharged particle in nucleus with mass slightly larger than that of the proton.

Newton's first law—If there is no net force on an object, then if it was a rest it will remain at rest. If it was moving, it will continue to move at the same speed and in the same direction.

Newton's second law—If a net force acts on an object, it will accelerate in the direction of the force. The acceleration will be directly proportional to the net force and inversely proportional to the mass. That is, $a = F_{net}/m$ or $F_{net} = ma$.

Newton's third law—When two objects, A and B, interact the force of A on B is equal in magnitude but opposite in direction of the force of B on A.

Newton (N)—Unit of measure of force. In the English system, force is measured in pounds (lbs).

NEXRAD—Next-generation weather radar, relies on the Doppler Effect to calculate the position and the velocity of precipitation.

Node—Locations in a standing wave where there is no amplitude caused by destructive interference.

Noise—Sound intensity at a wide variety of different wavelengths.

Noise, white—Roughly equal sound intensities at all frequencies.

Normal force (*N*)—Force pushing two surfaces together.

Nuclear mass defect—Difference between mass of a nucleus and sum of the masses of the protons plus the sum of masses of the neutrons. Larger the defect, the larger the energy needed to pull nucleus apart or energy released when nucleus is assemble.

Nuclear model—Extremely small positively charged particle, the nucleus, surrounded by electrons.

Nuclear physics—Study of the properties of the nuclei of atoms and the protons and neutrons of which they are composed.

Nuclear reactor—Method of producing electric power where energy released by nuclear fission is used to boil water producing steam that turns generators.

Nucleon—Generic name of protons and neutrons.

Object, light—Point from which light leaves into many directions.

Opaque—An object that allows no light through it.

Optical fiber—Strands of glass that use the principle of total internal reflection to transmit information near the speed of light.

Optics—Area of study within physics that deals with the properties of and applications of light.

Overtone—Mode of vibration are is not a whole-number multiple of the fundamental frequency.

Parabola—Path taken by an object acted upon by gravity that is given an initial horizontal velocity.

Parallel circuit—Circuit allows the charges to flow through different branches.

Paramagnet materials—Caused by unpaired electrons in atoms. Material is attracted by magnetic fields.

Parasitic drag—Force when an airplane wing, automobile, or any other object moves through a fluid.

Pascal (Pa)—Unit of measure of pressure. One pascal is one newton per square meter.

Pascal's principle—A fluid, like water exerts the same pressure in all directions.

Penumbra—Area of a shadow where light from only part of the source is blocked.

Period—The time it takes for a an oscillating object or wave to complete one full vibration; the inverse of the frequency.

Permanent magnet—Domains remain aligned after the polarizing magnetic field is removed, resulting in a permanent magnet.

Perpetual motion—Forbidden by the second law of thermodynamics.

peta (P)—10^{15}.

Pewter—An alloy of tin with copper, bismuth, and antimony.

Photon—Packet of light energy light quantum by Einstein in 1905. Given name photon in 1926. Has no mass or charge, but carries angular momentum. Always moves at the speed of light.

Photon energy—$E = hf$, where f is frequency.

Physical optics—Division of optics that depends on the wave nature of light, polarization, diffraction, interference, and the spectral analysis of light waves.

Physics—The study of the structure of the natural world that seeks to explain natural phenomena in terms of a comprehensive theoretical structure in mathematical form; from the Greek *physis*, meaning nature.

Physics education research—Study of how people learn physics and how best to teach them.

pico (p)—10^{-12}.

Pinna—Outer ear. A cartilage flap that forms a transformer to match the impedance of the sound wave in air to that at the end of the ear canal.

Pitch—Like loudness, a description of how the ear and brain interpret sound frequency.

Planck's constant—$h = 6.6 \times 10^{-34}$ J/Hz.

Plasma—State of matter consisting of electrically charged particles.

Plasma physics—Study of the properties of large numbers of electrically charged atoms.

Plastic deformation—Object's shape changed after deforming force is removed (see elastic deformation).

Plum pudding model—Swarm of electrons in a positively charged sphere.

Pneumatics—Use of a compressed gas rather than a liquid in a hydraulic machine.

Polarized wave—Transverse wave, like light. with amplitude in one direction.

Position—The location of an object. Requires a reference location (vector).

Positron—Antimatter form of electron with same mass but positive charge. Emitted in decay of radioactive nucleus or as result of collision of high-energy gamma ray with the nucleus.

Positron emission tomography (PET)—Method of forming three-dimensional image of human body. Uses a short-lived positron-emitting isotope inserted into a biologically active molecule.

Potential difference—Difference in electric potential or energy difference divided by charge, also called voltage.

Pounds per square inch (psi)—Measure of pressure in English system.

Power (P)—The change in energy divided by the time taken.

Power, electrical—Given by $P = I \times V$ (power equals current times voltage).

Precession—The rotation of the axis of rotation of a rotating object due to a torque on the object.

Precision—How well the results can be reproduced.

Pressure (P)—Force divided by area over which force is exerted.

Pressure of air—At sea level about 101 kPa.

Principle of equivalence—A central tenant of Einstein's General Theory of Relativity. The laws of physics in an accelerating reference frame or a gravitating frame are indistinguishable. You can't tell the difference between falling or being acted on by gravity. Requires that inertial mass equals gravitational mass.

Principle of relativity—The laws of physics are the same in two coordinate systems moving at constant velocity with respect to each other.

Proton—Positively charged particle in nucleus.

Pulley—Simple machine. A fixed, or unmovable, single pulley can be considered a wheel and axle where both have the same radius. The mechanical advantage is one, but the pulley changes the direction of the force. If the pulley is allowed to move you achieve a mechanical advantage of two.

Pulse width modulation (PWM)—Method of digital modulation in which a narrow pulse represents a 0, a wider pulse represents 1.

Quad—Unit of energy used for measuring energy use by nations. One quad is one thousand billion BTUs.

Quantum chromodynamics (QCD)—The theory that describes the interactions between quarks and gluons.

Quantum mechanics—Study of atoms by themselves, as well as in molecules, liquids, and solids in which the wave nature is important.

Quark—Particles that make up protons and neutrons and other short-lived particles. Fractionally charged, that is their charges are either 2/3 or –1/3 the charge of the proton and are called "up" and "down" respectively.

Quark generations—Each generation has two quarks: up and down in first, strange and charm in second, top and bottom in third.

Qubit—Quantum bit, the basic piece of information for a quantum computer.

R-value—Measure of resistance of material to heat transfer. R is the inverse of conduction: $R = 1/U$.

Radar—An acronym for "RAdio Detection And Ranging."

Radiation—A means of heat transfer when infrared waves areemitted by hotter objects and absorbed by colder ones.

Radio astronomy—Uses VHF, UHF, and microwave portions of the electromagnetic spectrum to study planets, stars, and galaxies.

Radioactive nucleus—Unstable nucleus that emits particles in decay to another nucleus.

Rainbow—Spectrum of light formed when sunlight interacts with water droplets. Angle between entering and leaving is 40° for blue light, 42° for red.

Rainbow, secondary—Reversed color spectrum caused by an additional reflection of the light inside water droplet.

Ramp—Example of an inclined plane $MA = F_{ourput} / F_{input}$ = (length of ramp) / (height of ramp).

Rarefaction—Region of lower pressure in a longitudinal wave.

Reflection—Light or other wave bounces off a surface or interface between two media.

Reflection, diffuse—Light rays reflect into different, random directions, as from paper.

Reflection, specular—Light rays reflect into a single direction, as from a mirror.

Refraction—Bending of light as it goes from one medium to another.

Refractive index (*n*)—Ratio of speed of light in a vacuum to that in a medium.

Resistance (*R*)—Friction that electrons encounter due to collisions with the atoms in a wire. Causes electric charges to lose electrical energy. Equal to potential difference divided by current.

Resistor—Device used in electrical and electronic circuits to put a definite resistance in a circuit.

Resonance—Occurs when an external oscillating force is exerted on an object that can vibrate. When the frequency of the external force equals the natural frequency.

Retina—Screen on back of eye. Has cells that detect light and a neural network that does preliminary processing of the information received.

Reverberation time—Time for a sound to diminish by a factor of 60 dB, that is, to 1/1,000,000th of its original intensity.

Rod—Rod-shaped nerve cells on the retina that sensitive to low light levels. Responsible for a general image over a large area, but not fine details.

Rolling friction—The result of deformation of either the rolling object or the surface.

Rotational inertia—See moment of inertia.

Rotational kinetic energy—Equivalent of kinetic energy for rotational motion.

Rydberg formula—Wavelength of the emitted radiation from hydrogen atom is given by $1/\lambda = R(1/m^2 - 1/n^2)$. The numbers m and n are the quantum numbers of the two energy levels. For example, the red line would be $m = 2, n = 3$. The Rydberg constant $R = 0.01097$ nm^{-1}.

Saturation—Extent to which other wavelengths of light are present in a particular color.

Scalar quantity—A quantity that has only magnitude (size).

Schroedinger's equation—Equation of the wavefunction of a particle. Wavefunction is related to probability of finding particle at a specific location.

Screw—Simple machine like an inclined plane wrapped around an axle.

Second (s)—The time it takes for 9,192,631,770 periods of microwave radiation that result from the transfer of the cesium-133 atom between lower-energy and higher-energy states.

Second law of thermodynamics—It is impossible to convert heat completely into work in a cyclic heat engine; there is always some heat output. Also: Heat doesn't flow from cold to hot without work input.

Second-class lever—Simple machine with effort force exerted at one end or bar, the input force applied at the other end, and the pivot between the two. *MA* less than, equal to, or greater than one.

Series circuit—Electrical devices arranged in a single line allowing only one path for the charges.

Shadow—Areas of darkness created by an opaque object blocking light.

Shear force—Force exerted that parallel to a plane.

Simple machine—Device that matches human capabilities to do work to tasks that need to be done. There are four major groups, the lever, the wheel and axle, the pulley, and the inclined plane.

Snell's law of refraction—Describes how light behaves at a boundary between two different media.

Solar eclipse—Occurs when the moon casts its shadow on Earth. The moon must be directly between the sun and Earth.

Solid—State of matter. Solids retain their shape because strong forces hold the atoms in their places.

Sonar—Acronym for "SOund Navigation And Ranging," a method of using sound waves to determine the distance between an object and a sound transmitter.

Sonic boom—Caused by a shock wave produced when an aircraft reaches the speed of sound.

Sound intensity—Power in a sound wave divided by the area it covers.

Sound wave—Created by some type of mechanical vibration or oscillation that forces the surrounding medium to vibrate.

Sparticle—A supersymmetric partner particle. The sparticle for a boson is a fermion; the sparticle for a fermion is a boson. Including sleptons, squarks, glunios, and photinos, sparticles are theoretical particles that are (or have been) proposed as part of an alternative to the Standard Model of sub-nuclear particles.

Special relativity—Study of the descriptions and explanations of the motion of objects moving near the speed of light.

Specific heat capacity—Amount of energy needed to increase the temperature of one kilogram of a substance by one degree Celsius.

Spectrum, absorption-line—Continuous spectrum with dark gaps produced by a continuous spectrum source viewed through a cool, low-density gas.

Spectrum, continuous—Light of all wavelengths within a range produced by a hot solid or a hot, dense gas.

Spectrum, emission-line—Light of a few, distinct wavelengths produced by a hot, low-density gas.

Speed (v)—Distance moved divided by the time needed to move (scalar).

Speed of light in a vacuum (c)—= 299,792,458 m/s.

Speed of sound—In air about 340 m/s or 760 mph (miles per hour) at 20° C. Increases by 0.6 meters per second for every degree Celsius increase in temperature ($v = (331 \text{ m/s}) (1 + 0.6\ T)$ where T is measured in Celsius.

Sphygmomanometer—Device used to measure blood pressure.

Stainless steel—A non-rusting alloy of iron and chromium and nickel. May also contain silicon, molybdenum, and magnesium.

Standard model—Presently accepted description and explanation of sub-nuclear particles and their interactions.

Standing wave—Two waves of same frequency moving in opposite directions produce a standing wave that appears to stand still.

Static—Not moving. All the forces and torques acting on a body must sum to zero the net force on the body is zero so the object does not move or rotate.

Steel—An alloy of iron and a small amount of carbon. Other metals may be added.

Stimulated emission—Atom in excited state struck by photon with correct energy is stimulated to emit another photon with same energy and drop to lower energy level.

Streamlines—Lines that represent the flow of a fluid around an object or through another fluid.

String theory—Proposed theory to replace the standard model. Each particle is a string, a tiny closed loop. Requires a ten-dimensional spacetime rather than the normal four-dimensional one.

Strong nuclear force—Force within nucleus that acts between protons and protons, protons and neutrons, and neutrons and neutrons, all with the same strength.

Sublimation—Change of state from solid directly to gas.

Superconductor—Allow electrical current to travel without resistance, and therefore no voltage drop across them or energy loss.

Superhigh frequency (SHF) waves—Frequency 3 Ghz–30 Ghz.

Superposition—Two or more waves at the same location pass through each other without interaction. The amplitudes of the two waves add or subtract.

Supersymmetry—Proposed theory to solve difficulties with the standard model in which each boson has a supersymmetric fermion partner particle (or sparticle) andeach fermion has a supersymmetric boson sparticle.

Suspension bridge—Bridge using cables from which wires are attached that suspend the roadbed.

Sustainable energy source—Primarily wind, water, and solar energy, ultimately receive their energy from the sun, and therefore will be available for billions of years.

Synthesizer—Electronic device that generates, alters, and combines a variety of waveforms to produce complex sounds.

Système International (SI)—The International System of Units. Based on the meter-kilogram-second (MKS) or metric system.

Teleportation—Communicating the state of an atom involving entangled photons.

Telescope, reflecting—Uses a concave mirror to gather light and focus it. Uses a secondary mirror to direct light to eyepiece.

Telescope, refractor—Uses one lens to gather, refract, and focus light toward an eyepiece that contains one or more lenses that create an image that they eye can see.

Temperature—A quantitative measure of hotness. The more thermal energy an object has the higher its temperature.

Tensile strength—Maximum tension force that can be applied without breaking object.

Tension force—Force that pulls an object apart. See tensile strength.

Tera (T)—10^{12}

Terminal velocity—Constant velocity attained by a falling object affected by drag in a gas or liquid.

Tevatron—Very high-energy synchrotron at Fermilab, near Chicago, Illinois.

Theory—An explanation of a large number of observations.

Thermal energy—The sum of the kinetic and elastic energy of the atoms and molecules.

Thermal physics—The study of objects warm and cold, and how they interact with each other.

Thermodynamics and statistical mechanics—Studies how temperature affects matter and how heat is transferred. Thermodynamics deals with macroscopic objects; statistical mechanics concerns the atomic and molecular motions of larger numbers of particles.

Thermograph—Picture that shows the temperature of every location in the picture.

Thermometer—Device to measure temperature.

Thermostat—A device that is part of a system that maintains a constant temperature.

Third law of thermodynamics—Absolute zero can never be reached.

Third-class lever—Simple machine with pivot at one end or bar, the effort force exerted at the other end, the input force between the two (*MA* less than one).

Threshold of hearing—Minimum sound intensity that can be heard (0 dB).

Timbre—Quality of sound. Sound spectrum produced by an instrument, characterized by relative amplitudes or harmonics.

Time division multiple access (TDMA)—Method of sending cell phone calls that splits up three compressed calls and sends them together.

Time standard—The cesium-133 atomic clock is the standard for the second.

Torque (τ)—Equivalent of force for rotational motion. If forces is at a right angle to the radius then torque equals the force times the distance from the axis of rotation.

Torr—Measure of pressure using measurement of height of mercury in a glass tube that would create a pressure that balances the pressure of the fluid. Used to be called millimeters of mercury (mmHg).

Torsional force—Force exerted to twist an object.

Torsional wave—Material is not only displaced vertically, but also twists in a wavelike fashion.

Total internal reflection—Occurs when a ray of light in a medium with a higher index of refraction strikes the interface between that medium and one with a lower index of refraction.

Transformer—Impedance matching device.

Translucent—Media that allow light to pass through, but that bend closely-spaced rays into different directions.

Transparent—Media that allows light to pass through the material. Rays of light are either not bent or closely-spaced rays are bent together.

Transverse wave—Wave and its energy move perpendicular to the direction of oscillations.

Trough—The lowest point of a transverse wave.

Tsunami—Series of waves in water caused by underwater earthquakes and volcanic eruptions.

Turbulent—Flow of fast-moving fluid or object moving fast through a fluid. Causes usually circular or helical motion of fluid.

Ultrahigh frequency (UHF) waves—Frequency 300 Mhz–3 Ghz.

Ultrasonic sound—Frequencies above human hearing, 20kHz.

Ultrasound—A method of looking inside a person's body to examine tissue and liquid-based organs and systems without physically entering the body.

Umbra—Area of the shadow where all the light from the source has been blocked.

Uncertainty principle—Limitation to simultaneous knowledge of position and momentum of particle: $\Delta x \times \Delta p \geqslant h/4\pi$. Equivalent equation for energy and time.

Universal gravitational constant—Proportionality between gravitational force between two objects and GMm/r^2.

Van Allen belt—Doughnut-shaped regions of charged particles around Earth. Created by charged electrons and protons from solar wind and cosmic rays trapped by Earth's magnetic field.

Vad de Graaff generator—Device using static electricity to produce very large potential differences.

Vector quantity—A quantity that has both a magnitude and a direction.

Velocity (v)—Displacement divided by the time required to make the change (vector).

Very high frequency (VHF) waves—Frequency 30 Mhz–300 Mhz.

Very large array (VLA)—Group of dozens of radio telescopes spaced as much as 13 miles apart from each other in Socorro, New Mexico.

Vocal cords—Or vocal folds. Vibrating source of the human voice.

Volt (V)—Unit of measurement of potential difference or voltage.

Voltaic pile—Or battery in which chemical energy is converted into increased energy of electric charges.

Watt (W)—Unit of power. One watt is one joule (J) per second (s).

Wave—A traveling disturbance that moves energy from one location to another without transferring matter.

Wave velocity—Depends upon the material or medium in which it is traveling.

Wavelength—The distance from one point on the wave to the next identical point; the length of the wave.

Weak nuclear force—Cause of beta decay. Weaker than strong nuclear force. Shown to be an aspect of the electroweak force. Carried by the massive W^+, W^-, and Z° particles.

Wedge—Simple machine used to force two objects apart. Examples are the knife, hatchet or axe, and plow.

Weight—Force of gravity on an object ($F = mg$).

Wheel and axle—Simple machine. If the radius of the axle is a and the radius of the wheel is w, then if the input force is applied to the wheel, the output force is given by $F_{output} = F_{input} (w/a)$.

White light—The combination of all the colors in the visible spectrum.

Work—Energy transfer by mechanical means.

X ray—Electromagnetic wave of very short wavelength or very high-energy photons. Can be emitted by an atom with many electrons that is disturbed by the collision with a high-energy particle.

Young's modulus—Applied pressure (called stress) divided by the ratio of the change in length to the original length (called strain).

Zeroth law of thermodynamics—If objects A and B are in equilibrium and B and C are in equilibrium, then A and C are also in equilibrium.

γ—In Einstein's Theory of Special Relativity a ratio that depends on the speed of an object. It is 1 when the speed is zero and infinite when the object moves at the speed of light. Affects distance, time, momentum, and energy.

Index

Note: (ill.) indicates photos and illustrations; f indicates figure; t indicates table.

398